Outbreaks, Epidemics, and Health Security

Outbreaks, Epidemics, and Health Security

COVID-19 and Ensuring Future Pandemic Preparedness in Ireland and the World

Authors

Sebastian Kevany, MA, MPH, PhD

Aoife Kirk, MB BCH BAO (HONS) MPH

Senior Editorial Contributor

Ruchika Gallani, BDS

Editorial Contributors

Eilis Conneally, MA, PgDip

Maisie M. Jones, BSc, MPH

Eimear Bourke, BCL, DipLaw

ELSEVIER

ACADEMIC PRESS

An imprint of Elsevier

Academic Press is an imprint of Elsevier
125 London Wall, London EC2Y 5AS, United Kingdom
525 B Street, Suite 1650, San Diego, CA 92101, United States
50 Hampshire Street, 5th Floor, Cambridge, MA 02139, United States
The Boulevard, Langford Lane, Kidlington, Oxford OX5 1GB, United Kingdom

ISBN 978-0-323-91705-6

For information on all Academic Press publications
visit our website at https://www.elsevier.com/books-and-journals

Publisher: Stacy Masucci
Acquisitions Editor: Elizabeth Brown
Editorial Project Manager: Tracy Tufaga
Production Project Manager: Punithavathy Govindaradjane
Cover Designer: Matthew Limbert

Typeset by STRAIVE, India

To Shandy and in memory of Dr. John Kevany.

Contents

About the authors

Sebastian Kevany, MA, MPH, PhD, is Associate Professor at the Daniel K. Inouye Asia-Pacific Center for Security Studies, Honolulu, Hawaii. He has conducted more than 100 global health and development missions to Africa, the Middle East, the South Pacific, Eastern Europe, and Asia with a focus on monitoring and evaluation, health security, international relations, human rights, health diplomacy, and international governance. Dr. Kevany holds BA and MA degrees from Trinity College Dublin, Ireland; an MPH degree from the University of Cape Town, South Africa; is a former adjunct assistant professor at Trinity College Dublin; is an affiliate of the University of California, United States; and holds a doctorate by life research from the University of Westminster (United Kingdom). He is the author of *Barefoot Global Health Diplomacy: Field Experiences in International Relations, Security, and Epidemics*, published in December 2020 by Elsevier/Academic Press, 9780128186817. He is also a US Coast Guard Auxiliarist; a research fellow at the United Nations University International Institute for Global Health; and a visiting scholar at the University of Leeds School of Politics and International Studies (POLIS).

Aoife Kirk, MB BCh BAO (Hons) MPH, is Medical Doctor. After qualifying from medicine in NUIG Ireland, Dr. Aoife Kirk spent the last few years working as a junior doctor in Ireland, New Zealand, and London, United Kingdom. During the COVID-19 pandemic, she worked within Public Health in Ireland, Respiratory and ICU departments in a Dublin hospital, and has witnessed a range of the widespread effects the pandemic has created. She was introduced to the concept of Planetary Health whilst undertaking MPH in London School of Hygiene and Tropical Medicine and is one of the co-founders of Irish Doctors for Environment (IDE). Dr. Aoife is the international link for IDE through work with Planetary Health Alliance, Health Care Without Harm, and WONCA Working Party on the Environment. She is passionate about the opportunities a carbon-zero future provides for the health of global citizens, biodiversity, interdisciplinary teamwork, research-driven policy, and providing safe, efficient healthcare services.

About the contributors

Annette L. Aguon is the administrator for the Bureau of Communicable Disease Control at the Guam Department of Public Health and Social Services where she has served in that capacity since 2019. She began her public health career in 1994, immediately following graduation working in the TB programme prior to joining the immunization programme 2 years later. Annette received her Bachelor of Science degree in medical technology from Seattle University, her Master of Public Administration degree from the University of Guam, and her Postgraduate Certificate in field epidemiology from the Fiji National University.

German Andres Alarcon Garavito is a Colombian nurse, with a master's degree in global health and development from University College London (UCL) with experience in qualitative methods, mental health research, and clinical trials. He is also a research collaborator at the Rapid Research Evaluation and Appraisal Lab (RREAL) from UCL, the London School of Hygiene and Tropical Medicine (LSHTM), and Uniandes Colombia. He is a member of Sigma Theta Tau International Honor Society of Nursing.

Coral Andrews graduated from the University of Southern California Sol Price School of Public Policy with a Doctor of Policy, Planning, and Development Degree (2021). She holds a Master of Business Administration degree from New Hampshire College (1989) and a Bachelor of Science degree in nursing from the University of South Alabama (1983). She has served in executive roles in the public, private, and plural sectors. Within the US Department of Defense, her work as a senior officer in the US Navy culminated with joint service working alongside global partners in international relations and diplomacy. In the private and plural sectors, she has served at the forefront of domestic health policy design and implementation.

Joelson Anere is a willing learner, creative intellectual, developmental builder, and creative leader with political acumen, political instinct, and political foresight. He is critical-minded and analytical. There're moments when Mr. Joelson Anere can be very philosophical. Mr. Joelson Anere believes in building from ground zero upwards and refining the many initiatives of his late parents.

Joelson Anere writes up a wide variety of abstract ideas on paper and shares them with a wider global audience to help geminate a new puritan spirit of positive change, positive development, that is inclusive and that the benefits of this change are equitably shared with all citizens. And you read them.

Mr. Joelson Anere holds a Bachelor of Arts degree in PNG studies and international relations, with a major in international relations and diplomacy. He is the winner of the Mineral Resources Development Corporation (MRDC) Academic Excellence Award in International Relations and Diplomacy at Divine Word University in 2005. Mr. Joelson Anere has also contributed written papers for a wide range of academic journals abroad.

Lyndsay S. Baines has a background in medical sociology and psychology and specializes in global health, well-being, and the psychosocial aspects of civilian and military trauma, oncology, solid organ transplant, and global health diplomacy. Currently, she is the lead for an innovative short course in global health diplomacy at the University of Oxford. The course attracts students internationally and across disciplines including career diplomats, military personnel, lawyers, economists, scientists, and healthcare professionals.

Rear Admiral **Michael Baker** (Retired) is a general surgeon and trauma specialist. He served 30 years in the US Navy, including combat deployment with the US Marine Corps, and retired with the rank of Rear Admiral. He has authored numerous articles on combat casualty care, strategic medical threats, triage and disaster response, and the use of military in foreign policy which he teaches for the University of California Berkeley Osher Lifelong Learning Programme.

Eimear Bourke is a final year medical student at the Royal College of Surgeons in Dublin, Ireland (RCSI). She worked as a solicitor before returning to medical school. She is particularly interested in public health, healthcare equity and lifestyle medicine. In her spare time she likes to be outdoors, hiking or running.

Ruairí Brugha qualified as a medical doctor in 1980, spent 6 years working in Africa in the 1980s and 1990s, and completed public health training in the United Kingdom. He spent 10 years at the London School of Hygiene and Tropical Medicine, as Head of the Health Policy Unit and Coeditor of Health Policy and Planning. In 2005, he became Professor of Public Health and Epidemiology at the Royal College of Surgeons in Ireland. His interest is in health systems and policy research, in Ireland, Africa, and globally. He transitioned to an Emeritus Professor role in January 2020 and in April 2020 to media work in Ireland on COVID-19. He has published 150 articles in international peer-reviewed journals.

Laura E. Buback has more than 10 years of experience in public health programme monitoring, evaluation, research, and training in sub-Saharan Africa, the Caribbean, and Southeast Asia. She is currently a Technical Lead at the UCSF Institute for Global Health Sciences (IGHS), where she provides technical assistance in HIV, malaria, and global health security research and training programmes. Laura led Case Investigation and Contact Tracing trainings and workflow management throughout the COVID-19 pandemic in San Francisco, California, and also supported the University of Guam and Guam Department of Public Health and Social Services to strengthen their COVID-19 response in 2020. Prior to IGHS, Laura spent 6 years working in West Africa working on monitoring and evaluation of nutrition, immunization, and maternal and child health programmes, as well as polio and Ebola outbreak response. Laura holds an MPH from Tulane University and a BA in sociology/anthropology from Lewis and Clark College.

Conor Burns is a surfer and a political and social observer.

Deon Canyon is a well-regarded international scholar in security cooperation, crisis management, biosecurity, Oceania, and grey-zone gaming. His cross-disciplinary research focuses on understanding, managing, controlling, and preventing complex and dynamic security threats with innovative approaches. Canyon has authored more than 240 publications during his 29 years at DKI-APCSS and other US and Australian universities.

Ciara Conlan is a medical doctor currently working in clinical virology. She co-founded Access to Medicines Ireland (AMI) in 2016 and Doctors for Vaccine Equity (DVE) in 2021. AMI and DVE have been leading voices in Ireland calling for global COVID-19 vaccine equity. She is interested in the intersection between social exclusion and infectious diseases and in the problems caused by the current commercial model of drug development.

Eilis Conneally is currently working in the Irish civil service. She received her undergraduate degree in Politics and Sociology from University of Aberdeen and Postgraduate Diploma in Global Health from the University of Glasgow. Her interests include Reproductive Health in Post-conflict societies.

Teofila P. Cruz, School of Health, University of Guam, Mangilao, Guam.

Marisa Dallabetta has worked as a PE teacher in a secondary school of art and music in Trento, Italy. Marisa's love for yoga, nature, a curiosity for life, and a healthy lifestyle led her to study this ancient philosophy in an Indian ashram, and now she spreads it and the practice of meditation to students with special needs, teenagers, and adults for a better and peaceful world.

Sarah O'Dwyer is an emergency doctor currently based in Hawke's Bay, New Zealand, having spent the first lockdown in Nelson, South Island of New Zealand. She graduated from medical school at the National University of Ireland, Galway, in 2015. She has a certificate in women's health and has a strong interest in global health, minority populations, and migrant inequalities.

Luisa Enria is Assistant Professor in the Department of Global Health and Development at the London School of Hygiene and Tropical Medicine. Her research draws from political anthropology to study postwar political violence, citizenship practices, and responses to health emergencies. She is a member of the COVID-19 Clinical Research Coalition Social Science Working Group.

Eamonn Faller is a doctor specializing in infectious diseases. He was involved in the response to COVID-19 whilst working at Cork University Hospital and has published six papers on the topic, including a report of the first documented community transmission of COVID-19 in Ireland. He previously worked with Medecins Sans Frontieres (MSF) on an HIV project in 2017 in Nyanza Province in Kenya and also participated in MSF's response to postelection unrest. He has written several articles and appeared before a Dail Committee in various patient advocacy areas including HIV and direct provision in the context of the COVID-19 pandemic. He is also currently completing a master's degree in healthcare economics, policy, and management at the London School of Economics.

Jeremy Fowler worked as a clinician and medical director at a tuberculosis and respiratory diseases treatment centre in northern Jordan from September 2010 to July 2021. He currently resides in California with his family.

Sabine Iva Franklin earned her PhD from the School of Organisations, Economy and Society at the University of Westminster. Her expertise is in international development, global health governance, and qualitative field research methods. She is a Visiting Researcher at the Westminster Development Policy Network and a Postdoctoral Researcher at the National University of Ireland, Galway.

Selma Furtado has a keen interest in strategy, ICT, and management in health care. She holds a strategy-focused Master of Business Administration and a master's degree in strategic management. In her spare time, Selma enjoys photography, sightseeing, reading, and indoor gardening.

Viveka Guzman, MBBS, MSc, is a SPHeRE PhD Scholar at the Royal College of Surgeons in Ireland. Her current research focuses on socio-ecological determinants of health and well-being. She is also the projects team lead at the Global Health Next Generation Network (GHNGN).

Margaret Hattori-Uchima is the Dean of the University of Guam School of Health. She was the 2019 Guam Nurse of the Year awarded by the Guam Nurses Association. In 2020, she was inducted into the American Academy of Nursing as the first Nursing Fellow from Guam.

Hattori-Uchima has provided leadership in the study and alleviation of health disparities and had a significant impact in the arenas of education, practice, and health care. She secured significant funding for the HRSA-funded Guam/Micronesia Geriatrics Workforce Enhancement Programme which aims to fill gaps in care for the elderly and those with Alzheimer's disease and related dementias through workforce development, family/caregiver training, and enhancement of practice environments.

Hattori-Uchima is the contact PI for the NIMHD U24: A cohort for Studying the Burden of Cardiometabolic Diseases in Guam and Pohnpei. She is one of the Guam co-MPIs in the NCI U54 Pacific Islands Partnership for Cancer Health Equity and Guam lead investigator in the NIH BUILD EXITO (Building

Infrastructure Leading to Diversity Enhancing Cross-Disciplinary Infrastructure and Training at Oregon). She led the UOG School of Health response during the first 2 years of the COVID-19 pandemic, assisting in vaccination, testing, and community outreach activities for homeless and marginalized communities at risk. Her professional and academic collaborations target capacity building and research in health disparities, elder care, with emphasis on the needs of patients and families coping with Alzheimer's disease, cancer disparities, and the biomedical training of underrepresented students in Micronesia.

Hattori-Uchima has a BSN from the University of Hawaii, an MSN from the University of Phoenix, and a PhD in nursing from Villanova University.

David L. Heymann is a medical epidemiologist and Professor of Infectious Disease Epidemiology at LSHTM. From 2009 to 2017, he was Chair of Public Health England, and during this period, he also led the Centre on Global Health Security at Chatham House (London). From 1989 to 2009, Heymann held various leadership positions in infectious diseases at WHO and in 2003 headed the WHO global response to SARS in his role as executive director of communicable diseases. In 1976, after spending 2 years working in India on smallpox eradication, Heymann was a member of the CDC (Atlanta) team to investigate the first Ebola outbreak in DRC and stayed on in sub-Saharan Africa for 13 years in various field research positions on Ebola, monkeypox, Lassa fever, malaria, and other tropical diseases. Heymann has published more than 250 peer-reviewed articles and chapters, is the Editor of the Control of Communicable Diseases Manual, and is an elected member of the UK Academy of Medical Sciences and the US National Academy of Medicine. In 2009, he was named an Honorary Commander of the Most Excellent Order of the British Empire for services to global health.

Zachary Hitchcock is a surfer and an environmentalist based in Oahu, Hawaii. He is the former lead of the Surfrider Foundation Hawaii Chapter and is a trained yogi.

Rita Issa is a medical doctor and clinical research fellow in climate change, migration, and health at University College London. In 2021, at the height of the COVID-19 pandemic, she spent 3 months aboard the Greenpeace Ship 'Arctic Sunrise' as the on-board medic.

Rahul M. Jindal is Professor of Surgery, Global Health at Uniformed Services University, Bethesda, United States. Jindal has authored more than 200 manuscripts, has been funded by the National Institute of Health, and is on the steering committee of 'The Quality of Cancer Care Committee'—a transgovernmental agency committee of Federal Departments. Jindal's research work has been cited in textbooks and editorials. Jindal has played a crucial role in the team that conducted groundbreaking surgery at Walter Reed on Thanksgiving Day in 2009. A 21-year-old senior airman, Tre Porfirio was shot three times by an insurgent in Afghanistan; he received the first-ever pancreas islet cell transplant after trauma.

In addition to his clinical activities, Jindal co-authored a textbook "The Struggle for Life: A Psychological Perspective of Kidney Disease and Transplantation." Jindal set up the first kidney dialysis and transplant programme in Guyana in 2007. His team visits Guyana and Suriname four times a year to perform kidney and corneal transplants. Dr. Jindal is working towards replicating this programme in the 15 CARICOM countries and in Botswana. He is the Co-Chair of SEVAK Programme in India and Guyana in which his team trains students in good preventative measures and diagnosis of diabetes and hypertension. Dr. Jindal continues his outreach as a Visiting Professor to several institutions in India, the Caribbean, and the United Kingdom. Dr. Jindal lectures at the University of Oxford and Anglia Ruskin University, United Kingdom, on their global health diplomacy courses. Jindal teaches in-class and distance learning global health modules in the MPH curriculum of Uniformed Services University. The Governor of Maryland appointed Jindal as Commissioner, Office of Service and Volunteerism, and Commissioner to the Human Rights Commission, Montgomery County. Jindal was awarded the Fulbright-Nehru Distinguished Chair to carry out research and teaching in India (2016 and 2022) and the Ellis Island Medal of Honor, 2015. The Carnegie Foundation, New York, named Jindal as the 'Great Immigrant—Great American', 2019. Every Fourth of July since 2006, Carnegie Corporation of New York honours Andrew Carnegie's legacy by celebrating the wide-ranging contributions of immigrants who enrich our communities and culture, strengthen our economy, and invigorate our democracy. Jindal was also awarded the faculty mentor of the year (2018), Alpha Omega Alpha (AOA) chapter of the Uniformed Services University.

Maisie M. Jones is a graduate of the Masters of Public Health (MPH) from University College Cork, Ireland, where her research thesis focused on the barriers and facilitators to asylum seekers accessing sexual health services within the direct provision system. She previously studied for her BSc in health and society at Dublin City University, during which time her interest and passion for global health peaked. Maisie has experience in many areas of health care including disability services, COVID-19 contact tracing, health research, and health communications and has interned with the Irish Global Health Network and ESTHER Alliance for Global Health Partnerships. She has been involved in several health research projects including studies related to cancer genetics services, dementia policy, and maternal, newborn, and child health in sub-Saharan Africa.

Kiran Kamble has a medical degree (MBBS) from Mumbai University and a Master of Public Health from Harvard University and is currently pursuing a PhD in Health Systems and Policy at the University of Memphis. He is a health services researcher with a focus on access, equity, quality, and cost issues associated with primary care. Kiran has 28 years of progressive experience in the health sector spanning across primary care practice, pharmaceutical manufacturing and retail, occupational health, to culminate into global public health systems and policy. He has worked with the World Health Organization (WHO) in the tuberculosis control programme in India and has provided technical support to governments of Abu Dhabi, India, Tajikistan, and South Korea in health systems strengthening, establishing and managing public-private partnerships, and revising national communicable disease programmes. In addition, he has provided short-term technical assistance (STTA) to the United States Agency for International Development (USAID), European Union (EU), and several other international health organizations in programme management and monitoring and evaluation of programmes and projects in communicable disease control, maternal and child health, community-based health programming, public-private partnerships, health policy analysis, and health sector reform. He has worked in all five continents, including fragile/transitioning nations such as Afghanistan, Central African Republic, Haiti, Iraq, Solomon Islands, and Yemen.

Enku Kebede-Francis holds a PhD in agronomy and plant science, an MS in environmental science, an MA in education, a BA in sociology, and a certificate in management. She has worked for the United Nations (UNESCO, UNDP, UNFPA, and UNDPI), is Assistant Professor at Tufts University Medical School/Department of Public Health, a Visiting Scientist at the USDA's Center for Human Nutrition Research Center for Aging, and Visiting Fellow at the Australian National University Medical School. She has designed and implemented preventive health programmes promoting women's health and tobacco cessation programmes. Her professional interests include preventing scurvy and childhood blindness in developing countries using micronutrients and advocacy in primary health care, primary education, and access to clean water and improved sanitation for all as rights. She has published a textbook titled *Global Health Disparities: Closing the Gap Through Good Governance* (2010).

Regina Keith is a health practitioner with 46 of experience in Europe, Africa, and Asia. She holds a PhD in Nutrition and is a senior lecturer at the University of Westminster in the United Kingdom. She is the course leader for an innovative MSc in global public health nutrition and is on the executive committee of the World Public Health Nutrition Association. She has worked in more than 39 counties at all levels of the healthcare system, focusing on strengthening timely access to equitable health services. Her research demonstrates that equity, voice, and governance through an intersectoral approach are essential for achieving universal health coverage by 2030. Her evidence suggests these services need to be free at the point of access. Her research focuses on exploring strategies to address barriers to timely health service utilization. She has extensive experience in maternal and child health, infectious disease control, strengthening health systems in low-income and fragile contexts, and bringing voices of marginalized groups into national and global health, nutrition policy, and planning cycles. She has worked for the NHS, private health services, NGOs like Save the Children, VSO, and World Vision and is the Director of Health Nutrition and Social Development for Cara International Consulting Ltd.

Ina Kelly graduated in medicine in University College Galway, trained in family medicine (general practice) in Perth, Australia, followed by Higher Specialist Training in public health medicine in Faculty of Public Health Medicine/Royal College of Physicians of Ireland and in the WHO, Geneva. She is Specialist in public health medicine/Medical Officer of Health, working mainly in Health Protection, in Ireland's Health Services Executive, Midlands region, and nationally is Chair of the HSE Public Health Medicine Environment and Health Group. Her special interest areas of work include public health medical law, waterborne infectious diseases, and health impacts of climate change, which of course are relevant in many aspects of health protection and inform the need for climate change adaptation. Since the advent of COVID-19, Ina has been very involved in the statutory investigation and control of cases and outbreaks in the Midlands region and in the interpretation of Medical Officer of Health legislation to establish and explain the legal basis of investigation and control of COVID-19 in Ireland in line with the EU GDPR.

Professor Ilan Kelman is Professor of Disasters and Health at University College London, England, and a Professor II at the University of Agder, Kristiansand, Norway. His research interest is in linking disasters and health, including the integration of climate change into disaster research and health research. This covers three main areas: (i) disaster diplomacy and health diplomacy; (ii) island sustainability, involving safe and healthy communities in isolated locations; and (iii) risk education for health and disasters.

Swee Kheng Khor is a Malaysian physician currently specializing in health systems and policies and global health, with fellowships at Chatham House, the United Nations University (UNU-IIGH), and ISIS Malaysia. Previously, he held progressively senior practitioner roles in clinical medicine, refugee and disaster relief, clinical research, and pharma anticorruption. In these roles, he was based in Malaysia, Singapore, Dubai, Shanghai, and Paris, covering more than 90 countries in Asia, Africa, Europe, and the Middle East. He holds postgraduate degrees in internal medicine (Royal College of Physicians of the United Kingdom), public health (Berkeley), and public policy (Oxford) and has published more than 170 articles in international academic journals, think tanks, and media.

James Larkin is a PhD student in Health Economics and holds a master's degree in global health. He has published several studies examining the influence of the pharmaceutical industry on health care. He is also a member of Access to Medicines Ireland (AMI), which has been a leading voice in Ireland calling for global COVID-19 vaccine equity.

Yodi Mahendradhata is currently the Vice Dean at the Faculty of Medicine, Public Health and Nursing, Universitas Gadjah Mada, Indonesia. He is a medical doctor by training, and he obtained a master's degree in International Health from the Institute of Tropical Medicine, Charite Medical Faculty, Humboldt University, Berlin, Germany. Dr. Mahendradhata completed his doctorate in medical science at the Institute of Tropical Medicine, Antwerp, Belgium and

Ghent University, Belgium. He is also currently leading the South East Asia Regional Training Center for Health Research supported by TDR, the Special Programme for Research and Training in Tropical Diseases. He has led developments of a national strategy for implementation research, a national strategy for tuberculosis control, and country proposals to the Global Fund to Fight AIDS, TB, and malaria. His research interests are in global health, implementation science, disease control, health policy, and health programme management. Dr. Mahendradhata was an Alexander von Humboldt Foundation Visiting Scholar at the Institute of Public Health, Heidelberg University, Germany. He has authored and co-authored numerous peer-reviewed articles published in reputable international journals. He is serving as an editorial board member of PLOS Global Public Health, Frontiers in Tropical Diseases and BMC Health Services Research journals. He has been awarded the Scopus Young Scientist Award by the United Nations University, Elsevier, and Alexander von Humboldt Foundation. Dr. Mahendradhata has been elected as a Fellow of the Royal Society for Public Health since 2018.

Serena Masino is Senior Lecturer (Associate Professor) in International Development and Economics at the University of Westminster's School of Organisations, Economy and Society. Her current field of research is in the political economy of development, and her research agenda looks at issues of global health policy alongside the interaction of globalization and labour standards in transnational production settings.

Deena Mehjabeen is currently a higher degree research candidate researching migrant and refugee mental health at the Translational Health Research Institute, Western Sydney University, Australia. She has a background in global public health, having completed her Master of Global Health (MGH) and Master of Public Health (MPH) degrees at the University of Sydney, Australia. Previously, she has undertaken her Bachelor of Science and Master of Science degrees in microbiology from the University of Dhaka, Bangladesh. She has substantial research experience in academia and nonprofit organizations. Her research interests lie in global health diplomacy, health promotion, global mental health, and women's health.

A Chevening Scholar and Social Researcher by profession, **Swati Mishra** has more than 9 years of experience working in the domain of public health in India. She is a graduate of the London School of Hygiene and Tropical Medicine and Tata Institute of Social Sciences. Her core expertise lies in researching access to health care for Dalits, Adivasis, and other vulnerable populations in India. She has worked in Chhattisgarh and Andhra Pradesh with Adivasis and Dalits to ensure quality health services. Currently, she is working on social and financial linkages of seasonal migrants from Ganjam (Odisha) engaged in the power loom industry of Surat (Gujarat).

Sourav Neogi is an alumnus of the London School of Hygiene and Tropical Medicine (LSHTM) and specializes in maternal and child health and hospital-acquired infections. Currently with Jhpiego, Dr. Neogi has previously worked with Norway India Partnership Initiative (NIPI) and supported the government of Bihar during COVID-19 pandemic with a vaccine roll-out focus.

Kim Nordmann, MGH, BSc, is a medical student at the RWTH Aachen University in Germany. She is part of the projects team and the internal affairs team of the GHNGN.

Sandra Chukwudumebi Obiora is a Researcher in the School of Management and Economics at the University of Electronic Science and Technology of China. Her work focuses on empirical research in finance, economics, carbon mitigation, alternative financing methods, strategic management, entrepreneurship, and other economic and development studies concerning sub-Saharan African and other developing, emerging, and developed economies. She takes particular interest in environmental sustainability and how economics can expand sustainably.

Yvette C. Paulino is an epidemiologist and Professor of Health Science at the University of Guam. She is a Pacific Islander dedicated to achieving health equity through research, training, and community collaborations. Professor Paulino's work comprises research in the Pacific on areca nut/betel quid chewing, including the epidemiology, oral precancer screening, effects of chewing on the oral microbiome and liver cancer development, and betel quid cessation. She is passionate about training and mentoring students in population health. Early in the pandemic, Professor Paulino co-led COVID-19 contact tracing training efforts in Guam. She continues to encourage the leveraging of community collaborations to address health disparities and promote public health.

Nichol Randall is currently pursuing a doctorate in global health at the UCLA Fielding School of Public Health and was formerly at the UCSF Center for AIDS Prevention Studies.

Michael J.A. Reid currently serves as the Associate Director of the Center for Global Health Diplomacy, Delivery and Economics, and Chief Medical Officer for UCSF's Pandemic Initiative for Equity and Action (UPIEA). He oversees research and capacity building initiatives in Latin America and sub-Saharan Africa related to enhancing HIV service delivery and strengthening global health security. He is an HIV clinician and sees patients at San Francisco General Hospital's Ward 86. He is also co-chair of UCSF's Decolonizing Global Health working group.

Mehr Muhammad Adeel Riaz, MBBS, is a junior doctor working in Pakistan. He is currently serving as part of the international team in the projects team at the GHNGN.

Maryam Rumaney is a freelance scientific consultant. Her formal training is in molecular science, where she holds a master's degree in human genetics. She has also worked in the field of public health as a laboratory advisor for Médecins Sans Frontières (MSF), where she was involved in multicountry project management and broad stakeholder engagement. During her time with MSF, she was involved with the roll-out of HIV and TB operational research projects in various countries. She has also served on a few consortia. Thereafter, she held a senior research assistant position at the African Cancer Institute at Stellenbosch University. Since 2018, she has been a self-employed scientific consultant offering laboratory advisory work, research support, and fulfilling editing assignments to a broad range of industries.

Danny Scheinmann is a writer, storyteller, and film-maker. He has told stories in over 30 countries, worked with street children in some of the most challenging cities in the world including Manila, Bogota, and Phnom Penh, and was one of the first UK performers into Siberia after glasnost in 1991. His novel, Random Acts of Heroic Love, was a Sunday Times best-seller, reaching No.1 on Amazon UK in fiction and being translated into 21 languages. In 2020, Danny worked alongside his wife Sarah Sutcliffe to make a film about change makers in health services around the world, which was funded by WHO. They have also made two other films in the health sphere about psoriasis and children born with HIV.

Ori Solomon, MGH, is a PhD candidate in Microbiology at McGill University. He is currently serving as the Secretary-General of the GHNGN.

Kathylene Castillo Southard, MSW, LCSW, is a clinical social worker and Buddhist who began practising in the Shambhala tradition, established by Chöygam Trungpa Rinpoche, in 2001. She is a former resident of Gampo Abbey where she studied under Pema Chödrön, the former HR Director for Dechen Chöling, a residential Buddhist Meditation Centre in France, and the former Head of Programmes for the London Shambhala Meditation Centre. She is now in private practice as a psychotherapist in Honolulu, Hawaii.

Carlos R. Taitano, Global Learning and Engagement, University of Guam, Mangilao, Guam.

Gladson Vaghela is studying Bachelor of Medicine and Bachelor of Surgery at GMERS Medical College, Gandhinagar, India. He is a Policy Fellow of The Network—Towards Unity For Health (a nongovernmental organization in an official relationship with the World Health Organization) and a Country Representative at Healthcare Information for All (a global human-rights-based movement, comprising 20,000 members from 180 countries) and serves at the Executive Committee of Asian Medical Students' Association International (a student-led organization having representation in more than 30 countries). Gladson's ultimate goal is to contribute to achieving equity in accessing healthcare services, especially in low- and middle-income countries. For one of his attempts to address inequity in accessing suicide prevention services, Gladson was awarded the Royal Society of Tropical Medicine and Hygiene Small Grant 2021 in partnership with the National Institute of Health Research UK, to conduct his study. As an aspiring physician-scientist, Gladson looks forward to working in the domain of global health and research.

Foreword

For those with a biblical perspective on the world, the COVID-19 pandemic echoes the plagues visited on ancient Egypt—though, in fact, successive waves of COVID-19 more closely resemble the tsunamis that devastated ancient Eastern Mediterranean civilizations as well as the coastal communities of the Indian Ocean in modern times. The myth of Noah's Ark reveals the importance of protective, anticipatory action and demonstrates that a tsunami (or pandemic) is not an equalizer or leveller. Instead, there are many losers, and few winners.

When the hatches are battened down, access to resources and the tools for survival are kept to the few. As the flood waters recede, which happened with COVID-19 after the short shock of the less devastating Omicron variant in early 2022, a changed landscape is revealed. However, actions to pick up the pieces and build back better, which is a term that first caught international attention after the 2004 Indian Ocean Tsunami, will be ultimately fruitless if lessons have not been learned.

The perspectives and analyses contained in "*Outbreaks, Epidemics, and Health Security: COVID-19 and Ensuring Future Pandemic Preparedness in Ireland and the World*" make an important contribution to the critical task of lesson-learning. One sees an irony in the title, as this collection of essays is a conscious effort to push against the tide of the 24-h news cycle and the attention span of national decision-makers, which in turn contribute to history repeating itself. The chapters are always worthy, and often successful attempts to identify, analyse, and report what countries and communities did well—and, in some cases, did badly.

Through such lessons, local and global communities can be better prepared for the next pandemic; be more aware and resilient in their responses; and be better able than the world was in 2019–22 to protect the vulnerable and those who lacked access to resources. But this book also speaks to a deeper need, which is to find meaning when life snatches away our way of life, and sometimes our loved ones.

"*Outbreaks, Epidemics, and Health Security: COVID-19 and Ensuring Future Pandemic Preparedness in Ireland and the World*" has as its initial lens a small two-state island-nation on the Western edge of Europe, which like many small nations can bring a more informed perspective on the responses of larger nations. It starts with an analysis by one of the book's authors and editors, Sebastian Kevany. The initial 'take' of public health specialists might be that COVID-19 has revealed, and brought to centre-stage, the critical role of public health for population protection. However, this closer look shows how the public health measures that are essential for the protection of the health of populations have opened up political fracture-lines and fissions—perhaps a metaphorical transition from floods to earthquakes.

Clearly, to a public health specialist, countries need better information and surveillance systems to enable speedier responses to future pandemics. However, public health responses have been politicized, as Kevany points out. Also, science is ignored by many politicians and interest groups, and information is prostituted by and to social media. Interestingly, one likely determinant of Ireland's often belated successes, although its delayed responses to its first, third, and fourth waves resulted in avoidable deaths, was the informed coherence of national mainstream media, and the comparative paucity in Ireland of social media misinformation.

Of note, "*Outbreaks, Epidemics, and Health Security: COVID-19 and Ensuring Future Pandemic Preparedness in Ireland and the World*" also contains a contribution from David L. Heymann, an epidemiologist who has had an illustrious career with the US Centers for Disease Control in Africa, the World Health Organization, and the London School of Hygiene and Tropical Medicine and Swee Kheng Khor. Khor and Heymann provide a historical perspective on epidemic controls, noting the importance of protecting borders and the use of quarantines. The latter's more recent origins are found in the actions of the city states of Dubrovnik, Croatia, and Venice, Italy, in the 14th and 15th centuries, in the face of ship-borne plague. Foreword by Swee Kheng Khor and David L. Heymann is framed around the three interlocking functions for health security at the national level: resilient healthcare systems, resilient public health-core capacities for epidemic control and investments and policies for healthy populations.

In between these chapters is a rich tapestry of contributions illustrating, and learning, from the different national and regional responses—from India, China, Africa, Cuba, and Papua New Guinea, amongst other places. Cross-country comparative analyses abstract lessons on the political and socio-cultural factors that determined national responses to COVID-19. Most of us hunkered down for much of the first 2 years of the pandemic, focused on the nightly updates from our own national experts—such as Dr. Tony Holohan in Ireland and Dr. Anthony Fauci in the United States. Opportunities to learn from other countries were often missed.

Several contributions summarize generic lessons from the pandemic regarding how society is structured. Infectious diseases are not the great levellers of society. They discriminate between those who can protect themselves through better housing, and those who cannot—the single most important factor, along with good nutrition, in the 19th-century victories against tuberculosis in Ireland, Europe, and the Americas. In this context, most notorious in the COVID-19 pandemic has been the inequitable access to—and to the means of producing—vaccines. Health inequity has for long been determined by income at the macro (national) and micro (individual) levels, and by the insidious nature of racism: comparative poverty and racism act as determinants of health, and of our health-care and socioeconomic responses to disease.

Many of the chapters rightly deal with the nuts and bolts of COVID-19 responses. The basics of control of an airborne infection—physical distancing, masks, and face and hand hygiene—continued to be overlooked in Ireland and elsewhere up to the end of 2021, as people clutched at other (often counterproductive) ways of restoring normality to their lives; or as they saw opportunities to advance extreme—usually right-wing—political agendas. The importance of governance responses is also recognized in this book.

"*Outbreaks, Epidemics, and Health Security: COVID-19 and Ensuring Future Pandemic Preparedness in Ireland and the World*" will help build the foundation of lesson-learning that all countries, without exception, need to do to be better prepared for the next pandemic. Its strength lies both in the diversity of examples and in the overview of the different dimensions of national and regional responses to the pandemic. Its lessons also extend beyond the COVID-19 pandemic in that the fundamental inequities revealed will also undermine how the world responds to issues such as climate change.

Some difficult analyses need to follow, in particular on the political and social factors that underlay the delayed, ineffective, and incoherent responses of many long-standing democratic states, such as the United States and the United Kingdom. And then there were the chaotic responses of the democratically elected populist governments of Brazil and India—yet it would be an oversimplification to assume that autocratic states are best designed to respond quickly to pandemics. China's hubris, as a result of its successful initial control of COVID-19 in 2020, undermined its response to the changing dynamics of the infection in 2022. Living with the omicron variant, through a combination of efficacious vaccines and natural infection-induced immunity, proved to be a more effective strategy than China's original zero-COVID strategy. As infection dynamics evolve, so must control strategies, guided by public health principles, not political hubris. There are also cultural factors to be teased out in comparing Asian and 'Western' responses. But, either way, we do need to shine an analytical light on the inability and failures of many democratic states at protecting their populations.

Whilst such political analyses of the global and national responses to the COVID-19 pandemic need to follow, what this book has achieved is to make the case for countries to invest in a strong public health competency, based on principles of equity and health as a human right, at the centre of political decision-making. We have learned enough to conclude that a *cordon sanitaire* response to pandemics will ultimately break down and that, as a global community, we cannot be safe until we are all safe.

Ruairí Brugha
RCSI University of Medicine and Health Sciences, Dublin, Ireland

Preface: Three interlocking functions for stronger health security

The underlying theoretical approaches and strategies to global health security constantly evolve in response to actual pandemics, and understanding this evolution will help us prepare for the next pandemic. The first part of this preface traces the evolution of global health security, with each recent pandemic adding new facets to global health security. The second part of this preface proposes the three interlocking functions for stronger health security in response to any pandemic, such as COVID-19, monkeypox or other emerging infections.

The history of global health security is the history of added facets

In 1951, the 4th World Health Assembly approved the International Sanitary Regulations (ISR). The ISR consolidated several international sanitary conventions then in place, which were mostly colonial instruments [1]. The ISR required all countries to notify the World Health Organization (WHO) within 24 h of a local area becoming infected with six diseases (plague, cholera, yellow fever, smallpox, typhus, and relapsing fever).

In 1969, ISR was changed to the International Health Regulations (IHR) [2], as scientists and health leaders realized that preventing infectious diseases is much more complex than merely ensuring sanitary conditions. Broadly speaking, the ISR and IHR depended on border controls to prevent the international spread of outbreaks caused by cholera, plague, smallpox, and yellow fever.

In 2002–03, the world was hit by the severe acute respiratory syndrome (SARS) epidemic, at a time when the world economy was increasingly interlinked through the forces of globalization, global trade, globalized supply chains, and cheap mass air travel. Although most of the infections occurred in Asia and the final count was approximately 8000 infections and 800 deaths worldwide over the 9-month period [3], the outbreak led to a fundamental rethinking of health security around the world.

The lesson from SARS was that border controls were unable to stop infections in a globalized world. Therefore, global health security strategies added another facet—one of strong national public health systems to prevent, detect, and respond to outbreaks at the source in order to prevent their spread. This requires strong public health surveillance, epidemic intelligence, laboratory networks, and a command-and-control infrastructure.

The timing of SARS during the IHR revision process (which began in 1995) led to several major changes before the revised IHR was approved in 2005. First, the strategy shifted from 'controlling diseases at borders' to 'detecting and containing diseases at source', recognizing that infectious diseases could cross international borders in a globalized world whilst still in the incubation period. For this, the IHR established required national core capacities for disease detection, surveillance, confirmation, control, and communication. Second, the list of diseases requiring mandatory reporting was enlarged from the historical focus on cholera, plague, and yellow fever (smallpox had been removed from the IHR after its eradication) to include all potential public health emergencies of international concern (PHEICs). And third, the Joint External Evaluations (JEE) mechanism was created to help countries monitor progress towards these core capacities.

The 2014–16 Ebola epidemic tested the IHR 2005. Similar to SARS, Ebola was concentrated in one region (West Africa) and had relatively few cases compared to COVID-19 (although it was deadlier than SARS with approximately 29,000 cases and 11,000 deaths over 3 years) [4]. The Ebola outbreak in West Africa also saw a collapse of health services, causing an increase in death from Ebola and also from other infections such as malaria because populations could not access health care when needed. The West African Ebola outbreak therefore had wide-ranging impacts on global health security, leading to the understanding that there needed to be one more facet to global health security, that of resilience and surge capacity in healthcare services including hospitals, intensive care units, and primary care along with the necessary human capital.

In summary, the historical IHR focused on border controls, SARS added the facet of strong public health infrastructure, and Ebola added the facet of resilient healthcare delivery infrastructure.

COVID-19 and individual health security: Healthy populations

SARS and Ebola were short-lived and limited in geographic scope. The prolonged duration, the disproportionate serious illness amongst those with comorbidities and the elderly, and the global scale of COVID-19 demonstrate that strong public health and resilient healthcare delivery infrastructure alone are insufficient to respond to pandemics. Therefore, consensus is developing that preparedness must also be concerned with the health status of individuals as a basic building block of health security: healthy populations can, for pandemics such as COVID-19, better resist serious illness after infection.

Healthy populations require an all-of-government approach to build, implement, and sustain Health in All (Public) Policies. This is especially crucial as populations age, and it requires strong attention to the social determinants of health. The sum total of Health in all Public Policies is a healthy population with proper nutrition, a minimum of noncommunicable disease, and robust mental health.

Therefore, as countries continue to grapple with COVID-19 and imagine a postpandemic health architecture, three interlocking functions appear to be required for stronger health security at the national level. One, countries must build resilient healthcare systems with built-in surge capacity (including for primary health care). Two, countries must build resilient public health core capacities that meet IHR standards. And three, countries must deliver proactive investments and policies to enable healthy populations.

Each of these three interlocking functions has strong conceptual frameworks and large bodies of evidence [5–7]. They are also familiar to health agencies at the national, regional, and international level and backed by strong international commitments like the United Nations High-Level Meeting on Universal Health Coverage and the WHO's Global Action Plan for Healthy Lives and Well-Being for All. They also mirror the WHO's Triple Billion strategy [8] and Sustainable Development Goal (SDG) #3 of Good Health and Well-being and provide a pragmatic 'middle path' for countries without forcing them into false choices between public health, healthcare delivery, and healthy populations.

Measuring the three interlocking functions could be accomplished by consolidating their metrics and indicators into existing frameworks, like IHR, WHO's UHC Service Coverage Index [9] and World Health Statistics [10], or the SDG progress reports by the United Nations Department of Economic and Social Affairs (UNDESA). This will help institutionalize the three interlocking functions into familiar existing frameworks without having to reinvent the wheel. Implementing the three interlocking functions also requires a broad coalition of health and nonhealth stakeholders, backed by high-level political commitment at the United Nations General Assembly, with or without a pandemic treaty.

To implement the three interlocking functions, countries should first hold themselves accountable with strong enforcement mechanisms. The JEE, for example, can converge with other global enforcement frameworks, such as country obligations for human rights, labour rights, international trade or finance, or climate change [11]. Enforcement could be a mix of self-reviews (like the Human Rights Council Universal Periodic Review [12] process), peer reviews (like the World Trade Organization Trade Policy Review Mechanism [13]), and external reviews (like the Paris Agreement for Nationally Determined Contributions to reducing carbon) [14].

Second, countries should ensure sustainable funding and capacity-building through government engagement and funding, matched in low- and middle-income countries (LMICs) by donor agencies and backed by systematic ways to build capacity for leadership, governance, and systems strengthening. In other words, we must support all countries, especially LMICs, to strengthen their public health, healthcare delivery, and population health through adequate and sustainable financing and capacity-building.

Third, countries should build a strong partnership to address One Health issues. Formal collaborations between public health, medical, veterinary, environmental health and other relevant specialties in One Health are crucial for pandemic preparedness in particular and population health in general. This integrated approach will allow for a more robust approach to prevent, detect and respond to threats "at source".

COVID-19 will not be the last large-scale and global pandemic, and countries must ensure their health security with interlocking healthy populations, resilient health systems, and strong public health. It will take political will, skill, and speed if the world is to move faster than its pathogens.

Swee Kheng Khor
David L. Heymann

References

[1] International Sanitary Regulations, World Health Organisation Regulations No 2, World Health Organisation, 21 May 1951. Retrieved from: https://apps.who.int/iris/bitstream/handle/10665/101391/WHA4_60_eng.pdf?sequence=1&isAllowed=y. (Accessed 16 July 2021).

[2] D.P. Fidler, From international sanitary conventions to global health security: the new international health regulations, Chin. J. Int. Law 4 (2) (2005) 325–392, https://doi.org/10.1093/chinesejil/jmi029.

[3] https://www.cdc.gov/sars/about/fs-sars.html. (Accessed 16 July 2021).

[4] https://www.cdc.gov/vhf/ebola/history/2014-2016-outbreak/index.html. (Accessed 16 July 2021).

[5] M.E. Kruk, M. Myers, S.T. Varpilah, B.T. Dahn, What is a resilient health system? Lessons from Ebola, Lancet 385 (9980) (2015) 1910–1912.

[6] H. Kluge, J.M. Martín-Moreno, N. Emiroglu, et al., Strengthening global health security by embedding the International Health Regulations requirements into national health systems, BMJ Glob. Health 3 (2018), e000656.

[7] D.L. Heymann, L. Chen, K. Takemi, D.P. Fidler, et al., Global health security: the wider lessons from the west African Ebola virus disease epidemic, Lancet 385 (9980) (2015) 1884–1901.

[8] World Health Organization, Triple Billion Dashboard. https://www.who.int/data/triple-billion-dashboard. (Accessed 4 July 2021).

[9] World Health Organization, UHC Service Coverage Index. https://www.who.int/data/gho/indicator-metadata-registry/imr-details/4834. (Accessed 4 July 2021).

[10] World Health Organization, World Health Statistics. https://www.who.int/data/gho/publications/world-health-statistics. (Accessed 4 July 2021).

[11] S. Lehtimaki, A. Reidy, K. Nishtar, S. Darehshori, et al., Independent Review and Investigation Mechanisms to Prevent Future Pandemics: A Proposed Way Forward, United Nations University International Institute for Global Health. April, 2021.

[12] United Nations Human Rights Council, Universal Periodic Review. https://www.ohchr.org/en/hrbodies/upr/pages/uprmain.aspx. (Accessed 4 July 2021).

[13] World Trade Organization, Trade Policy Review Mechanism. https://www.wto.org/english/tratop_e/tpr_e/tp_int_e.htm. (Accessed 4 July 2021).

[14] United Nations Framework Convention on Climate Change, Paris Agreements, Nationally-Determined Contributions. https://unfccc.int/process-and-meetings/the-paris-agreement/nationally-determined-contributions-ndcs/nationally-determined-contributions-ndcs. (Accessed 4 July 2021).

Acknowledgements

We acknowledge Professor George Rutherford, Professor David L. Heymann, Professor Ruairi Burgha, Dr. Deon Canyon, the Eng family, the Kevany family and Dr. Jeremy Emmet-Booth.

Part I

Big picture lessons

Chapter 1

The irreversible global politicization of health

Sebastian Kevany and Aoife Kirk

Chapter outline

1.1 Health and politics

In the light of the events of 2020 and 2021, is it possible that the concepts of public health and health security, rather than fading away, will become more and more an issue for 21st-century Ireland and many other countries. In recent decades, the green and environmental movements have become powerful forces in modern politics [1]. In this way, related concerns regarding climate and planetary health have managed to permeate, with good reason, all aspects of public policy. If this is the case, such developments stand to be potentially divisive ones, ones on which political parties and reputations may stand or fall.

No longer is an issue such as air pollution the sole responsibility of select government departments. Instead, it is a consistent consideration and tangible presence across sectors and within government departments, for example, ministries of housing, health, foreign affairs, finance, and many others. Such divisions of departments within the civil service and in the political arena may also reflect attitudes towards public health, epidemic control, or health security. Perhaps these attitudes will not be on the scale of Brexit in the United Kingdom—but, nonetheless, these are considerations that threaten to split public opinion into extreme and possibly polarized camps.

Outbreaks, Epidemics, and Health Security. https://doi.org/10.1016/B978-0-323-91705-6.00001-0

During the lockdowns across Europe, there were those who lobbied for airline profitability and tourism; those who lobbied for pub and restaurant openings; and those who lobbied for a return to office and workplaces. There were policies in certain countries to charge citizens who broke the COVID-19 national lockdowns or precautions; controversies over benefits to those who became unemployed during lockdowns; and political missteps regarding ignoring public health advice at various stages.

On the other hand, there were also those who were primarily concerned with children's education; and those who would prefer efforts and resources across all government departments to have placed public health on a pedestal first, rather than gross national product (GNP), although the two invariably interrelate. How do these conflicting matters relate and interrelate in the context of global health security [2]?

Such trade-offs are not easy to predict. Those who claim that the economy has to be put before public health may have made short-sighted gambles at that time. Should epidemic conditions develop in certain specific ways, the reverse may, in fact, be true: countries such as New Zealand, that managed to keep their epidemics under close control in 2020–21, profited from a significant public health dividend as the restaurants, pubs, and sporting events remained open for most of the year, whilst the rest of the world looked on.

Suppose this effort, which was exemplified, can be combined with efficient and effective border testing, a public health workforce, and healthcare staff and properly funded services, low-risk countries may become increasingly attractive for tourism—without putting home populations at risk. GDP and economic performance can only improve if built on a healthy society to start with—healthy countries result in healthier economies.

Perhaps, Anthony Fauci had the best answer of all. Throughout his dealing within matters of politicization and polarization, his view remained: "You must stay completely apolitical and non-ideological. I'm a scientist, and I'm a physician. And that's it." Fauci was the lead of National Institute of Allergy and Infectious Diseases in the United States over the last 36 years and a scientist-public health activist who also worked with US administrations on various public health crises, including HIV, SARS, the Avian flu, Zika, and Ebola.

Yet, even though he remained high in the approval ratings of the American public for his position on advising the Trump administration, he has not been immune to critics. Media, journalism, and general public attitudes have all played a remarkable role during this pandemic crisis—a position which has often led to the highlighting of misinformation, which is a lesson we may wish to ensure does not repeat itself [3].

Yet, even with such seemingly straightforward demarcations in attitudes—many of which will likely become more entrenched—further considerations are at play. Other elements of epidemic and infection control debates raise the stakes even further, as related events and policy decisions play into arguments about immigration, hard borders, diplomacy, international relations, surveillance societies, and even housing—issues which can now all be traced back to health and health security issues, in some way.

PICTURE 1.1.1 **Walks and water during lockdown.** Aoife Kirk.

1.2 Guest essay: Global health inequalities, crises, and governance

German Andres Alarcon Garavito

Globalization has incited a vast number of interconnected problems that affect the world as a whole, crossing international frontiers and generating large-scale risks. One example of this is in the domain of infectious diseases, which requires both improved local health systems and cooperation between nations to address outbreaks. Although many risks are associated with globalization, there are also new opportunities to achieve worldwide partnerships and collaboration. Whilst countries may undertake significant efforts to improve the health of citizens, political and practical capacity is affected by globalization. Global governance seems to be a key answer to manage poor quality and ineffective local governance actions [4]. Additionally, though the current globalization growth should represent a proportional crisis for all the countries in a connected world, there is still a socio-economic and development division between the northern and southern hemispheres. The unequal distribution of mortality, morbidity, burden of diseases, and the socioeconomic consequences of these conditions are highly correlated with the colonial foundations of the international sanitary institutions that reinforce hemispheric disparities [5].

This essay will define global governance, its relationship with global health, and global health's evolution. It will discuss how global governance is excessively influenced by global power nations and multinational companies located in countries of the northern hemisphere, such as the United States or the European countries; how this domination has expanded the social inequalities in low- and middle-income countries (LMICs); and in what manner some of the main features of global governance, such as leadership, participation, and the rule of law could represent a key solution or could aggravate local and global health governance issues.

Global health background

According to Hoffmann and Ba (2005), global governance can be defined as a phenomenon that seeks to manage global problems [6]. Some authors claim that there is a need for dealing with collective problems through global institutions such as the United Nations (UN) and specifically in the area of health, the World Health Organization (WHO). The WHO's command is valuable because of its almost universal recognition by sovereign nations. Hence, international coordination regarding health threats via the WHO's programmes and prearranged mechanisms are facilitated more easily than individual interventions regarding similar hazards [7]. International cooperation is thus a key element of the WHO's capacity.

However, others argue that the importance of organizations like the WHO is limited because their agenda is not shaped equally, but mostly by the members who offer funding. Another criticism levelled at the WHO is that whilst it tries to intervene in the negative consequences of globalization in LMICs, it does not take coordinated actions to decrease the northern-southern disparities, which commonly intensify globalization's consequences in the southern hemisphere [6]. These regional differences are not new considering that since the first international sanitary conferences in the 19th century, a strong dominance of industrialized countries over developing ones had existed [8]. The evolution of global health has had an evident influence on global health policies [9]. The entire story of global health has been a challenge to decouple itself from its colonial background, from tropical medicine's attempts to preserve the lives of White people and the workforce in colonies established in developing countries, to international health's objective of studying health problems in LMICs and not in high-income countries (HICs) [10]. Historically, colonialism's influence on global health has increased the gap between HICs and LMICs and, specifically, the gap between northern and southern hemispheres. The 'rich North' where HICs are located has been building a large part of its economy in the extraction of natural resources from the 'poor South' (LMICs) [11]. This phenomenon may be attributed to massive flows of capital moved transactionally and the influence of multinational corporations, not globalizing wealth but rather globalizing poverty [5]. Global poverty is linked to deficient governance and socio-economic inequalities, which are boosted by globalization and capitalist processes [12].

Health in global governance

There is an overlapping division within global governance, which includes global environment, health, and economic governance. This highlights the complexity of multisectoral governance and explains some of the difficulties governments may encounter in allocating funding [12]. However, separating sectors that are interconnected generates a lack of coordination during the execution of strategies, decreasing efficiency and effectiveness. In an attempt to combat this, activities that in the past were considered to be part of other sectors, such as climate change, violence and conflict, or international trade, are now under the health sector scope too, for example, and can be seen within the UN's structure. Every day, institutions such as the World Bank are more committed to health-related problems as well [4].

The WHO, as the major global health governance player, collaborates with diverse participants like nongovernmental organizations (NGOs) (which had not been deeply consulted before 1990) or other associated institutions, for example, UNICEF. Though UNICEF and other UN organizations have been participating more in global health governance, their scope and participation are still limited and, on occasion, not coordinated with the WHO's actions. Alternative but less known actors are the civil groups, like research organizations (e.g. universities) or collective movements (e.g. HIV/AIDS activism), which influence health policies either formally or informally. Similarly, collaborative efforts between the private and public sectors are frequent [4].

Though having many different groups trying to improve global health is important to accomplish the goal, their disjointed efforts can result in less efficient change and a redundant use of resources.

Missing global health leadership

There is also a bigger controversy regarding the role of the WHO as the principal stakeholder of global health governance. Whilst the WHO has played an important historical role in global health, recent arguments against the WHO's leadership have been increasing [13]. Crises like the AIDS pandemic, Zika virus pandemic, and even the current coronavirus (COVID-19) pandemic have caused global concern regarding how the WHO responds to crises. Aside from political clashes between global players like the United States and China, the WHO has been accused of being strictly hierarchical and closed to the power of networks and innovation during the COVID-19 crisis [14]. Its critics are requesting significant reforms, some of which are calling for institutional change and governance centred on networking.

A notable example of networking and institutional innovation is the Open Source Drug Discovery (OSDD) in 2008 [14]. The initiative proposes public and open access to collaboration spaces to discover and develop new affordable drugs to treat neglected tropical diseases (NTDs) in LMICs [15]. Innovation activities like this provide beneficial opportunities to encourage open development of new medicines without private interference and represent a deviation from typical drug development, which is often sponsored by multinational companies [14].

Although the global economy is shaped by these companies, the WHO can reassume its leadership role as the principal global health representative and confront this problematic structure. Running strategies, such as open-source projects, are just one example of how to deal with this structure. This will require determination from the WHO and possible extra funding from international agencies and governments. However, it will generate public access to transparent innovations in the treatment of significant diseases such as malaria or tuberculosis, which are by their very nature a good example of the global inequality in health. Malaria cases in the last 5 years have been primarily located in the African region, which accounts for 94% of all malaria cases and deaths [15]. Lack of access to antimalarial medicines due to reduced accessibility and financial constraints are signs of how unequal health can be in LMICs.

Global participation

Participation in decision-making is vital not only for health policy processes, but also for global health governance. If citizens are not heard or empowered to defend their collective interests and rights, policies will not be beneficial for them, or effective for the governments responsible for them. Besides, no significant participation could be accomplished if there are no ways to communicate. Information governance is a crucial strategy to increase the channels of discussion and thus achieve considerable participation from not only citizens but also other participants such as pharmaceutical companies and NGOs [16].

Some notable strategies to enhance participation include consensus conferences, national health forums, and formal consultations during legislature drafting, amongst others [17]. Another example is global public-private partnerships (GPPPs). GPPPs are an ideal approach to include more actors in global health initiatives and reduce inequalities in deprived populations. This is exemplified in numerous accesses to medicine programmes like the Mectizan Donation Programme. Since 1897, the American pharmaceutical company Merck has donated millions of doses of Ivermectin, used to treat onchocerciasis and other NTDs in LMIC in South America and Africa. As a result, onchocerciasis was eradicated in four countries in Latin America and in one country in Africa [18]. This was achieved through a collaboration between this company, health ministries, local NGOs, and the WHO.

On the other hand, some detractors from NGOs, LMICs, and even UN agencies have expressed concerns about GPPPs. It has been reported that private institutions' objectives rarely match with the UN's or NGOs' interests [4], which implies a risk of corruption in consequence of the misuse of these partnerships.

To recap, trust in government appears to be higher when participatory practices are included across all government actions [19]. Participation is one of the most important processes to achieve good governance. Strategies such as GPPPs are vastly useful to enhance involvement in global health governance, but they may require better planning and oversight to ensure that all the actors are focused on one common goal, namely, to address global health inequalities, and not in particular interests to gain profit.

The rule of law

The rule of law is defined as the mechanism controlling how the state and its citizens are held responsible under consistent and uniformly implemented laws that follow international human rights guidance, procedural and legal transparency, and equality [20]. The rule of law plays an important role in attempting to achieve global health equity [21], and it can be used in attempts to reduce inequities in the most vulnerable communities, thus helping in the pursuit of global health justice [22].

Regrettably however, the rule of law as a main feature of governance has been problematic, considering the fact that even the law itself might be insufficient and can be easily avoided by wealthy individuals [23]. Performance standards of law such as accessibility or transparency are minimal; thus, the most vulnerable individuals are the ones whom the law fails the most [24].

Countries with a strong and well-structured rule of law tend to have better public health systems. It has been demonstrated that the higher the rule of law index, the higher the life expectancy, and the lower the rates of maternal deaths and chronic diseases [25].

When the rule of law is actively achieved, it is a determining factor to ensure citizens in LMICs are not denied access to healthcare services. The way to accomplish this is with active legislation [26]. National constitutions should enshrine the right to health and access to essential features like water, food, and housing. This not only protects the most vulnerable individuals but also helps to reduce the social gap between populations.

The rule of law should be enhanced in LMICs to implement macro measurements that generate high impact solutions. One example of this occurs in Colombia, where the use of legal actions as principal tools to expand the judicialization of health has developed the Universal Health Coverage (UHC) panorama. UHC increased from 24% in 1993 and 80% in 2007 [23], to 96% in 2015 [27].

Hence, it is important to highlight the key role of law and justice in LMICs to guarantee access to health care and quality of delivery. The Colombian example shows a structured pathway to request compliance with the right to health, whilst other countries also combine this strategy with constitutional legislation to generate a higher impact in terms of coverage.

Conclusion

An interconnected world is often cited as a benefit of globalization. Unfortunately, the more connected the world is, the more divided it apparently becomes. Considering that global disparities can be presented not only between countries but also between entire regions, it is no surprise that the inequalities between the northern and southern hemispheres are chiefly caused by globalization processes.

Firstly, some multinational corporations' practices are important components of this divided and unequal world. Developing their businesses in LMICs for extractive purposes, generating cases of corruption, excessive accumulation of wealth, and sometimes human rights violations are causes of inequality [28].

Secondly, global health evolution is undoubtedly linked to the colonialism process, which has been another consistent factor of globalization. It is a current challenge to start the decolonization of global health and thus think of global health problems as truly global and not just problems in LMICs that some HICs try to solve.

Thirdly, the WHO has been recognized as the leader in global health governance since its foundation. However, due to some unreliable performance indicators during regional and global health crises, its manner of functioning and even its existence have been questioned. The WHO is accused of having lost its leadership role due to processes that are not opened to innovation but attached to unnecessary bureaucratic practices. This has caused a perception of missing authority. Open-source strategies, GPPPs, and participation of citizens can help the WHO regain its directive role.

Overall, this essay argues that good global governance can be achieved with structural change and strategic measures to enhance participation. Strategies to connect nonstate, state, and civil actors appear to be efficient. Whilst the GPPPs' results are proof of this, it is vital to ensure sufficient supervision to prevent corruption. Finally, the rule of law should be promoted, protected, and reinforced in some LMICs, but it has shown to be a significant legal approach to increase UHC and remove institutional barriers to healthcare services.

1.3 Changing political stances

One of the greatest ironies in this new political *milieu* was the new alliances that have been forming under this time of strain—often unusual partnerships, to say the least. For example, those more to the left of the political spectrum—who may have traditionally advocated for increased public spending on health, but who might also hold more liberal views on surveillance or borders—were forced, based on outbreaks, to reconsider their positions. This ironic switch of partnerships between public health and conservatives played out in the United Kingdom, for example, where politicization and polarization perhaps often undermined the argument of the importance of expert objective knowledge in early 2020 [29].

The political battlefield in the United Kingdom, one which has been even further polarized since the infamous Brexit referendum, may result in a national investigation in future. The question remains—would this investigation conclude that political decisions resulted in delayed decisions of enforcing lockdowns? An example was the national announcement of cancelling the planned Christmas relaxations of lockdown for London and South East England by the Prime Minister on 20 December 2021—merely 5 days before Christmas—which arguably resulted in more severe restrictions than would have been necessary.

PICTURE 1.1.2 COVID airports. Eimear Bourke.

Future epidemic control may thus, ironically, be served better in future by the interests of those conservative political elements which are traditionally less supportive of public health. Infection control may be associated with less movement of human populations and therefore more rigid borders. Perhaps, public health may be served by a society with more surveillance, one in which personal data would be fed back to the general public with details regarding precise outbreak numbers and locations? The question remains if several political realignments will have to be made instead of public health actions at the forefront of the political sphere.

But, as we have seen worldwide and in our local environments, both public health and security considerations have also been used—sometimes rightly, sometimes wrongly—to advance other agendas. Councils wanting to limit access or change parking protocols have, in many cases, leveraged the opportunities presented by the epidemic. Cyclists petitioned for further bicycle lanes; local businesses began demanding pedestrianization and an increase of public space for outdoor dining. Using some of these insights, certain cities (e.g. Paris and Barcelona) decided for the interim of summer 2021 following the first two waves of the COVID-19 pandemic to encourage the motto of 'build back better', which entailed putting actions in line with the SDGs at the forefront of their agenda [30]. Other places, however, remained eager to return to commercial, corporate, and capitalist interests as soon as possible.

PICTURE 1.1.3 **We all became familiar with the distance of 6 feet, or 2 m.** Aoife Kirk.

1.4 Guest essay: COVID-19 opinion piece on the Irish experience of COVID-19

Ina Kelly

As a specialist in public health medicine who has worked in investigation and control of infectious diseases for years, including during the SARS (2003), H1N1 (2009), and Ebola (2014) pandemic periods, I had thought that globally we were prepared for early detection, effective control, and elimination of 'pandemic potential' viruses. When I saw the early warning and response system (EWRS) report on 31 December 2019 of the possibility of such a virus, I expected effective elimination action as there were early indicators of concern—R_0 appeared to be greater than 2, and mortality and morbidity were of percentages likely to incapacitate even the highest functioning health services, given the exponential spread likely.

Instead, there were delays and downplaying of the severity of the emerging virus, which undermined its timely and effective control. This appears to have been a preventable pandemic, and it remains to be measured what the magnitude of the mental, physical, and social health impacts on the global population will turn out to be.

Within Ireland, the adverse impact has been very severe in many ways, with 5609 deaths to date affecting primarily our older citizens, including more than 90% of deaths in the 65 years or older age group, and many outbreaks in residential care facilities. We have had significant numbers of people hospitalized, with a proportion of them requiring ventilation in intensive care units. Then, we also have ongoing impacts from 'Long COVID' (unmeasured as yet) as well as the mental and physical health impacts due to the challenges of living a healthy and happy life during a pandemic.

Ireland has done well in some things during the pandemic—we have had a lot of community solidarity, limited division amongst experts, and there appeared to be a high level of trust in public health advice, which came through the national Chief Medical Officer. The health service has provided a very impressive coordinated vaccination programme, as well as massively enhanced testing and contact-tracing services. Ireland also brought in pandemic unemployment payment and additional business support that in many situations assisted the control of COVID-19 and supported our people through these almost unprecedented times.

However, COVID-19 has affected our already overstretched health service that prior to COVID-19 had long waiting lists for many nonurgent diagnostics and treatments. Waiting lists have worsened, and our healthcare workers, like in every other country, are exhausted, without the hope of respite at present.

From a public health perspective, one of our most understrengthened and underresourced services prior to the pandemic was public health medicine (PHM). We in PHM have been in a process since early 2019 of trying to negotiate more attractive terms and conditions to bring this neglected branch of medicine up to a state of equality with other branches of medicine with the help of our union, the Irish Medical Organization. The pandemic raised the public and media profile of our essential but largely invisible service, but it was frequently demoralizing to have to fight an industrial relations battle when we were so busy fighting the COVID-19 virus.

The pandemic has been transformative in many ways as we now have an agreed reform programme for PHM and health protection; the other domains of PHM are more widely valued than before. We have had increased multidisciplinary staffing which will allow us to widen our remit and increase our efficiency and effectiveness.

Whilst we have quite comprehensive statutory functions to protect health, their implementation has been very patchy—our legislation is old, and this suggested to many that our laws did not need to be implemented. However, these laws were demonstrated to be essential to control COVID-19, while also providing the legal basis for related areas of the General Data Protection Regulation (GDPR).

Health inequalities came to the fore during our efforts to control COVID-19 outbreaks as we found that essential workers, who needed to work in congregated settings throughout the pandemic, were often low-paid, living in crowded or shared accommodation, with few choices to protect themselves. Many in these groups are migrant workers, so language differences and the influence of international misinformation have been barriers that we have been striving to address. We have seen increased risks of infection in many other congregated settings such as asylum seeker accommodation, which stimulated greater efforts to reduce the density of residents in such settings. Disenfranchised groups such as Irish Travellers and Roma[a] were more affected by large outbreaks in their communities, having less information about COVID-19 and vaccines, and possibly having less trust and engagement with health services. All of the above have offered opportunities for public health services to work with these groups and try to provide more accessible and equitable services.

At this time, Ireland is enduring a fourth wave of COVID-19, with most hospitals having no spare ICU capacity. We have high rates of adult vaccination, and many of those vaccinated are now getting booster doses, but it is less clear how to get this wave under control, as we have opened up society and there appears to be less public support for further lockdowns. In addition, the international context of inequitable vaccine access is a global health security problem, apart from being an ethical obligation that remains unfulfilled.

Prior to COVID-19, my public health role included contributing to climate action and it is clear that addressing both this pandemic and climate change is going to require significant global transformation—political, economic, social, and technological. Both of these global public health issues require strong global public health leadership including to advocate for the need for global empathy and solidarity; drive integrated and equitable action internationally; highlight the health costs of delays, division, and misinformation; and advise on and monitor the investment in effective health security in every country in the world.

This is a huge challenge, but there is also an opportunity to address some of the global unfairness that, because it has been less visible, has continued. There is a spotlight on it, so now is the time for public health in every country to call for and support integrated action to address these existential issues for many populations in our world.

[a] Minority groups in Ireland.

1.5 Health, politics, and the economy

How do we begin to examine a country's productivity? How can we cross-compare? Gross domestic product (GDP), a concept that was coined less than a 100 years ago following the Depression in the 1920s, was created to measure manufacturing productivity. Yet, today, perhaps with GDP, we are not optimizing the effects of society, in particular as the GDP metric often has an unconscious bias against the elderly, women, and lower socio-economic classes. Whilst GDP metrics have their purposes, perhaps a proposal that includes biodiversity as a form of GDP, or measures of health—or maybe even a metric that measures the inclusion of sick pay, unpaid care work, or intersectionality—is more appropriate for the 21st century.

A further set of examples of advancing agendas relates to housing and rental rates: should remote working become commonplace in future, even if only partially, logic suggests that a significant amount of commercial real estate will be freed up in a way that, with some effort, might help to address housing issues—thereby, in turn, driving down rents and property prices. Unfortunately, a combination of increased costs and low supply has resulted in other cities worldwide experiencing a massive surge in housing prices [31].

Some will welcome this, others not—the construction sector may slow down; those who bought houses just before the epidemic may end up paying more than those who do so in the aftermath. Moreover, health and economic consequences have shown us throughout this pandemic that perhaps not all are equal—raising questions about whether we should return to 'normality' at all. But, in all cases—whatever the rights or wrongs about how the situation plays out over the forthcoming few months—the theme remained the same: public health is an increasingly political issue.

A final consideration in this regard relates to populism—the election of candidates and parties on single-issue tickets such as domestic economic growth, national reunification, or leaving the European Union (to name but a few). Again, the political implications of the pandemic in this regard are clear: single-issue parties are not always to be trusted on other issues.

Our elected governments thus perhaps need to back away from extremism and overspecialization to include a broad range of skills, expertise, and agendas in their DNA. But, perhaps, this too requires an examination of the governments we are electing—are they diverse? Are they inclusive? Do they represent the local citizens' views? What are they driving forward on their agenda—is health on the table? How can cross-sector governments work to improve 'the new normal'?

Even scientists, who swear by fact and reason, have often been pawns on the battlefield of pandemic politics. Wearing masks, vaccines, and questioning the scientific data have all been elements within the pandemic which were questioned and often scorned. The more significant questions remain in post-pandemic landscapes as to how will societies build back from COVID—will it be progressive, improving societies in a green and sustainable manner, or are we to return to a society with deepening inequalities [32]?

PICTURE 1.1.4 Ireland's mask-wearing guidelines. Aoife Kirk.

Perhaps we need to reconsider Fauci's statement—and return to the understanding that it may require depoliticization and depolarization for public health to be at its most successful. Public health must remain apolitical and nonideological to continue fighting against future pandemics, prolonged sickness, and socioeconomic issues. Yet, is that in itself too idealistic? A road forward is looking towards building that new system, where we ensure universal access to health; fairer working conditions and workers with a living wage; social justice and challenging ageism; racism; climate action; and economic rights which may align for a stronger 21st century to continue public health gains.

1.6 Key points

- Both the left wing and right wing, in Ireland and around the world, have adopted entrenched political positions on public health as a result of the pandemic.
- Future global health crises should, wherever possible, be managed in a bi-partisan way, mindful of the divides they create.
- Only through political consensus and cooperation can effective public health responses, such as vaccination campaigns, be effective.

References

[1] L. Elliott, The Global Politics of the Environment, Macmillan International Higher Education, 2004.
[2] B. Williamson, R. Eynon, J. Potter, Pandemic politics, pedagogies and practices: digital technologies and distance education during the coronavirus emergency, Learn. Media Technol. (2020) 107–114.
[3] V. Von Wyl, Challenges for nontechnical implementation of digital proximity tracing during the COVID-19 pandemic: media analysis of the SwissCovid app, JMIR Mhealth Uhealth 9 (2) (2021), e25345.
[4] R. Dodgson, K. Lee, N. Drager, Global health governance, a conceptual review, Glob. Health 1 (2018) 439–461.
[5] O. Aginam, Global Health Governance: International Law and Public Health in a Divided World, University of Toronto Press, Toronto; London, 2005. 1–202 p.
[6] M.J. Hoffmann, A.D. Ba, Contending Perspectives on Global Governance: Coherence and Contestation and World Order, Routledge Taylor & Francis Group, 2005. 1–278 p.
[7] J. Frenk, S. Moon, Governance Challenges in Global Health, N. Engl. J. Med. 368 (10) (2013 Mar 7) 936–942. [cited 2021 Feb 16]; Available from: http://www.nejm.org/doi/10.1056/NEJMra1109339.
[8] C. Clinton, D.L. Sridhar, Governing Global Health: Who Runs the World and Why?, 2017. 282 p.
[9] J. Rey, P. Koplan, C. Bond, M.H. Merson, S. Reddy, M.H. Rodriguez, et al., Towards a common definition of global health, Lancet 373 (2009) 1993–1995. Available from: http://globalhealthedu.org/Pages/GlobalvsInt.aspx.
[10] S. Abimbol, On the meaning of global health and the role of global health journals, Int. Health 10 (2018) 63–65.
[11] S. Chatwood, P. Bjerregaard, Y.T. Kue, Global health—a circumpolar perspective, Am. J. Public Health 102 (2012) 1246–1249. [cited 2021 Feb 18]. Available from: http://ajph.aphapublications.org/.
[12] C. McInnes, A. Kamradt-Scott, K. Lee, A. Roemer-Mahler, S. Rushton, O.D. Williams, The transformation of global health governance, Acta Univ. Agric. Silvic. Mendelianae Brunensis 53 (2014) 1689–1699. Available from: http://publications.lib.chalmers.se/records/fulltext/245180/245180.pdf%0A. https://hdl.handle.net/20.500.12380/245180%0A. https://doi.org/10.1016/j.jsames.2011.03.003%0A. https://doi.org/10.1016/j.gr.2017.08.001%0A. https://doi.org/10.1016/j.precamres.2014.12.
[13] L.O. Gostin, E.A. Mok, et al., Br. Med. Bull. 90 (2009) 7–18. [cited 2021 Feb 16]. Available from: https://academic.oup.com/bmb/article-lookup/doi/10.1093/bmb/ldp014.
[14] R. Smith, K. Lee, Global health governance: we need innovation not renovation, BMJ Glob. Health 2 (2) (2017) 2016–2018.

[15] GMP/WHO, World Malaria Report 2019, WHO Regional Office for Africa, 2019. 1–232 p. [cited 2021 Feb 18]. Available from: https://www.who.int/news-room/fact-sheets/detail/malaria.

[16] I. Mikkelsen-Lopez, K. Wyss, D. De Savigny, An approach to addressing governance from a health system framework perspective, BMC Int. Health Hum. Rights 11 (1) (2011 Dec 2) 13. [cited 2021 Feb 16]. Available from: https://bmcinthealthhumrights.biomedcentral.com/articles/10.1186/1472-698X-11-13.

[17] E. Barbazza, J.E. Tello, A review of health governance: definitions, dimensions and tools to govern, Health Policy 116 (2014) 1–11.

[18] Mectizan Donation Program, Achievements I Mectizan Donation Program, 2015, [cited 2021 Feb 18]. Available from: https://mectizan.org/what/overview/.

[19] M. Andrews, The good governance agenda: beyond indicators without theory, Oxf. Dev. Stud. 36 (4) (2008) 379–407.

[20] U. Points, Good governance in the process of public health law reform, in: Public Health Law Reform, Good Governance and Human Rights, 2018, pp. 73–78. Washington, (Chapter 5).

[21] J.P. Ruger, Global Health Justice and Governance, Oxford University Press, 2018. 1–394 p.

[22] K. Lee, A. Kamradt-Scott, The multiple meanings of global health governance: a call for conceptual clarity, Glob. Health 10 (1) (2014) 1–10.

[23] M. Barendrecht, Rule of law, measuring and accountability: problems to be solved bottom up, Hague J. Rule Law 3 (2) (2011 Jun 4) 281–304.

[24] T. Besley, M. Kudamatsu, A. Merlo, B. Olken, N. Nunn, Health and Democracy, American Economic Review, 2006, pp. 313–318.

[25] World Justice Project, The Twin Crises of Public Health and the Rule of Law, 2020, Available from: https://worldjusticeproject.org/our-work/publications/policy-briefs/twin-crises-public-health-and-rule-law.

[26] L.O. Gostin, K.O. Cathaoir, S.A. Roache, R.S. Magnusson, R. Krech, S.A. Solomon, et al., Advancing the right to health—the vital role of law, Am. J. Public Health 107 (2017) 1755–1756. [cited 2021 Feb 11]. Available from: https://www.hhri.org/publication/advancing-the-right-to-health-the-vital-role-of-law/.

[27] R. Guerrero, S. Prada, A.M. Pérez, J. Duarte, A.F. Ahuirre, Universal health coverage assessment: Colombia, in: Global Network for Health Equity (GNHE), 2014. Available from: http://gnhe.org/blog/wp-content/uploads/2015/05/GNHE-UHC-assessment_Indonesia.pdf.

[28] E. Giuliani, Why multinational enterprises may be causing more inequality than we think, Multinatl. Bus. Rev. 27 (3) (2019 Sep 16) 221–225.

[29] D. Fancourt, A. Steptoe, L. Wright, The cummings effect: politics, trust, and behaviours during the COVID-19 pandemic, Lancet 396 (10249) (2020) 464–465.

[30] H. Clark, A. Gruending, Invest in health and uphold rights to "build back better" after COVID-19, Sex Reprod. Health Matters 28 (2) (2020) 1781583.

[31] D. Rogers, E. Power, Housing policy and the COVID-19 pandemic: the importance of housing research during this health emergency, Int. J. Hous. Policy 20 (2020) 177–183.

[32] S.L. Greer, E.J. King, E.M. da Fonseca, A. Peralta-Santos, The comparative politics of COVID-19: the need to understand government responses, Glob Public Health 15 (9) (2020) 1413–1416.

Chapter 2

Fortress Ireland

Sebastian Kevany and Aoife Kirk

Chapter outline

2.1 Would closing borders have worked?

During late November 2020, it became increasingly apparent in Ireland and abroad that public and government consensus was moving back towards prioritizing normal social functioning for schools, social functions, and sporting fixtures over ease of tourism and travel. There appeared to be a direct trade-off here—easing one exacerbates the public health issues of contact tracing and curbing the infection rate. However, stricter border efforts—such as those employed by Iceland, New Zealand, or indeed at times the United Arab Emirates—whilst resulting at that time in significantly lower imported infections, were also a significant help to the functioning of many aspects of local, rather than international, society [1].

Throughout late 2020, public opinion in Ireland swayed between the two extremes—keeping society open, similar to the neighbouring United Kingdom and attempting to 'live with the virus' [2], versus attempting a

Outbreaks, Epidemics, and Health Security. https://doi.org/10.1016/B978-0-323-91705-6.00007-1

'Fortress Island' approach, for example, New Zealand, opting for 'Zero-COVID'. But despite the chance of eliminating infection, Ireland failed to reach zero new cases over the summer months. Perceptions were changed again following this, as the risk of dramatic second waves over the winter suddenly seemed real by the time autumn had arrived, which threw the plans for the normal functioning of schools into question. Yet ethically, not to mention in terms of practicalities, there could have been no possibility of an Ireland in which restaurants and pubs operated, but schools were closed [3].

PICTURE 1.2.1 Scary headlines were part of our daily lives at a very difficult time for us all. Sebastian Kevany.

Thus, the situation was perhaps no longer related to the considerations of trade-offs and cost-effectiveness. Instead, more rightly, the focus was around the epidemic control and containment, if not the elimination of the virus. Like many dynamic and emotive issues in Ireland, such as national reunification, the intensity of feeling attached to infection control is (rightly) ingrained in the national consciousness and perhaps defied economic analyses. Instead, the cries of COVID-Zero became something more profound: a crusade and a rallying point, a matter of national pride and determination. But, perhaps, a sense too far? A return lockdown occurred later, but not necessarily because Irish society allowed its government to create the conditions for the epidemic to return to overwhelming strength.

The construction of a proverbial 'Fortress Ireland' on this basis appeared—in which, like rabies, the epidemic threat is carefully and thoroughly excluded in an orchestrated, publicized, and cross-sectoral fashion. If this approach was accompanied by port-of-entry testing and aggressive and intensive contact tracing efforts, perhaps zero-COVID-19 would have been feasible at one stage during 2020 [4]. This, of course, would have relied on close cross-border co-operation with Northern Ireland; an adherence to European Union regulations; secure management of trade (particularly with the United Kingdom); and—by far most importantly—automatic testing at all points of entry, along with the facilities to enable strictly monitored quarantine enforcement, on the entire island of Ireland.

PICTURE 1.2.2 Covid airport signage. Maisie M. Jones.

2.2 Guest essay: COVID-19 lessons from Cuba: Isolation and diplomacy

Ilan Kelman

The COVID-19 pandemic simultaneously brought about isolation and connectedness. The former emerged through restrictions on movement, whilst the latter came through an increasingly online world for those with the technology. Islands exemplified both these characteristics—as stereotypes and in reality—leading to the importance of exploring the intersections between island studies and global health [5,6].

Cuba is an island country generating contentious debates swirling around these topics, notably for the country's period of communist leadership. It started when Fidel Castro overthrew a US-backed dictator in 1959, and has continued to the present day, even as the country's governance has begun to start to show cracks. Two key endeavours across these decades provide COVID-19 lessons: dealing with isolation and aiming for connectedness through global health diplomacy.

Dealing with isolation

Cuba under communist rule became used to dealing with isolation. Most prominently, the US government imposed various forms of trade restrictions with and sanctions against Cuba and others who wished to trade with Cuba. Cuba had allies through the USSR and its bloc which disappeared in 1991 as the Iron Curtain collapsed, making US restrictions more potent.

Supply shortages became common in Cuba, especially for oil, and soon extended into mechanical and electronic products, including spare parts for vehicles and handheld mobile devices. With US law often targeting non-US businesses and governments conducting business with Cuba, the island country developed domestic capacity and improvised as much as possible.

This experience served Cuba to some extent during the COVID-19 pandemic, which started in 2020. Being used to striving for high service levels with limited supplies, Cuba in 2020 appeared to have done better than most countries at keeping the disease under control whilst aiming for its own vaccines [7,8]. The main challenge in assessing Cuba's response to COVID-19 has been finding credible and verifiable data, particularly of deaths and hospitalizations, but also reflecting on case numbers, since testing must be conducted and reported in order to record a positive case.

It is now clear that, as with many other countries, any initial success in Cuba did not translate into continued success. As Cuban vaccines started to be approved and used in the country in mid-2021, confirmed cases and deaths skyrocketed. Protests followed food and medicine shortages, leading to both a government crackdown and increasingly vociferous social media criticism of the Cuban government by Cubans [9]. Some medical staff publicly complained that Cuba's healthcare woes were less about the pandemic and the US trade embargo and were more about domestic mismanagement. That is, they did not accept isolation as an excuse for the difficulties.

The events in Cuba during the COVID-19 pandemic further demonstrate how isolation can be overcome. Even with countries closing down internal and external travel, goods and information continued to flow. People with good Internet access gained connections by spending time at home online, especially being able to order anything they needed for delivery, where available. Those lacking good Internet access or the technology to use it became even more isolated than they were before COVID-19 manifested.

Unsurprisingly, the COVID-19 lesson for dealing with isolation is that—as with almost all disasters—inequities are exacerbated, not necessarily created anew [10]. Those with more opportunities and resources before the pandemic tended to find ways to gain from isolation, whilst those already behind ended up further behind. Isolation is not inherent, not even to islands such as Cuba, nor is isolation inherently bad or inherently good. The pandemic demonstrated yet again the known lesson that pre-existing conditions play a major role in determining the adverse or beneficial impacts of isolation.

Global health diplomacy

Measures for connecting can tackle isolation, with diplomatic endeavours amongst the techniques. Aspects of 'soft diplomacy' and 'public diplomacy' offer goods and services to raise a donor's profile, hoping that the recipients garner a positive view of the donor. Health diplomacy has been exemplified throughout history: cross-country efforts to control plague; the International Sanitary Conferences starting in 1851; and smallpox, polio, and rinderpest vaccinations in conflict zones.

Soon after Castro took power in Cuba, medical diplomacy became one of his foreign policy mainstays [11]. After Chile was rocked by an earthquake in 1960, Cuba sent a handful of medical staff to assist, soon ballooning into tens of thousands of Cuban doctors dispersing around the world to support health clinics and training. When an ideological ally led Venezuela from 1999 to 2013, Cuba exchanged medical services for oil. After Hurricane Katrina struck the United States in 2005, Cuba offered humanitarian medical assistance which was declined.

The main lesson from these examples is that health diplomacy—and its various forms including medical diplomacy, vaccine diplomacy, disease diplomacy, global health diplomacy, ecohealth diplomacy, one health diplomacy, and planetary health diplomacy—can succeed with respect to health care, but has limited impact on diplomacy [12]. Cuba-Venezuela health diplomacy worked due to pre-existing conditions of ideological similarity. Cuba-US health diplomacy failed due to pre-existing conditions of ideological differences.

This basic premise of health diplomacy ought to have been applied for COVID-19. Doing so would have recognized the importance of long-term direct international cooperation regarding pandemic prevention. When that failed, then international cooperation regarding pandemic management could have been implemented, without expectations of political changes or new friendship emerging.

Countries supplying medical equipment around the world near the pandemic's beginning included Russia, China, and Taiwan, without discernible impact on long-term relations between the donating and receiving locations. Countries ostracized from the world receiving numerous offers of assistance were Iran and North Korea, with limited acceptance of this aid. Cuba also proffered doctors and medical assistance to deal with COVID-19 in 2020 and then recalled some Cuban personnel in 2021 as the situation worsened on the island.

In the meantime, vaccine distribution became contentious. Richer countries purchased and distributed vaccines amongst their own populations, making donations to poorer countries by using mainly excess supplies. Cuba continued to hope that its vaccines would support Latin America and the world, inhibited by efficacy and supplies [8]. The United States suggested sending vaccines to Cuba, resulting in the usual diplomatic dancing between the two countries vying, or pretending to vie, for an agreement. In tandem, several others friendly to Cuba provided help, such as Bolivia, Canada, Mexico, and Russia.

COVID-19 diplomacy for Cuba provides the same lessons as previous centuries of health diplomacy: diplomatic outcomes should be neither expected nor assumed from international health cooperation. Instead, the most successful intertwining of health and diplomacy happens when the diplomacy already exists, as in pre-existing links or cooperation, permitting health-related cooperation.

What's next?

Irrespective of pandemics, US presidents, communism, and other disasters (especially referring to the first three points), the one-party state of Cuba cannot last forever. Will the transition be smooth, with the main power brokers shifting to more openness and democracy? Will the change be unwilling, through violence, sudden death, or a reluctant ceding of power?

With a preference for a peaceful destiny, two main points from Cuba are lessons which could and should have been applied to COVID-19. They remain relevant for future crises, health-related and beyond.

First, isolation from the world benefits no one, further illustrated by North Korea [13]. Even where borders are legitimately closed for most travel, connections contribute to surviving and thriving. Links include open Internet, legal exchange of goods and currencies, and virtual services. This statement is not endorsing unrestricted trade, unmonitored wealth transfer, or extensive travel. Instead, it seeks smooth flow with mutual gains, without incurring unsustainable human and environmental costs. Neither entirely open nor entirely closed borders are optimum. A balance of choice and restrictions is required.

The second lesson is that this balance yields diplomacy, facilitating global health successes much more than global health diplomacy could do so alone. Focusing on health outcomes, such as vaccination, can achieve plenty regarding health and diplomacy separately, yet it does not link the two topics for long-term, positive results for both. For full health success, diplomacy and connections have thus far been an important prerequisite.

Cuba's historical and current challenges have much to teach the world, with Cuba having plenty to learn as well, so that it becomes a mutual exchange. Ultimately, it is about everyone—sovereign states alongside governance at supranational and subnational levels—supporting each other for health and diplomacy in tandem.

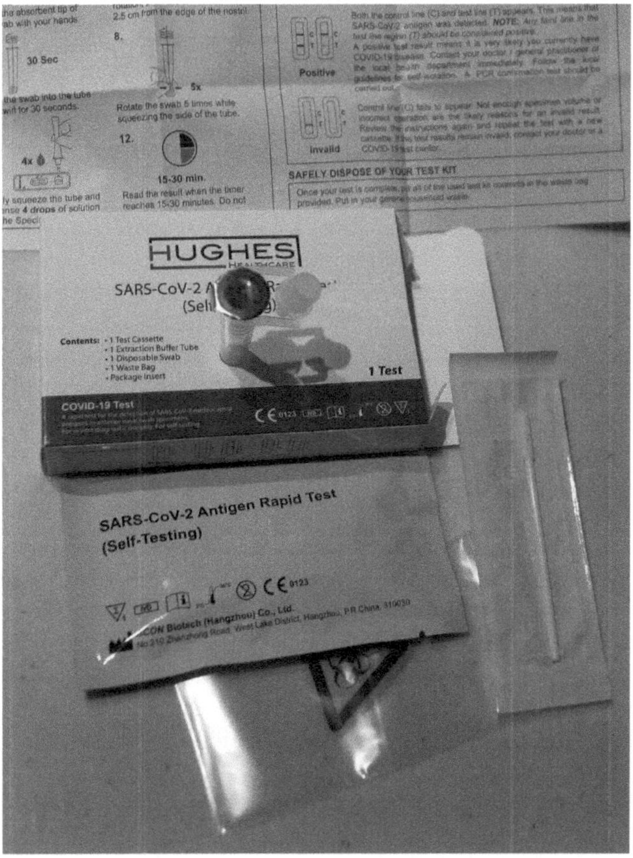

PICTURE 1.2.3 Testing kits were hard to attain at first, then became widely available. Ilan Kelman.

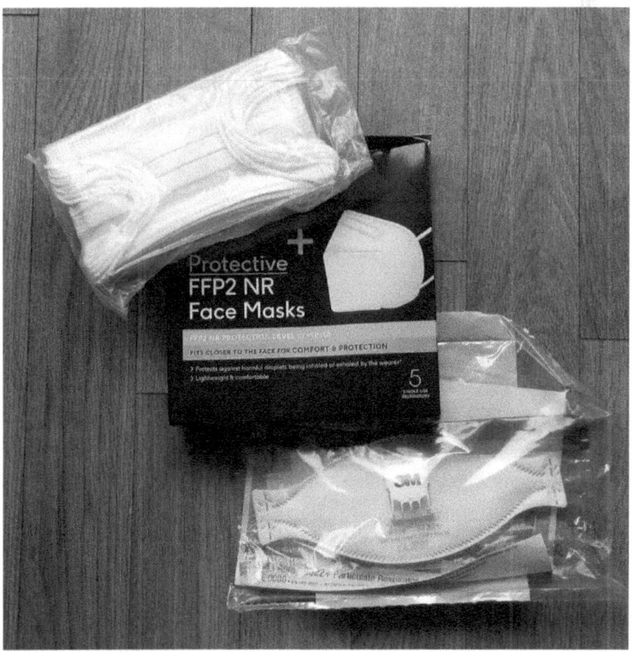

PICTURE 1.2.4 Face masks became part of all of our daily lives. Ilan Kelman.

2.3 The costs of isolation

So far, so good—until one begins to legislate for the downsides, which perhaps explain why a 'Fortress Ireland' approach never took root. Health security can too easily leverage for neo-isolationism and xenophobia [14]. Similarly, tourist and travel sectors, which contribute heavily to Irish employment and the financial industry (understandably), continued to lobby for the relaxation of quarantine restrictions [15]. Whilst it was difficult to find an easy solution to the airline issue, as the industry knows the risks of their businesses in the free market—the hospitality sector successfully instead encouraged domestic tourism. This domestic tourism provided both holiday options for those who required them and a much-needed revenue and income stream for Irish hotels, restaurants, and guesthouses [15].

Yet even these limited-success scenarios, which were perhaps the best Ireland could have hoped for, required careful management for optimization. In this regard, primary importance was using the tools of fields such as health diplomacy and security to manage the effects of the international relations of a 'Fortress Ireland' position. Moreover, as the country considered a longer-term recalibration of border and travel policies, such a scenario also required vigilance to ensure that Ireland, or other countries, did not become (or be perceived as) neo-isolationist states.

An example of a 'Fortress Ireland' approach requires a look back at history, with an example of the 2007 'Foot and Mouth' outbreak across the United Kingdom and Ireland [16,17]. Certain features were reminiscent of citizens during COVID-19, such as the reminder of hand-washing rigorously through widespread general public health information and campaigns. In addition, a 'One Health' approach was taken (the link between zoology, animals, and people sharing infectious disease and risk of spread).

At that time, a strict ban was implemented and agreed upon by the Northern Irish and Republic of Ireland governments to discontinue the importation of cattle, sheep, pigs, and goats from Britain into the island of Ireland until the outbreak was deemed 'under control' by the Department of Agriculture and health officials [17]. Perhaps that previous outbreak was a stark reminder to those who argued for the 'one island' approach with stricter border controls during the pandemic.

Perhaps, in such future scenarios, the role of Ireland's diplomats would be, amongst their other efforts, to work on managing the consequences of a 'Fortress Ireland' policy. This policy would ensure that commercial and different perceptions are protected and that the country continues to function as a small and open economy in as many ways as possible. Increasing global health expertise at the embassy level and within the Department of Foreign Affairs can only strengthen Ireland's position and increase the chances of success of such a strategy.

2.4 Guest essay: The biggest threat to global health security: COVID-19 and neoliberalism

Sabine Iva Franklin

In early 2021, I wrote an op-ed for a British media website asking whether the novel coronavirus (COVID-19) was responsible for the hundreds of thousands of American lives—or could it be poor governance? [18]. In this, I drew upon my research of local emergency management during the 2014 Ebola Epidemic in Liberia and Sierra Leone and made comparisons with the response in the United States. Here, I elaborate further on how a past neoliberal ideology may have impacted health governance in the United States.

Granted, in both disease outbreaks, we experienced fear and misinformation, such as rumours and conspiracies, hindering the effectiveness of social distancing mandates. Public health and government officials also struggled to gain trust from local communities as denials about the outbreak and severity of the disease spread as fast as the virus itself. Thus, the evidence pointed to poor governance as primarily responsible, particularly when we look at how countries that were at the epicentre of the Ebola outbreak responded to COVID-19, compared to the American response to COVID.

Age and the economy

Indeed, I argue that, in certain geographical areas, the COVID-19 governance in the Western world is largely neoliberal, in that economic growth and production were at times placed above human lives. If this seems like a bold statement, then see this quote from the Republican Lieutenant Governor of Texas, Dan Patrick, during an interview on FOX News. He says that older Americans, who were most at risk for severe disease and death from COVID-19, would rather sacrifice their lives than have the economy shutdown due to stay-at-home orders [19].

I am hard-pressed, however, to find any data to substantiate claims that a significant number of older Americans would die to preserve the nation's economy. However, if many had felt this way, then their sacrifice would have been in vain—precisely because the pandemic itself is the market failure [18]. And for as long as the pandemic continues, so will economic decline.

Race also plays a role, as by March 2020 data showed that African Americans were nearly three times more likely to die from COVID-19 than White Americans [20]. Thus, it is perhaps not a coincidence that the anti-lockdown and anti-mask protests were often largely made up of far-right White protestors, carrying confederate flags and Nazi insignia, who both felt that the public health restrictions were overblown and downplayed the threat of the novel coronavirus [21,22]. Nonetheless, neoliberalism is not entrenched in the right wing only; there was also a bipartisan call to 'mitigate' instead of eradicating the virus from our soil.

According to the US Centers for Disease Control and Prevention (CDC), community mitigation strategies are actions that people and communities can take to slow the spread of a new virus with pandemic potential [23]. This results in the goal of living with the infectious disease, which is a departure from eradication or elimination strategies to stop the viral outbreak altogether [24]. I certainly do not feel that COVID-19 has been an Ebola-style epidemic, which has seen death rates up to 90% in some outbreaks [25]. Rather, because data suggest that 99% of COVID-19 deaths were of people over the age of 50, with many having preexisting conditions or obesity [26], it was perhaps all too easy to write off these casualties as members of society who are less productive to the economy.

This type of ageism is not new, as previous studies found that employers and employees perceive older workers (age 50 and older) to be less productive than younger ones [27]. Of course, this stereotype is often not true, and in the United States, this discrimination may have cost the economy hundreds of millions of dollars [28].

Delta and the economy

At the time of writing, there were fears that the B.1.617.2 variant, also known as the Delta variant, will unleash a third wave of disease around the world [29]. The United States once again has the highest cases, but the demographics have shifted. This wave was touted as the 'pandemic of the unvaccinated', and even former COVID-19 deniers in the far-right media and political spectrum are now encouraging people to get vaccinated [30].

This is not a coincidence: the rise of hospitalizations and deaths is in younger people, aged 15 to 45 [31], the same age demographic that makes up the majority of the labour force. Moreover, across the country, we are also experiencing a labour shortage [32]. There were over 9 million job vacancies, and employers claimed that they could not find workers. However, this is a complex problem that cannot solely be attributed to the rise in COVID-19 cases.

Low-wage jobs, such as in the restaurant and bar industry, have seen the biggest shortages of workers. For non-American readers, it is important to understand this context, because jobs in this industry are notoriously exploitative with pay as low as $2.13 an hour. This is the lowest legal minimum wage in some states for 'tip employees' like waitresses and bartenders [33]. Food service workers are expected to receive enough tips to supplement their salaries, which is often unfair to the employees and the customers.

It is also rare that a bar or restaurant would provide benefits such as healthcare insurance or a retirement package. Furthermore, this industry was largely affected by public health restrictions, and many businesses reduced services or closed down during the pandemic. Therefore, many workers left the industry altogether. Other factors, such as workers desiring more flexibility and fearing for their safety because of disease spread, also play a role in the shortage. However, for politicians who are desperately trying to tout the end of the pandemic and a return to normalcy, a wave of the Delta variant that is threatening younger Americans may just undo that. Therefore, with the availability of vaccines, far-right pundits and politicians are now more openly expressing the deadliness of the outbreak and the need to stop the spread.

Lessons for the future

So, what can we learn from our experiences from the COVID-19 pandemic—and how can we apply them to global health security? First, there is the false dilemma of choosing to save either human lives or the economy. Undoubtedly, quarantine restrictions and stay-at-home orders have an initial impact where movement of goods and trade decline. However, these are temporary and with today's technology can often be mitigated. The world went mostly virtual for a year, and this kept some economic production going. However, there is a technological disparity between low-income and high-income countries, where broadband Internet access is less available.

The main policy that assisted with preventing more poverty in the United States was the pandemic unemployment assistance (PUA) benefits, which provided $600 a week to workers or freelancers that lost work due to the pandemic [34]. The PUA benefits were reduced to $300 a week in January 2021, mainly because of pushback from those in the legislature that these were incentivizing the unemployed to stay home and not look for work. However, over 13 million people in the United States would have fallen into poverty without the PUA, creating a long-term disaster for the economy.

The second lesson is that there is a lack of public health infrastructure in the United States, despite the country being ranked as the 'most prepared' for public health emergencies in the Global Health Security Index in October 2019 [35]. This ranking is relative, as it stated that no country is fully prepared to handle an epidemic or pandemic; however, current case and death rates show that the United States performed poorly compared to many other nations. We have 4% of the world's population, but as of January 2021, about a quarter of the world's COVID-19 cases [36]. Therefore, we need to rework our understanding of the global health security and public health infrastructure nexus.

Global health security as practised in the Western world is limited to building public health systems to detect and contain an infectious disease outbreak [37]. The United States works with other nations to help build systems around identifying and containing infectious diseases that may also threaten homeland security [38]. However, securitizing health in this way is limiting and may not be effective in stopping outbreaks of novel pathogens or containing them when they hit our soil.

For instance, the novel coronavirus was initially detected in China, and there have been some reports that the Chinese government was not forthcoming with vital information in the early stages. 'We're going on very minimal information', said American epidemiologist Maria Van Kerkhove, now WHO's technical lead for COVID-19, in one internal meeting. 'It's clearly not enough for you to do proper planning' [39].

Furthermore, after the 2014 Ebola Epidemic, the United States scaled up its global health security programme to help train disease detectives in other countries, including China. However, this was scaled down in 2018, as the funding was scheduled to end in 2019 [40]. Thus, many CDC technical experts were leaving various places, including China, which raised uncertainties regarding the sustainability of the programme. Therefore, these two examples highlight the political nature of global health security: it only works if there is close and sustainable international cooperation.

The American Public Health System

The second part of this lesson is that the lack of public health resources in the United States contributed to the spread of COVID-19. Thus, although we were able to identify our first coronavirus case, we were unable to contain the disease spread. Governance is neoliberal—as discussed above—but so is the current public health infrastructure. Health care is, by and large, a market institution and is dominated by firms, such as pharmaceutical and insurance companies. Thus, the goal is not the human right to healthy living but to maximize profits.

Healthcare providers also set up practices in locations where they are likely to sustain their businesses. Therefore, rural counties and ethnic minority neighbourhoods in urban areas are underserved. A lack of access to primary care was thus one of the facilitators of disease spread in 2020 [41,42]. Because these practices operated like businesses, and many saw a loss of patients (i.e. customers) between February and May 2020, this forced many offices to close. And closed primary care practices meant more people visit the emergency room, which was a further source of COVID-19 transmission.

Furthermore, this may also hamper the roll-out of vaccines, as many of these areas do not have access to pharmacies. Again, pharmacists would not be incentivized to operate in areas that are less profitable under market conditions. Hence, whilst there has been a lot of media coverage regarding vaccine hesitancy amongst conservative voters [43–45], many of these voters are also from rural areas that are underserved. Thus, a more complete picture of the problem would include the challenges to access the vaccines in underserved communities. However, historically, the default narrative has been to demonize rural constituents as ignorant or not caring about their health, rather than to question why the richest country in the world cannot provide universal primary healthcare and public health resources [46,47].

In my opinion, neoliberal governance of the COVID-19 pandemic in many countries is to blame for the country's high case and death rates. For future global health security, we can formulate policies that are not within this ideological paradigm; however, neoliberalism is entrenched in the American healthcare system and other sectors that long predate the pandemic. Just as some government officials were willing to write off the casualties of older Americans in the beginning of the outbreak, we have also been writing off the lives of poor people who did not have access to adequate health care.

Therefore, the pandemic lessons are actually revelations of unjust health and labour policies that did not support working Americans. It is a view of the path that led us here. Now, the road forward is to build a new system that provides universal access to health; workers with a living wage; fairer working conditions; social justice; and economic rights that are all part of the foundation for 21st-century strong democracies and global health governance [48,49].

PICTURE 1.2.5 Links between Covid and social unrest were evident in many parts of the world. Sabine Iva Franklin.

2.5 One island, two countries

Dr. Gabriel Scally, lead epidemiologist and public health consultant, spoke on numerous occasions to the media during 2020–21 on the different approaches between the United Kingdom and Ireland, emphasizing the importance of an 'all-island approach' [50]. In one instance, he insisted on the significance of an all-island approach that would have been taken if it were affecting one of the main economic factors of agriculture [50]. This claim indeed was backed up by the political path taken for the foot-and-mouth outbreak.

Fortunately, there were other short-term solutions. Technology made many redundant of the needs of commercial travel; international relations could often be maintained just as easily online as in person [51]. Events such as international conferences, business trips, and weekend trips abroad have become more complicated and perhaps altogether unnecessary, with companies and technologies stepping in to maintain efficiency and networking through online events.

Economically, plenty of arguments increasingly favour Ireland aspiring to become a no-virus zone. Although these discussions had many shortcomings regarding this goal, the longer term proved that such countries might profit from both an earlier return to normal societal functioning and significant health security dividends, in much the same way as zero-case countries rapidly became desirable places to live and work.

2.6 Ireland versus New Zealand

The argument remains—no country is an island—from PPE supply to vaccine equity in what has become a very interlinked, international, and globalized world. However, how long does it last before borders break down, either politically or locally? New Zealand remained an outlier in the 'Fortress Island' approach, perhaps because geographically, it is located at least a 3- to 4-h flight away from the nearest mainland and therefore remains a self-sufficient island, for the most part [52]. However, economically, New Zealand and Australia rely heavily on tourism. Unfortunately, a mandatory quarantine approach has proven unpopular as a tourism vacation option, with vaccination roll-out and vaccine passports still only on the horizon for the countries with the strictest border controls.

The decisive leadership approach shown by Jacinda Arden, the then Prime Minister of New Zealand, remained exemplary during many dithering and delayed responses worldwide. Taking early action, listening to public health guidance, closing borders, early contact tracing and testing, and strict (albeit shorter) lockdowns resulted in New Zealanders returning to relatively normal everyday life as quickly as possible throughout 2020–21, despite the restrictions on travel abroad. There were many associated benefits to public health and protection of the health system.

For example, one could explore the clear contrast between New Zealand and the UK's approach. Three days after the pandemic announcement by the World Health Organization (WHO) on 30 January 2020, New Zealand began introducing disease prevention measures and strengthening them in subsequent weeks. In comparison, the United Kingdom had its first cases in hospital by this stage, with no contact tracing in place and the main airports still open for international travel until June 8 2020—5 months after the global declaration of a pandemic. Still, this was at least three months after the United Kingdom went into complete lockdown. The United Kingdom thus dismissed the island approach from the offset [53].

Thus, Ireland partially became a fortress in a quasi-physical and public health sense—but not in virtual or otherwise significant terms. The country attempted to maintain its 'Thousand welcomes—Céad Mile Failte[a]' policy. Unfortunately, without any question, this tourism-friendly approach was not without the risk of school and other closures, nor the breakdown in law and order that such events precipitate—nor at the expense of public health, as was on occasion in the earlier lockdown in March 2020. In the current global context, the price of such an image was simply too high.

---

[a] Céad Mile Failte literally translates to 'a hundred thousand welcomes'.

2.7 Guest essay: The New Zealand response—A medic abroad

Sarah O Dwyer

The New Zealand (NZ) experience of COVID-19 has been quite different from that of the rest of the world. Having taken the initially successful route of an elimination strategy, this enabled open travel within the nation between the two islands; a lack of mask wearing; and has allowed for concerts, festivals, and gatherings without number restrictions—essentially, normal day-today life, although in our own NZ-based nationwide bubble. At times, this contrast felt almost like living outside some alternative dystopian universe when affronted with the difficulties of loved ones abroad in prolonged lockdowns, with news of the challenges their healthcare systems faced daily.

Whilst I believe the New Zealand government acted wisely with this initial strategic decision, using to its advantage the remoteness and island nature of the country, and despite its own difficulties—as a foreigner myself, I have found quite difficult the uncertainty of when I will be able to return home given the Managed Isolation and Quarantine (MIQ) protocols. I also fear and do not think this COVID-free time over the last 18 months has been efficiently used to ensure adequate measures for ongoing protection of the nation. The eventuality of needing to open borders to the rest of the world is looming, and elimination cannot and will not be permanent at this stage.

Whilst countries in Europe were rolling out vaccines before the New Year, New Zealand has been slow to implement this, with the programme only taking full steam now that we have a further outbreak. The elimination strategy led to a slow vaccine uptake initially, partially as some felt it was an unnecessary step given COVID-19 did not exist here. There has been a 'wait and see what happens' attitude amongst many, in addition to vaccine scepticism and concerns. This was slow to be addressed and acted upon, and now, we face a race to vaccinate, whilst simultaneously trying to control an outbreak. I feel that perhaps the time we had without the effects of active COVID-19 should have been utilized better in organizing clinics, providing education surrounding vaccination, and targeting those vulnerable communities most at risk.

Research carried out on the NZ population during the outbreak in 2020 identified the Māori and Pasifika populations as being at highest risk from contracting the COVID-19 illness—both in terms of being more likely to have an underlying illness and more likely to have poorer access to health care, and based on the increased morbidity and mortality seen in these groups during the H1N1 outbreaks [54]. With this in mind, many advisory groups had called for prioritized vaccinations of these ethnic groups as vulnerable populations.

Despite this advice, this was not included in the roll-out strategy. I am concerned that this has led to a failure to protect some of our most at-risk populations, especially when the data from this current outbreak indicates again that groups suffering the majority of the infection burden are overwhelmingly the Māori and Pasifika. Whilst I am no expert in this field, as a medic working in the emergency department, I do question at times, why expert advice was not followed regarding vaccination and information sessions, which now are leaving these vulnerable populations behind and putting them at risk again, leading to further scepticism of a health system which has been accused repeatedly of systemic racism.

This pandemic has been challenging for all governing bodies, and we have all been learning as we go and from mistakes made. New Zealand has been held up as one of the better-managed nations regarding outbreak response. The strong work of the government and public health teams is also reflected in boasting one of the lowest death rates in the developed world. However, I wonder if this prolonged global positioning on the leader board has led to a false sense of security, leading perhaps to an element of complacency. We are now, at the time of writing in September 2021, facing a reality check that was only a matter of time.

Whilst we have the benefit of the experience of other countries' management progress, we do not yet have the first-hand experience ourselves of healthcare strain and I worry that we remain inadequately prepared regarding resources and staffing. Whilst many systems have been set up for contact tracing, vaccine roll-out strategy, and even saliva-based rapid testing, New Zealand has often opted to develop its own devices, perhaps not benefiting from the experience of those who have gone before. As countries worldwide begin to open up again, we remain static, and interestingly, a recent article ranking the best countries to be with COVID-19 has ranked Ireland as number one with New Zealand coming in at 37 [55].

As a healthcare worker, I feel incredibly fortunate (if not a bit guilty) to have avoided the trauma that other countries have faced; however, it is only a matter of time before these foreign realities become a reality for us too, as our Auckland colleagues are currently experiencing. With the borders bracing to open, it will be a challenge for us all to learn to manage this relatively new disease we have not yet truly experienced, although I hope that the current focus on vaccination and the benefit of education from our colleagues' experiences can help guide us through and mitigate the effects.

PICTURE 1.2.6 For those who like to get away from it all, Covid presented some unexpected advantages—if you could get there. Sarah O Dwyer.

2.8 Key points

- Health security at the border level was not effectively explored in the early pandemic stages and could have helped avoid health system overload.
- However, policies, processes, and other protocols were not in place to enforce or impose such restrictions to the degree necessary for their effectiveness.
- Land borders present a unique challenge, as demonstrated by the success of small island states in the South Pacific and elsewhere in controlling the pandemic.

References

[1] E.L. Sigurdsson, A.B. Blondal, J.S. Jonsson, M.O. Tomasdottir, H. Hrafnkelsson, K. Linnet, J.A. Sigurdsson, How primary healthcare in Iceland swiftly changed its strategy in response to the COVID-19 pandemic, BMJ Open 10 (12) (2020) e043151.

[2] A. McLysaght, We Can Live With the Virus? This Is Dangerous Wishful Thinking, 2021, August 28, Retrieved December 6, 2021, from https://www.irishtimes.com/opinion/we-can-live-with-the-virus-this-is-dangerous-wishful-thinking-1.4657794.

[3] A. Doyle, W. Hynes, S.M. Purcell, Building resilient, smart communities in a post-COVID Era: insights from Ireland, Int. J. E-Plan. Res. 10 (2) (2021) 18–26.

[4] ISAG, A Better Way Forward: COVID Elimination in Northern Ireland, a compassionate, effective and decisive approach, 2020, Retrieved December 6, 2021, from https://www.isag-covid19.org/home.

[5] Island Studies, Isl. Stud. J. [cited 2021 September 12]. Available from: https://islandstudies.ca/.

[6] I. Kelman, Pandemic and post-pandemic islandness: building and wrecking resilience, 2021, Islands Economic Cooperation Forum: Annual Report on Global Islands 2020, p. 105. Available from: https://projects.upei.ca/unescochair/files/2021/04/Annual-Report-on-Global-Islands-2020-LowRes.pdf.

[7] E. Morris, I. Kelman, Coronavirus response: why Cuba is such an interesting case, The Conversation (2021). [cited 2021 September 12]. Available from: http://theconversation.com/coronavirus-response-why-cuba-is-such-an-interesting-case-135749.

[8] L. Taylor, Why Cuba developed its own covid vaccine—and what happened next, BMJ (August 5) (2021). n1912.

[9] Havana Faces Severe Reputational Damage Over Unrest, 2021. Emerald Expert Briefings. January 1 [cited 2021 Sep 12]; oxan-es(oxan-es). Available from: https://doi.org/10.1108/OXAN-ES262988.

[10] P. Blaikie, T. Cannon, I. Davis, B. Wisner, At Risk: Natural Hazards, People's Vulnerability and Disasters, Routledge & CRC Press, 2021. Available from: https://www.routledge.com/At-Risk-Natural-Hazards-Peoples-Vulnerability-and-Disasters/Blaikie-Cannon-Davis-Wisner/p/book/9780415252164.

[11] J.M. Feinsilver, Fifty years of Cuba's medical diplomacy: from idealism to pragmatism, Cuba. Stud. 41 (2010) 85–104.

[12] I. Kelman, Do health interventions support peace through "disaster diplomacy?", Peace Rev. 31 (2) (2019) 158–167.

[13] North Korea Disaster Diplomacy, 2021. [cited 2021 September 12]. Available from: https://disasterdiplomacy.org/northkorea.html.

[14] S. Rushton, Global Health Security: Security for Whom? Security From What? 2011, November 07, Retrieved from https://onlinelibrary.wiley.com/doi/abs/10.1111/j.1467-9248.2011.00919.x.

[15] F. Altuntas, M.S. Gok, The effect of COVID-19 pandemic on domestic tourism: a DEMATEL method analysis on quarantine decisions, Int. J. Hosp. Manag. 92 (2021) 102719.

[16] D.T. Haydon, R.R. Kao, R. Paul Kitching, The UK foot-and-mouth disease outbreak—the aftermath, Nat. Rev. Microbiol. 2 (8) (2004) 675–681.

[17] Houses of the Oireachtas, Dáil Éireann debate—Friday, 2 March 2001, 2001, March 02, Retrieved December 6, 2021, from https://www.oireachtas.ie/en/debates/debate/dail/2001-03-02/3/.

[18] S.I. Franklin, COVID-19: is it the disease or poor governance that is killing us? 2021, Meeting of Minds UK. Available at: https://meetingofmindsuk.uk/realreads/covid-19-is-it-the-disease-or-poor-governance-that-is-killing-us/. (Accessed 26 July 2021).

[19] Interview, 2020. Tucker Carlson Tonight, Season 5, Episode 58. Youtube, 23 March https://www.danpatrick.org/tucker-carlson-tonight-march-23-2020/. (Accessed 25 July 2021).

[20] APM Research Lab, The Color of Coronavirus: COVID-19 Deaths by Race and Ethnicity in the U.S., *APM Research Lab*, 2021. Available at: https://www.apmresearchlab.org/covid/deaths-by-race. (Accessed 31 July 2021).

[21] M.N. Hoskin, The whiteness of anti-lockdown protests, Vox (2020). 25 April. Available at: https://www.vox.com/first-person/2020/4/25/21234774/coronavirus-covid-19-protest-anti-lockdown. (Accessed 31 July 2021).

[22] A. Marcotte, Anti-lockdown "movement": powered by racism, Salon (2020). 4 May. Available at: https://www.salon.com/2020/05/04/anti-lockdown-movement-powered-by-racism/. (Accessed 31 July 2021).

[23] CDC, Implementation of mitigation strategies for communities with local COVID-19 transmission, 2021. COVID-19: Work & School—U.S CDC https://www.cdc.gov/coronavirus/2019-ncov/community/community-mitigation.html. (Accessed 29 July 2021).

[24] M.G. Baker, N. Wilson, T. Blakely, Elimination could be the optimal response strategy for COVID-19 and other emerging pandemic diseases, BMJ 2020 (371) (2020), m4907.

[25] S.I. Franklin, The Local Governance of the 2014 Ebola Epidemic: A Comparative Case Study of Liberia and Sierra Leone (Doctoral dissertation), University of Westminster, 2019.

[26] L. Beckett, Older people would rather die than let COVID-19 harm US economy—Texas Official, The Guardian (2020). 24 March. Available at: https://www.theguardian.com/world/2020/mar/24/older-people-would-rather-die-than-let-covid-19-lockdown-harm-us-economy-texas-official-dan-patrick. (Accessed 27 July 2021).

[27] H. Van Dalen, K. Henkens, J. Schippers, Productivity of older workers: perceptions of employers and employees, Popul. Dev. Rev. 36 (2) (2010) 309–330.

[28] K. Terrell, Age discrimination costs the nation $850 billion, study finds, AARP (2020). 30 January. Available at: https://www.aarp.org/politics-society/advocacy/info-2020/age-discrimination-economic-impact.html. (Accessed 28 July 2021).

[29] A. Khan, How the Delta Variant Changed The Course of the COVID-19 Pandemic, *Al-Jazeera*, 2021. 27 July. Available at: https://www.aljazeera.com/features/2021/7/27/how-the-delta-variant-has-changed-the-course-of-the-covid-19-pandemic. (Accessed 28 July 2021).

[30] M. Sotomayor, J. Alemany, M. DeBonis, Growing number of republicans urge vaccinations amid delta surge, Wash. Post (2021). 21 July. Available at: https://www.washingtonpost.com/politics/growing-number-of-republicans-urge-vaccinations-amid-delta-surge/2021/07/20/52a06e9c-e999-11eb-8950-d73b3e93ff7f_story.html. (Accessed 25 July 2021).

[31] A. Joseph, How delta is pushing the U.S. into a new phase of the COVID-19 pandemic, STAT (2021). 19 July. Available at: https://www.statnews.com/2021/07/19/delta-vaccines-new-phase-pandemic/. (Accessed 27 July 2021).

[32] T. Spiggle, What does a worker want? what the labor shortage really tells us, Forbes (2021). 8 July. Available at: https://www.forbes.com/sites/tomspiggle/2021/07/08/what-does-a-worker-want-what-the-labor-shortage-really-tells-us/?sh=2f95c34f539d. (Accessed 28 July 2021).

[33] Department of Labor, Minimum Wages for Tipped Employees, Wage and Hour Division—U.S Dept. of Labor, 2021. Available at: https://www.dol.gov/agencies/whd/state/minimum-wage/tipped. (Accessed 27 July 2021).

[34] B. Zipperer, Over 13 Million More People Would be in Poverty Without Unemployment Insurance and Stimulus Payments, *Economic Policy Institute*, 2020. 17 September. Available at https://www.epi.org/blog/over-13-million-more-people-would-be-in-poverty-without-unemployment-insurance-and-stimulus-payments-senate-republicans-are-blocking-legislation-proven-to-reduce-poverty/.

[35] GHSI, Inaugural Global Health Security Index Finds No Country Is Prepared for Epidemics or Pandemics, *Global Health Security Index*, 2019. 24 October. Available at: https://www.ghsindex.org/news/inaugural-global-health-security-index-finds-no-country-is-prepared-for-epidemics-or-pandemics/.

[36] S. Ahluwalia, R. Abraham, Global COVID-19 cases surpass 100 million as nations tackle vaccine shortages, Reuters (2021). 27 January. Available at: https://www.reuters.com/article/us-health-coronavirus-global-cases/global-covid-19-cases-surpass-100-million-as-nations-tackle-vaccine-shortages-idUSKBN29W0L6. (Accessed 28 July 2021).

[37] CDC, Global Health Security, Centers for Disease Control and Prevention, 2019. 7 January. Available at: https://www.cdc.gov/globalhealth/healthprotection/ghs/index.html. (Accessed 29 July 2021).

[38] USAID, Global health security, Glob. Health (2021). 8 March. Available at: https://www.usaid.gov/global-health/health-areas/global-health-security. (Accessed 29 July 2021).

[39] Associated Press, China delayed releasing coronavirus info, frustrating WHO, The AP (2020). 2 June. Available at: https://apnews.com/article/united-nations-health-ap-top-news-virus-outbreak-public-health-3c061794970661042b18d5aeaaed9fae. (Accessed 29 July 2021).

[40] J. Greenberg, Fact Check on Presidential Candidate Joe Biden, Politifact—The Poynter Institute, 2020. 21 May. Available at: https://www.politifact.com/factchecks/2020/may/21/joe-biden/under-donald-trump-key-cdc-us-staff-china-fell-0/. (Accessed 29 July 2021).

[41] M. Dill, We Already Needed More Doctors. Then COVID-19 Hit, Association of American Medical Colleges, 2021. 17 June https://www.aamc.org/news-insights/we-already-needed-more-doctors-then-covid-19-hit. (Accessed 30 July 2021).

[42] J.P. Geyman, Beyond the COVID-19 pandemic: the urgent need to expand primary care and family medicine, Fam. Med. 53 (1) (2021) 48–53, https://doi.org/10.22454/FamMed.2021.709555.

[43] N. Asgari, "A form of brainwashing": why trump voters are refusing to have a vaccine, Financial Times (2021). 21 July https://www.ft.com/content/39ff87ce-57b7-4007-9504-7eb2c7bc911f. (Accessed 30 July 2021).

[44] D. Ivory, L. Leatherby, R. Gebeloff, Least vaccinated U.S. counties have something in common: trump voters, The New York Times (2021). 17 April. Available at: https://www.nytimes.com/interactive/2021/04/17/us/vaccine-hesitancy-politics.html. (Accessed 30 July 2021).

[45] T. Keith, How do you reach trump voters who say they don't want the vaccine? Try doctors, NPR (2021). 19 March. Available at: https://www.npr.org/2021/03/19/978407316/how-do-you-reach-trump-voters-who-say-they-dont-want-the-vaccine-try-doctors. (Accessed 30 July 2021).

[46] J. Ivin, I know why poor whites chant trump, trump, trump, STIR J. (2016). 1 April. Available at: http://www.stirjournal.com/2016/04/01/i-know-why-poor-whites-chant-trump-trump-trump/. (Accessed 20 August 2016).
[47] S. Smarsh, Poor teeth, Aeon (2014). 23 October. Available at: https://aeon.co/essays/there-is-no-shame-worse-than-poor-teeth-in-a-rich-world. (Accessed 3 March 2021).
[48] S.I. Franklin, Why Health Is a Human Right: Why Hasn't Our Democracy Evolved Into a Modern Social Contract to Protect the Health and Wellbeing of Its People?, Healthcare Section, Tarbell, 2020. 14 July. Available at: https://tarbell.org/2020/07/why-health-is-a-human-right/. (Accessed 3 September 2020).
[49] L.O. Gostin, B.M. Meier (Eds.), Foundations of Global Health & Human Rights, Oxford University Press, New York, NY, 2020.
[50] G. Scally, Gabriel Scally: North and Republic Must Harmonise Covid-19 Response, 2020, March 31, Retrieved December 6, 2021, from https://www.irishtimes.com/opinion/gabriel-scally-north-and-republic-must-harmonise-covid-19-response-1.4216073.
[51] D.S. Ting, L.C. Wei, V. Dzau, T.Y. Wong, Digital technology and COVID-19, Nat. Med. 26 (4) (2020) 459–461.
[52] S. Jefferies, N. French, C. Gilkison, G. Graham, V. Hope, J. Marshall, C. McElnay, et al., COVID-19 in New Zealand and the impact of the national response: a descriptive epidemiological study, Lancet Public Health 5 (11) (2020) e612–e623.
[53] T. Colbourn, Unlocking UK COVID-19 policy, Lancet Public Health 5 (7) (2020) e362–e363.
[54] https://journal.nzma.org.nz/journal-articles/estimated-inequities-in-covid-19-infection-fatality-rates-by-ethnicity-for-aotearoa-new-zealand.
[55] https://www.bloomberg.com/graphics/covid-resilience-ranking/.

Chapter 3

Lessons for humanity

Sebastian Kevany and Aoife Kirk

Chapter outline

3.1 What have we learned?

After what will perhaps be regarded as the greatest crisis of the 20th century, what has humanity actually learned? What has nature taught us, and what have some of its bluntest possible instruments attempted to teach us? In all disasters, whether they are a product of human agency or purely natural, lessons are learned from the message that nature is trying to tell us. These lessons can include developing early-warning systems or adapting construction and city planning in earthquake-prone areas.

To ignore these suggestions would be arrogant in the extreme: humanity cannot assume that it is protected evermore from future epidemic threats because of our current response efforts. Even without further waves of infections in the near future, the pathways and mechanisms for the rapid spread of new forms of epidemic disease worldwide in very short periods are now clear [1]. Yet, by the time of writing this in late 2021, amnesia is a deadly ally to the virus. Will we see the repercussions of forgetting the lessons of 2020–21 in years to come?

Thus, on the most extensive picture level possible, what has humanity discovered that it has been doing wrong? The throwaway argument is that nature teaches us that there are too many of us: that this is a natural cull to control the human population. Yet, the realities are far more subtle and nuanced than such unfounded

Outbreaks, Epidemics, and Health Security. https://doi.org/10.1016/B978-0-323-91705-6.00012-5

and irrational messaging, which encourages us to do nothing. Some readers may recall some related arguments put forward in favour of the HIV/AIDS epidemic: as some commentators watched the death tolls mount, they were confident that, in the long term, it was the right thing—a lesser evil—for a greater good.

Instead, too much high-density living might be one fair and acceptable message we can all accept and work with. With growing population levels around the world and increasing urbanization and rural depopulation, humans live in closer quarters than ever before. Urban sprawl has been changing natural ecosystems and increasing contact between animal species and humans for decades. This, unfortunately, leads to a perfect scenario for the spread of zoonoses [2]. Issues such as air pollution, and high population density and mobility, now result in cities, which may also be seen as uniquely positioned to develop sustainable recovery solutions for COVID-19 and beyond.

For example, will the revolution in remote working, combined with more widespread recognition of the public health risks of urban life, thus spur a more even distribution of us between city and countryside in future? A world where 'working from home' becomes the norm, whereby commuting becomes more and more redundant?

PICTURE 1.3.1 Social distancing in Donegal, Ireland. Eilis Conneally.

In the same way, what has nature now told us about, for example, packed commuter trains and about intense flows of humans, whether in cars or on buses, converging with their schedules, at the same time of the day? These practices were inherently unhealthy in themselves—even without infectious disease concerns.

Yet, because perhaps of their gradual increase, such experiences became an accepted part of the status quo—despite our gradual realization of how unpleasant, occasionally degrading, and even inhumane such experiences of overcrowding could be.

Whilst public transportation plays a role in ensuring equity amongst those without access to private vehicles, the challenges throughout lockdown and perhaps beyond regarding capacity and compliance with physical distancing regulations may reinvent how we commute [3]. For most of the last 20 years, Asia has been using face masks when on commuter trains to avoid the risk of the spread of infectious diseases and decrease the risk of inhaling the higher air pollution in the larger cities.

Indeed, mask-wearing in China began from the beginning of the creation of the 'new republic', with Dr. Wu Lien-Teh, a Chinese doctor who won the Nobel Prize in Medicine in 1935, leading the response to the pneumonia pandemic of Manchuria and Mongolia, which eventually claimed 60,000 victims [4]. His contribution to public health at that time, which included the use of masks during the pandemic, resulted in international coverage of mask-wearing 'plague defenders' [4]. Dr. Wu is now often regarded as one of the first people to modernize China's medical services and education.

PICTURE 1.3.2 **Masks.** Sebastian Kevany.

The peak popularity of mask-wearing on public commutes was during the SARS epidemic in 2002–03. SARS, also a coronavirus (a cousin of COVID-19) reinforced the importance of mask-wearing. To wear a mask emphasizes not only the safety of oneself but also a consideration for the safety of others. This cultural shift towards pro-mask-wearing in Asian culture over the last decade may have assisted in certain countries controlling the first wave of COVID-19. Another interesting thread to explore is the culture of being a 'good citizen' [5], including mask-wearing, symbolizing the 'collective action' versus the individualism often experienced in more westernized countries [6]. Perhaps, we must also look back to history as a reminder?

One strategy to explore within the public transport sector is an evidence-based, interdisciplinary, interventional approach, with inputs from diverse stakeholders such as the government, local councils, private industry, city planners, transport authorities, and public health and infectious disease experts. International treaties, such as the Sustainable Development Goals Agenda or the global commitment to reduce greenhouse gases through the 2015 Paris Agreement, may provide a framework to deliver this vision. However, more importantly is the political will and commitment to prepare and account for future pandemics and other global crises on the horizon, such as climate change. Signing these treaties is one aspect; more crucially, it is vital to keep national governments accountable for the commitments.

Too much unnecessary and often careless international travel by the general population is a further possible lesson from nature that we perhaps cannot afford to ignore. These short-term travel trips and weekend breaks have implications for climate change, pollution and resource extraction, and consumption considerations—all of which have been a product of the most mobile and industrial period of human history. But, again, can technology, along with behavioural change, come to the rescue—have we seen the curtain coming down on an era in which air travel was almost always the first, rather than the last, option for international endeavours?

3.2 Guest essay: Social observations of a post-lockdown environment

Selma Furtado

With the imminent lifting of most of the remaining restrictions in Ireland on 22 October 2021 [7], a fatiguing plod through 18 months of varying degrees of the COVID-19 pandemic lockdowns is finally in sight. For the majority of us, it cannot come too soon. Articles which I read during the pandemic introduced me to terms like lockdown fatigue [8] and crisis fatigue [9]—terms which are disconcertingly descriptive of the angst [10] that has gripped much of the global populace since early 2020. The lifting of restrictions means that we will finally 'return to normality'. But, having lived through a 'new normal', can we even describe 'normality' any longer? Over the past month or so, as most of the restrictions were eased in Ireland, I have been observing my environment, my own reactions, and the actions and reactions of my friends and people around me. This is what I have seen:

The ubiquitous hand sanitizer

The start of the pandemic was marked by global hand sanitizer shortages [11]. Now, however, there is ubiquitous presence of hand sanitizer and an almost automatic gravitation towards hand sanitizer stations wherever they may be placed in buildings. This is not 'normality' as we once knew normality to be. I have seen people whip out mini hand sanitizers from pockets and purses every few minutes to vigorously cleanse their hands. They are not to be faulted—this is one of the preventative measures [12] which we have all learned during 18 months of lockdown and it is evidently—quite thankfully, I must admit—here to stay.

The refrain of 'too many people'

As lockdowns slowly eased and I resumed tentative ventures outdoors, the overwhelming feeling I had was that everywhere, there were 'too many people'. At first, the feeling was almost unconscious as I navigated the wide central streets that I had grown used to walking along unhindered by pockets of people standing around and talking. So accustomed had I become to the feeling of wide-open spaces uninterrupted by what now seemed to be 'crowds', that I found myself hurrying through my errands and darting back to the 'safety' of home.

I was not even immediately aware of this change in behaviour. Clouded by over a year of mainly two-dimensional interactions with friends and acquaintances over Zoom, my reactions to people reappearing on streets and roads, in shops and other buildings, going about their daily business, appeared to me at first to be the usual reaction of a naturally introverted person like myself. I was extremely surprised and interested therefore when the exact phrase was uttered by a supremely extroverted friend of mine without any prompting. My friend was describing her first post-lockdown lifting venture into a shopping mall and she said to me, 'I couldn't do it. There were too many people in the shop and on the streets'. Yet, another friend used the same phrase—at a different time—to describe her reaction to similar conditions. She, too, is far more extroverted than I have ever been.

The guarding of personal space

Even in an environment of over 80% of the adult population having been fully vaccinated [13], I see people making wide detours around those who are not wearing masks. Personal space has become a precious commodity where individuals back away from others who stop to speak to them—the reaction even more pronounced when the other is not wearing a mask.

Public transport capacity restrictions might have been lifted [14] and fellow travellers might all be adequately masked, but the unconscious refrain of 'too many people' appears to be large in the minds of the populace. Travelling on the Luas earlier this month, in a tram that no longer had the strategically placed 'do-not-sit' social distancing signs on designated seats, I tentatively sat next to a fellow female passenger after asking her whether I might take that seat. She shuffled away from me and I took this as acquiescence that I might. She got up and stood for the rest of the journey—over nine stops to where she finally alighted.

Handshakes have all but vanished. The merest suggestion of a handshake is met with a huge reflexive step backwards—so inured have we become in a mere 18 months to not shaking hands. There is a very real guilt associated with what now seems to be a huge faux pas of attempting to shake another's hand. A friend put her hand out for a handshake the other day and then apologized, saying: 'We aren't allowed to do that anymore'. This seems to be the common sentiment although there is some tentative debate about whether the handshake will be making a return [15]. The compromise of 'selective handshaking' [16], as this article seems to suggest, does not appeal to me and, from what I have observed, to many others. I scour the Internet and every article on handshake etiquette now seems to be superfluous. How soon will it be before we have articles on the etiquette of business elbow bumps, I wonder?

Our new 'comfort zoom'

Behavioural psychology states that we become conditioned through actions repeated several times over a period of time. My friends say that their comfort zones have changed and my casual observations in this 'lockdown emergence phase' support their remarks. Where they once wore masks from necessity and because it was mandatory, my friends have at various times said that they still plan to wear them once the restrictions and policies on mandatory mask-wearing have been eased—because they have grown 'used to them'. Zoom, our main 'support prop' for communication during the various phases of lockdown, does not seem to be something which any of us will discard soon, either. Unlike 18 months ago, today, I personally cannot think of a single person amongst my acquaintances and friends who will give me a blank stare if I mention a conversation over Zoom.

Cocooned in the comfort of our homes even if we have been outdoors earlier in the day, it is to use Zoom or similar communication software that we turn—even when we could step outside our homes and meet in person.

'The new normal' versus 'normality'

Informal discussions with my friends have led me to conclude that each of us detests the phrase 'the new normal' with comparable degrees of fervour. Where at the start of the pandemic, the phrase evolved to encompass all the actions of mask-wearing, hand-sanitizing, the backing away to 2 m whenever we met someone, it now induces a shudder in each of us. What was originally 'the new normal' has now become 'normal'—so habituated have we become to these things after 18 months of endless repetition. But, if this 'new normal' has become 'normal', the lifting of restrictions promises the return to 'normality'—which, for clarity's sake and the avoidance of confusion—I will refer to as the 'prenew normal'. This was when we comfortably shook hands on meeting someone, we gave hugs freely, and we did not think twice about opening doors using our hands or approaching to speak to a person—whilst standing well within 1 m of them. Is this the 'normality' to which we can expect to return?

Given the observations I have made, I have to say—not for some months, perhaps.

So where do we go from here?

If we cannot really expect normality to be the 'prenew normal', I must therefore—again for clarity's sake and borrowing from technological product development [17]—refer to it as the 'new "new normal"'. The effect of the global pandemic on our social lives has been significant—from social distancing to the 'bubbles' [18] which we have had to maintain over many, many months. The 'this, too, shall pass' [19] heard and seen many times during lockdown is now within our collective grasp, but we should not expect it to be the 'normal' we were all used to before all this began. If we are to ensure even a semblance of permanence to this 'new' 'normality' as we emerge from lockdown, our now inculcated new behaviours—especially those related to hand hygiene and coughing and sneezing—should become a part of our daily etiquette. Once the last of the restrictions are lifted, as the Irish government has repeatedly emphasized, the onus to ensure that we never again go into another lockdown now lies on each and every one of us.

PICTURE 1.3.3 Going out, pandemic. Selma Furtado.

3.3. Humans versus nature

From the individual level to societal impacts, nature is a force to be reckoned with. On the one hand, the pandemic has improved air quality in different cities across the world, reduced greenhouse gas emissions and noise and water pollution, and the burden of unsustainable pressure often associated with tourist areas. One example of a high footfall tourist area that received a lot of media attention was Venice. In the first lockdown, a 'fake news' viral tweet was shared of dolphins returning to Venice which turned out to be one of the many 'fake news' stories circulated and shared [20].

Yet, the residents of Venice in 2020 were left with an incredible 'restart' button on their values and how tourism may access the city in future. Since August 2021, cruise liners and ships over 25,000 tonnes or longer than 180 m have been banned from the shallow Giudecca Canal in Venice [20]. This is one of the many policies that cities have taken the opportunity to introduce, in this aspect using it as a moment to protect the environmental, artistic, and cultural heritage of Venice, declared a world heritage site by UNESCO [21]. However, this environmental win has its own sunken costs, including the risk of loss of livelihoods and economic loss to those who rely heavily on tourism in the area.

On the subject of ports, too little attention to health security has, perhaps, been a further lesson for humanity—an unwillingness to put in place those inconvenient measures, at air, sea, and other ports, that would routinely protect populations from the spread of epidemic diseases. Despite the efforts of some academic groups and United Nations bodies, the reality is that we have never appreciated, in our daily lives—until now—the importance of using borders to control epidemics. At best, such considerations previously arose in knee-jerk ways, for example, in times of Ebola outbreaks—an extemporaneous strategy that has let us all down badly in recent months.

To paraphrase another way, have our borders actually reached a stage where they have gone from being too hard to too soft? Is there, bizarrely and ironically, actually some logic to neo-isolationist thinking in the public health and epidemic context? Is there a sweet spot for health borders in which they are neither too rigid nor too loose? Or a situation in which necessary travel for personal, business, or leisure purposes functions adequately and safely—but a lot of unnecessary, dangerous, and wasteful movement does not?

There are many other considerations and possible lessons, as well. The pressing need for a more virtual existence for work and commerce—even for the daily means of survival, as supermarket shopping lists are transacted online—may also, in the long run, serve both to highlight and curtail our collective abuse and misuse of the Internet. When people are living (as we have been) in an increasingly virtual environment for practical purposes, the concept of spending more time online for recreational purposes becomes, perhaps, less enticing. Similarly, possibly, cars were once regarded as leisure instruments before they became mere workhorses.

We have also perhaps learned of another dimension of our Internet use that may need correction: not just the belief that the technological landscape is an all-powerful panacea, but that what we read online is not necessarily true. Never before has a public health—or perhaps any—issue garnered this much online attention. Communications, collaborations, and vital messaging have been facilitated. Still, the costs are equally high: the abounding misinformation, uninformed commentaries, conspiracy theories, and other malignant effects of instant, round-the-clock news saturation [22].

In truth, only academia (and even then only a subset of publications) and verified statistics can be trusted—yet, too often, the interpretation of such information is only possible by those experienced in the field. Even the respected broadsheet media perhaps succumbed to hype and sensationalism. Unquestioning trust in what we read has, perhaps for good, now been irreversibly questioned. The political layer adds to this, for example, the US elections were held amid the second lockdown in November 2020. The use of Twitter as a medium to broadcast political statements was used by presidential candidates, often before the campaign team or public officials would have information. Does it take this long, and thus far, before tech companies must intervene?

3.4 Guest essay: Pandemic reflection from WHO 'changemakers in health' filmmaker

Danny Scheinmann

I'm a storyteller and filmmaker, and for the past year, I have been making a film for the WHO, which seeks to amplify the voices of changemakers in health.[a] We interviewed stakeholders from all levels of health services from all over the world: be it doctors in Indonesia, India, or Argentina, hospital administrators in East Timor, nurses in Nigeria, patient advocates in the United States, community health workers in the remote parts of Liberia, or paediatricians in Canada. All of them had contributed in novel ways to the improvement of health outcomes. I have no wisdom of my own to offer in this field, except the stories of the wonderful people we spoke to. Interestingly, most of the interviewees did not speak directly about the impact of the pandemic on their work, but rather about issues they had been facing for years, which suggests that the roots of the problems we faced during the pandemic have been present for a long time.

In Liberia, Bendu Whykai and Moses Chealey are two community health workers (CHWs) plying their trade in very remote, rural areas. So remote that it took them the better part of a day to get to an office with an internet connection to speak to me. Neither Bendu nor Moses set out to be health workers. They were nominated by their communities because they were already respected role models and would be listened to. Moses left his teaching job to be trained by the charity Last Mile Health, and now every day he visits 35 local families whether they are ill or not. He has to walk long distances between households to reach them. Bendu works out in the fields where there is a constant flux of labourers. This is a country which has been on the frontline of HIV, Ebola, and malaria. They have high infant and maternal mortality rates and are now dealing with COVID-19. The communities which Bendu and Moses serve have seen dramatic improvements in health outcomes in comparison to other rural areas that are not served by CHWs. I asked them what they were doing when they visited families who were not even ill. They said they were spotting problems before they arose—stopping people drinking from a dirty creek, removing standing water near people's homes that was attracting mosquitos, checking on pregnant women and facilitating antenatal visits at the local health facility, educating about hygiene, disease spread, and basic medicines, helping with family planning and contraception, making sure people visited doctors before symptoms develop, promoting vaccines—the list goes on.

There are three things we can learn from them that could provide insight into how we handle a pandemic. The first is the idea that 'health' comes to your door every day in the form of someone you know and respect from your community. The second is that instead of all the resources being tipped towards people who are already ill, they are put towards preventing people from getting ill in the first place. And the third is health education from a peer which is vital in the battle

[a] Watch the WHO 'Changemakers in Health' film here: https://www.youtube.com/watch?v=-oppYBfKQwU.

against misinformation and conspiracy theories. The Liberians told me that before they were nominated, the countryside was rife with fake 'healers' selling all kinds of tinctures and herbs which promised to cure every ailment under the sun, but only made matters worse. Since the CHWs started visiting, many people have now stopped buying from them. Ignorance fans gullibility. The situation is no different from the vortex of online antivaccine conspiracy theories that engulf people in the privacy of their homes—if only their friendly neighbourhood CHW had set them straight before the panic had set in.

I asked Moses what happened when a person refused to take action and was consequently putting their neighbours at risk. He said that he would sit down with church leaders and village elders and they would agree on a plan of action to help the individual understand the consequences of their actions. I found this whole community approach touching, coming, as I do, from a place where so many people are isolated even from their neighbours.

In the United Kingdom, during the height of lockdown, people took to their doorsteps every Thursday to applaud the NHS workers for their incredible efforts. There was a brief period of extraordinary goodwill. On my street, we organized a WhatsApp group to help get food to those isolating. A neighbour set up a food bank for people who had lost their jobs. But, then the gestures fizzled out, the clapping stopped, and we retreated into our homes. Could we have made more of it and harnessed that goodwill in a more organized way? Could we have enlisted people to carry out basic tasks on behalf of their loved ones isolated in hospitals or in their homes? This might be especially relevant in countries whose health services are already short-staffed and underfunded.

We could learn from the experience of Dr. Arti Maria. She is a paediatrician working in a public hospital in India. They were desperately short-staffed. In the Neonatal Intensive Care Unit (NICU), they would often have only 1 nurse caring for 15 babies. It was almost impossible to keep the unit as sterile as it should be, change all the babies' nappies, clean them, and administer high-quality care. Every day she would walk past desperate mothers sitting on the floor outside, wanting to know how their ill babies were getting on. One day she had a revelation: what if she were to let these women into the NICU to help care for their babies? Dr. Arti Maria readily admitted that her initial intention was to solve a staffing problem, but what emerged was far greater than this. Of course, there were concerns amongst the staff at first about outsiders getting in their way and spreading infection. The solution was to teach the women how to wash their hands properly, how to put on PPE, and how to manage simple tasks in the NICU like changing nappies, cleaning, and alerting doctors if they saw changes on the monitors.

The constant training of mothers would add an extra burden to an already threadbare staff, but eventually, they created training videos and learned how to manage the process efficiently. What then unfolded was beyond all expectations. A mother's love is a very powerful and underestimated form of medication. They kept their babies clean, cuddled them, maintained very high standards of hygiene, and kept an eagle eye. The babies thrived, the nurses were able to focus on more skilled work and the mothers (many of whom were uneducated) were empowered. The active participation of parents actually led to fewer complaints because they witnessed first-hand how hard the staff worked. When, finally, the women returned home with their fragile babies, they were well equipped to look after them. Furthermore, there was a knock-on effect as they passed on their knowledge to their family and friends. The scheme was so successful that it was piloted across four states and now the government has embarked on a national roll-out. Dr. Arti Maria described the process as transformational for all the stakeholders.

During the pandemic, many hospitals were short-staffed and many patients were ignored because of those shortages, and yet, so much love and care were on hand. This speaks to a common theme that came up in our interviews about patient and family empowerment and, as one doctor said, 'love impacts outcomes'. Loved ones will do anything to help and are very careful to do things right. If the most basic tasks are done by family, then this leaves skilled professionals more time to do their jobs.

Dr. Arti Maria goes on to say, "Whom does health belong to? Health is a positive attribute of sickness. Sickness is a negative attribute. Sickness belongs to patients, and patients need doctors and nurses. Health is a positive attribute which belongs to people – healthy people. So, who should take charge of that? Not doctors – I think health belongs to the community and the community should look after it. If people receive preventative healthcare and communities are empowered, will I need doctors as much? Maybe not. I think that we should seek doctors and nurses only for sickness and we need to promote preventative health. That should be the focus at the school and preschool level and everywhere. Health education has to start much earlier. This is how we can impact the community to take charge - social accountability of health."

Many of our interviewees spoke about the need to teach children about basic hygiene, nutrition, mental health, and more in school. They wanted people to understand their own role in health and to become partners in administering health, and not always be reliant on health services to fix things after it's too late. Similarly, early education could teach us how to prevent the spread of epidemics instead of relying solely on government policy which can be politically motivated or corrupt.

I spoke to a hospital administrator in Indonesia who said that seriously ill people with COVID-19 were refusing to come to his hospital because of a ruling that, should they die, the family would not be able to wash the body. Whilst this policy makes total sense in a European context, it had a negative effect in his region because it is a Muslim tradition for the family to wash the corpse. As a result of the policy, patients who might have been saved in a hospital died unnecessarily at home. Furthermore, he feared that the unsupervised washing of corpses may have actually added to the spread. Would it have been better to talk to community leaders, provide PPE, and come to some kind of acceptable solution within the hospital?

I would like to leave it to others, more qualified than myself, to draw their own conclusions from these stories, but if there is one common thread it is that early education and more ambitious community engagement is the way forward.

Another theme which emerged time and time again was the need for a change in culture within health services. A survey of the NHS in the United Kingdom carried out just as the pandemic struck showed that 1 in 4 nurses and health visitors leave the NHS within 3 years of joining, and around a third intend to quit within the next 2 to 3 years, whilst there were already 40,000 vacancies. Similarly, 35% of GPs are intending to quit and there were 6000 GP vacancies before the pandemic struck. Many of our interviewees, who were facing similar issues in their countries, spoke of chronic excessive workload, burnout, and a culture of blame as contributing factors. According to a recent report, between 1 June 2020, and 1 June 2021, NHS workers missed more days of work because of mental health-related issues (2.5 million days) than because of COVID-19 (2.1 million days). The total combined cost to the NHS of mental health- and COVID-19-related absences was £682 million.[b]

In a culture of blame, staff hide their mistakes for fear of being criticized, sacked, or sued. If the mistakes are not known, then quality improvements cannot be made because there is no data on which to base an improvement. We heard that sometimes on a death certificate the cause may be attributed to 'complications' of a disease when in truth it might be because of an error like the administration of the wrong dose of a particular drug. One example was from a hospital in Nigeria where insulin doses of 40 iu and 100 iu were kept in the same-coloured boxes in the same drawer in the same fridge. Doctors, who sometimes worked 36-h shifts, could easily confuse the doses. Giving the wrong dose could be fatal. When a sharp-eyed head of QI noticed a high death rate amongst diabetic patients, he realized what was happening and the solution was as simple as changing the colour of the boxes to make it easier for the doctors to identify different doses.

Professor Michael West of the King's Fund and Dr. Rachna Chowla of King's Health Partners explained to us the need to move from a culture of blame towards a restorative culture. They went on to elaborate that instead of constantly looking for who was at fault, we should be asking who was hurt by this mistake (including the doctor or nurse), how can we help them and what can we learn. For this to happen, there needs to be a shift towards compassionate leadership. They identified four steps of compassionate leadership: listen, understand, empathize, and then take intelligent and appropriate action.

What does compassionate leadership look like in practice? Dr. Salhab el Helou and Dr. Enas El Gouhary, senior leaders from the Neonatology Division at McMaster Children's Hospital in Hamilton, Canada, did a survey that revealed that

---

[b] *Workforce burnout and resilience in the NHS and social care* report from the House of Commons' Health and Social Care Committee published on 8 June 2021.

80% of their trainee doctors had experienced burnout and nearly half said that they did not feel it was 'psychologically safe' to speak out about mistakes or raise concerns about some of the issues they faced. They were shocked and embarked on a culture change programme with a raft of initiatives including instituting pre-intervention huddles and postintervention debriefs where all the staff involved (not just doctors) were able to discuss what needed to be done beforehand, and then afterwards share what went well, what was missed, and how they felt. This has led to more consensus amongst staff and higher quality care. The environment is more supportive and caring, and patient outcomes have improved over the past 5 years. Dr. El Gouhary explained how they now teach trainee doctors how to have constructive, noncritical conversations. She also gave an example of when trainee doctors were asked what their top priority was for improvement, they agreed that it was spending more time with their families. A hot issue was that their shift would end at 6 p.m. and the absolute deadline for collecting their children from day-care and after-school clubs was 6 p.m. Almost every day they felt like bad parents as they arrived late for pickup and this contributed to a feeling of intense pressure and failure at home. They brought the shift pattern forward by half an hour and instituted a few other quick-win measures based on their feedback. Burnout reduced from 80% to around 45% without any extra cost. The process is ongoing, but the direction of travel is now clear.

The common thread from these stories of changemakers in the field was a need for deep-seated systemic change. In summary, there needs to be a shift towards well-being and vitality with greater investment in stopping people from getting ill in the first place. In the case of a pandemic, this may mean stockpiling PPE, but perhaps more importantly, training community health workers to engage with people before they fall ill. This should be accompanied by a campaign of early education in schools to teach children about infection, hygiene, and other health issues with the long-term goal being to empower individuals and the community to take more responsibility for a healthy lifestyle.

There needs to be a focus on patient- or family-centred care, but this can go much further than merely consulting patients on what they want. As Dr. Arti Maria's experience shows, it can be a transformative process which allows families to share in the administration of care where it can be shown to be beneficial or where health systems are under pressure.

Finally, there needs to be an overhaul of the leadership style in health services with the aim of alleviating the enormous pressure that health workers are under, especially in times of crisis. As one doctor put it; 'you have to look after yourself if you want to look after others'. The cost to the NHS from lost days and high attrition is a clear indicator that the current system is leading to a crisis in human resources and wasted investment. Compassionate leadership results in better outcomes for all stakeholders, patients, and workers alike.

3.5 Health is wealth

Inattention to personal hygiene might also be a lesson for the 21st-century. In the millennial era, we have had much less concern for the tradition of washing hands regularly than, perhaps, a generation ago. As we are now mostly freed from the spread of most epidemic diseases in developed countries, there was a corresponding decline in preventive measures. The lesson is simple: without personal hygiene once again instituted as an integral part of human behaviour, the epidemic risks increase.

And what of other concerns—what of racism? From the catastrophic overlap of epidemic and racial problems that arose in the United States, perhaps a different lesson may be derived: humanity cannot afford to live with racial tensions that can spill over into epidemic problems if time and place coincide. Such a series of events, each of which compounds the other, quickly turn public health emergencies into complex humanitarian crises; public health becomes exponentially harder to manage when accompanied by such tensions.

Indeed, racism has been defined by the American Medical Association as an "urgent threat to public health, the advancement of health equity, and a barrier to excellence in the delivery of medical care" [23]. This pandemic has not only highlighted the inequalities of access to care, care worker burden, and frontline workers, but it has also demonstrated that the COVID-19 virus seems to affect blacks and minorities more severely. Chronic diseases such as cardiovascular disease, obesity, hypertension, and diabetes are more frequent and severe in minority populations. The disproportionate impact amongst racial and ethnic minorities is yet another example and stark reminder of ongoing health disparities, which need to be addressed worldwide [24].

Turning to the political realm, should the heads of state that we elect now be evaluated based on their response to public health threats, rather than single-issue politics such as European Union membership or national geographical unity? Likewise, should critical positions such as the leadership of the World Health Organization and its related departments be determined based solely on expertise rather than nominations based on regional origin?

And the list of possible lessons that nature may be attempting to teach us goes on: a need to move away from the brain drain of scientists towards tech and finance; a need for robust public health systems. The more generalized risks of increasingly polarized societies may also be revealed based on a need to pay more attention to socio-economic patterns of disease. In addition, the need for the enforced end of the live animal trade, in the Far East and elsewhere, has unquestionably been highlighted.

We have learned that the term 'emergency' is inappropriate for climate change concerns compared to epidemic risks, yet we have also been shown the need to reconsider our frantic pace of life. Likewise, we perhaps have felt a need to move away from celebrity culture during lockdown times and reflect on the social congestion it creates: all of these are possible lessons for humanity to learn.

But perhaps the most important lesson that nature may be trying to teach us is that, despite our borders and health security measures, no country or community on earth is any longer an island. No matter how hard we try and how unpleasant it may be, we now live in an irreversible era of global coexistence—therefore demanding unprecedented levels of cooperation, if for no other reason than enlightened self-interest. In this way, nature has reminded us of our limited power over organic global forces despite all our industrial, agricultural, artistic, technological, and other achievements. Our sense of unrivalled domination and control has been rudely shocked: we have, as a famous football manager once remarked, been at least temporarily knocked off our perch.

A final related ethical-philosophical question is regarding what needs to be retained in our behaviour change from our individual and collective experiences. In Ireland, it was unfortunate to see rush-hour commuter traffic jams returning as the intensity of the epidemic decreased during the summer months. Should humanity revert to its prepandemic status quo, a repeat of the events of 2020/21 will inevitably occur: we have thus already seen that a wholesale return to old ways is inappropriate. Even with vaccines now primarily administered, the cliché that those who ignore the mistakes of history are destined to repeat them has never been more appropriate.

If humanity chooses to revert to the preepidemic status quo, the same events will inevitably recur. Fortunately, for many, the so-called 'new normal' is not such a bad thing: cutting down on the unnecessary or optional car and long-distance travel, thereby reducing pollution and increasing road safety; reducing consumption and consumerism. All the whilst reducing excessive shipping and

PICTURE 1.3.4 Dalkey Winter swims. Aoife Kirk.

air transport and improving ocean and air quality: the list goes on. In addition, we have seen not just threats to, but improvements in, physical and mental health as people stay out of pubs, exercise more, understand that smoking is more dangerous than ever, and appreciate nature more. These are perhaps just some of the behavioural changes that nature is urging us to—even insisting that—we maintain.

3.6. Key points

- Humans should, where possible, attempt to manage the behaviours that have led to this pandemic situation, including excessive movement, consumerism, and materialism.
- A wide range of 20th-century behaviours and standards should, where possible, be reevaluated—including a movement away from individualism to a more collective consciousness.
- The way that we treat our bodies, from excessive smoking and alcohol to unhealthy diets, should change—but only in a way that is fair and accessible to all members of society.

References

[1] S. Rousseau, N. Deschacht, Public awareness of nature and the environment during the COVID-19 crisis, Environ. Resour. Econ. 76 (4) (2020) 1149–1159.

[2] A. Córdoba-Aguilar, C.N. Ibarra-Cerdeña, I. Castro-Arellano, G. Suzan, Tackling zoonoses in a crowded world: lessons to be learned from the COVID-19 pandemic, Acta Trop. 214 (2021), 105780.

[3] L. Budd, S. Ison, Responsible transport: a post-COVID agenda for transport policy and practice, Transp. Res. Interdiscip. Perspect. 6 (2020), 100151.

[4] Z. Ma, Y. Li, Dr. Wu Lien Teh, plague fighter and father of the Chinese public health system, Protein Cell 7 (3) (2016) 157–158, https://doi.org/10.1007/s13238-015-0238-1.

[5] Discover Society, The "Good" Coronavirus Citizen, The "Covidiot", and the Privilege of #StayAtHome, 2021, Available from: https://archive.discoversociety.org/2020/04/01/the-good-coronavirus-citizen-the-covidiot-and-the-privilege-of-stayathome/#. (Accessed 5 December 2021).

[6] T.H. Manh, N.N. Anh, N.T. Duc, N.B. Minh, M.T. Huong, N.M. Hieu, Adherence to mask wearing on public transport during the COVID-19 pandemic and influential factors: the case of Hanoi, J. Transp. Sci. 72 (4) (2021) 486–497.

[7] C. Humphries, Ireland to Drop Almost all COVID-19 Restrictions in October, 2021, Available from: https://www.reuters.com/world/europe/ireland-drop-almost-all-covid-19-restrictions-october-2021-08-31/. (Accessed 30 September 2021).

[8] M. Young, L. Linehan, Have You Hit Lockdown Fatigue? Two Wellness Practitioners Have Some Advice for You, 2020, Available from: https://www.thejournal.ie/readme/phase-three-mindfulness-5143855-Jul2020/. (Accessed 30 September 2021).

[9] K. Philip, Opinion: What is Crisis Fatigue and Have We Reached That Point? 2020, Available from: https://www.thejournal.ie/readme/crisis-fatigue-5279755-Nov2020/. (Accessed 30 September 2021).

[10] M. Dolan, Column: Angst About Pandemic Without End, 2020, Available from: https://www.reuters.com/article/uk-global-tailrisk-idUKKBN28Q16Q. (Accessed 30 September 2021).
[11] S. Hegarty, Coronavirus: Where has all the Hand Sanitiser Gone? 2020, Available from: https://www.bbc.com/news/stories-52126847. (Accessed 30 September 2021).
[12] HSE, n.d. Protect Yourself and Others From COVID-19. Available from: https://www2.hse.ie/conditions/covid19/preventing-the-spread/protect-yourself-and-others/ [Accessed 30 September 2021].
[13] N. Towey, Covid-19 Vaccine Tracker: How Many People Have Been Inoculated in Ireland? 2021, Available from: https://www.irishtimes.com/news/health/covid-19-vaccine-tracker-how-many-people-have-been-inoculated-in-ireland-1.4481095. (Accessed 30 September 2021).
[14] L. Boland, Public Transport Services Return to Full Capacity Today for the First Time in Over a Year, 2021, Available from: https://www.thejournal.ie/public-transport-full-100-capacity-restrictions-bus-train-5536334-Sep2021/. (Accessed 30 September 2021).
[15] R. Chin, The Handshake Went on Hiatus During the Pandemic. Some Doctors Hope It's Gone for Good, 2021, Available from: https://www.startribune.com/the-handshake-went-on-hiatus-during-the-pandemic-some-doctors-hope-it-s-gone-for-good/600073105/?refresh=true. (Accessed 30 September 2021).
[16] R. Trent, Is the Handshake Back? How We're Greeting Each Other as the World Reopens, 2021, Available from: https://www.ctvnews.ca/health/coronavirus/is-the-handshake-back-how-we-re-greeting-each-other-as-the-world-reopens-1.5457828. (Accessed 30 September 2021).
[17] H. Takeuchi, I. Nonaka, The new product development game, Harv. Bus. Rev. (January-February) (1986) 137–146.
[18] C. Sherlock, Explained: What is a Support Bubble? 2020, Available from: https://www.rte.ie/news/coronavirus/2020/1019/1172534-extended-household-bubble-explained/. (Accessed 30 September 2021).
[19] M.E. Black, Covid-19—This Too Shall Pass, 2020, Available from: https://blogs.bmj.com/bmj/2020/03/18/mary-e-black-covid-19-this-too-shall-pass/. (Accessed 30 September 2021).
[20] RFI, Venice Ban on Cruise Ships Comes into Force Amid Mixed Reactions, 2021, Available from: https://www.rfi.fr/en/europe/20210801-venice-ban-on-cruise-ships-comes-into-force-amid-mixed-reactions-italy-economy-environment-culture-unesco-protests. (Accessed 5 December 2021).
[21] UNESCO, Venice and Its Lagoon Rest in Fragile Balance. Findings of New Study on the World Heritage Site, 2021, Available from: https://en.unesco.org/news/venice-and-its-lagoon-rest-fragile-balance-findings-new-study-world-heritage-site. (Accessed 5 December 2021).
[22] J. Roozenbeek, C.R. Schneider, S. Dryhurst, J. Kerr, A.L.J. Freeman, G. Recchia, A.M. Van Der Bles, S. Van Der Linden, Susceptibility to misinformation about COVID-19 around the world, R. Soc. Open Sci. 7 (10) (2020), 201199.
[23] American Medical Association, AMA: Racism is a Threat to Public Health, 2021, Available from: https://www.ama-assn.org/delivering-care/health-equity/ama-racism-threat-public-health. (Accessed 5 December 2021).
[24] C.E. Johnson-Agbakwu, N.S. Ali, C.M. Oxford, S. Wingo, E. Manin, D.V. Coonrod, Racism, COVID-19, and health inequity in the USA: a call to action, J. Racial Ethn. Health Disparities (2020) 1–7.

Part II

Lockdowns and politics

Chapter 4

Lockdown alternatives

Sebastian Kevany and Aoife Kirk

Chapter outline

4.1 The lockdown dilemma

In late 2020, Ireland faced the divisive and largely unwelcome prospect of an extended second lockdown—one that was, inevitably, going to be more difficult to enforce than the first. Though both the Republic of Ireland and the Northern Ireland governments unquestionably tried their best to control the epidemic during the summer months, the question of what progress Ireland as an island made in the meantime is an important one. But, perhaps, such a step may have been inevitable under any circumstance [1].

Outbreaks, Epidemics, and Health Security. https://doi.org/10.1016/B978-0-323-91705-6.00015-0

The lockdown concept is part of a delaying phase in epidemic control—a period to allow individuals, society, and government to adjust to new circumstances—to prepare. In this regard, Ireland could be said to have, in some ways, used this time wisely—equipping the country with time to ensure face mask and personal protective equipment (PPE) availability (among other necessities which were initially in short supply). The gains in knowledge, and associated behaviour change, have also been critically important [2].

According to one postlockdown analysis, Ireland had one of the harshest and most stringent lockdowns in Europe, with Italy and the UK also among the top five countries with the most rigid rules in place [3]. The 2-km restriction is one such example—which many argue was unfair in the context of an island with a rural population—resulting in many people becoming tourists within their city, street, and neighbourhood.

PICTURE 2.4.1 Dry COVID lockdown humour in Ireland. Eimear Bourke.

4.2 Guest essay: Experiences and policy lessons from the first wave of the COVID-19 emergency in Italy: The importance of the geo-political context[a]

Serena Masino[b] and Luisa Enria[c]

Italy was the first European country to be hit by the COVID-19 emergency that went on to turn into a global pandemic. Images of overwhelmed hospitals, military trucks carrying coffins, and tightly regulated lockdowns announced what was to come across the continent [4]. Transmission had already been underway in part of the country from at least the beginning of January 2020 [5]. When evidence of the sustained spread of the disease emerged weeks later, the Italian government moved to introduce a series of emergency lockdown measures [6]; but this was likely too little too late at that time. Indeed, in the months that followed, and for the most part of the Spring in 2020, the country's epidemic grew exponentially and so did its death toll [7].

Understanding how the country got where it did from an epidemiological point of view is imperative in order to be able to operationalize any lessons learnt for future epidemics. However, it is increasingly recognized that outbreak responses need to also take into account the social, political, and economic dimensions of epidemics [8–11]. The ability to put in place effective responses, in fact, differs across social and geographical contexts because social policies, political economies, and historical trajectories have significant effects on different countries and communities' experiences, as well as on their ability and modality of adherence to epidemic control measures. The COVID-19 pandemic has given rise to a wide range of such analyses highlighting the impact of different policy decisions on transmission, the relationship between marginalisation and risk of infection, and the historical roots of mistrust in epidemic response measures including vaccination [12–20].

In the Italian context, specifically, a divide between northern and southern regions has long been politically salient, and, during the first wave of the pandemic, it accounted for marked differences in COVID-19 caseloads and responses. Indeed, as mentioned earlier, transmission continued unchecked for weeks and grew exponentially initially in Lombardia, a wealthy region in the industrial north of the country. However, as work-related movements are frequent, the disease was quick to spread to all other regions within a matter of weeks. In spite of scores of return migrants attempting to free the lockdown in the north to head back to their family homes in the south [21], the first wave of the pandemic broadly spared southern regions the crisis that engulfed the health systems in the north. This can be largely attributed to the difference in the timing of lockdowns in southern regions, once it was realized that the virus was quickly spreading in the north. The time lapse between the occurrence of sustained transmission in the two parts of the country, combined with fears raised by reverse migration, allowed southern regions to prepare themselves and lock down preemptively.[d]

---

[a] This research received ethical approval from the University of Bath and the University of Westminster and consent was obtained for all interviews.

[b] Dr Serena Masino, s.masino@westminster.ac.uk, University of Westminster, 35 Marylebone Rd, NW1 5LS, London (Corresponding Author).

[c] Dr Luisa Enria, Luisa.Enria2@lshtm.ac.uk, London School of Hygiene and Tropical Medicine, 15–17 Tavistock Pl, London WC1H 9SH, United Kingdom.

[d] This differs markedly from the experience of the following second wave in the autumn of 2020, when sustained community transmission was already under way all across the national territory.

As beds for the hospitalized were not freeing up at pace with new case numbers in Lombardia, private and public bed spaces in Intensive Care Units (ICUs) rapidly filled up as early as mid-March 2020. Many other regions in Italy only had a fraction of the ICU availability of Lombardia, particularly in the south. Thus, aware of their specific socio-economic conditions, the risk, and the concern in southern regions was such that, if the spread of the disease went unchecked, they could have been completely unable to provide ventilation, resuscitation, and ongoing care for critical COVID-19 patients, as well as emergency care for other patients with significant comorbidities.

More specifically, a number of factors characterize living and socio-economic conditions in the south as very different from the other half of the country. To start with, medical conditions of chronic nature may be more prevalent among the low-income sections of the population given their impaired or diminished access to adequate healthcare [22]. This, combined with living and working conditions that do not allow for effective social distancing, makes this part of the population particularly vulnerable to COVID-19. Living conditions in some of the urban areas in southern Italy indeed made it almost impossible for social distancing and containment measures mandated by central and regional governments to work. These areas are mostly home to low-income households and have very high levels of population density, with the poorest often cohabiting in a one-room open space. In addition, they experience a high prevalence of informal work. Due to their inability to access formal social safety nets during the pandemic, informal workers were unable to socially distance as they had to keep on working whenever possible.

Indeed, informal work being widespread in Italy [23] meant that many did not qualify for social assistance upon losing their jobs as a result of the pandemic. Although informal work is prevalent throughout the country, the vulnerability deriving from it is higher in southern regions. Nonetheless, no concrete steps were taken to establish a government-run income supplementation scheme with universal coverage that attempted to level such inequalities. This led to widespread complaints that national support measures had been laid out in a way that favoured northern regions, while the specific reality of the south had been overlooked. This narrative became particularly powerful as it reinforced an already existing record of exclusion and discrimination between different areas of the country.

An important policy point that emerges is therefore that while an appropriately timed preemptive lockdown was effective at sparing southern regions the brunt of the health crisis in the first wave of the pandemic, what hit them far more heavily turned out to be the socio-economic consequences of the national lockdown. This was due to the markedly different conditions prevailing there, with higher levels of informality, vulnerability, and deprivation. While a lesson learnt is thus definitely around appropriate timing of lockdowns, another equally important lesson is that measures of epidemiological nature are hard enough in a well-rounded socio-economic response plan. In particular, the institutional failure to establish a universal income supplementation scheme that was independent of employment and residence status led to perceptions of a skewed and unequal institutional response. Because this reflected historical patterns of structural inequality in the country [24,25], politically relevant perceptions of response measures were quickly exacerbated. The consequence was ultimate that crucially important trust and cohesion around the institutional response were weakened, if not partially lost in the successive phases of the epidemic.

4.3 Lockdowns and societal control

Governments also, of course, have to be honest with themselves about the missed opportunities which resulted in challenging positions of 'yo-yo' lockdowns throughout the year. The mainstream communication of the epidemiological term of 'flattening the curve' took on a more personal, implicit tone when the state enforced restrictive movements on citizens.

Yet, the World Health Organisation (WHO) Director General announced in April 2020 that lockdowns, while a feasible tool to suppress the virus, are only one of the many parts of public health response to a pandemic or epidemic [26]. That response includes a functioning and satisfactory test and trace strategy, ensuring the breaking of the chain of the virus through border controls, quarantine facilities, and advice to those exposed to the virus [26].

It also confirmed that infection control measures are carried out, including (in the case of a respiratory illness) mask wearing, hand hygiene, and good ventilation. However, at the same time, the WHO also warned that a premature lifting of lockdowns would spark further infections, resulting in longer-term economic damage.

Yet, there still remained public health policy options that would have mitigated the need for the enormous social, economic, and law and order damage done by the lockdown policy and other opportunities to shift the debate and narrative into a more positive, proactive, nonlockdown direction at that critical time [27]. So what lessons, if any, can we take from this?

PICTURE 2.4.2 Waiting room social distancing in Jordan. Jeremy Fowler.

4.4 Guest essay: Invisible Indians—The pandemic, mass exodus, and never-ending pain of migrants of India

Swati Mishra

India is one of the fastest-growing major economies globally, with a projected GDP growth of 11% during the FY 2021–22 [28]. However, the nature of work continues to be stuck in the past. With over 90% of the entire workforce working outside formal contracts [29], India's economy continues to be predominantly informal. Despite decades of growth, the overall proportion of informal workers' total employment has remained relatively unchanged. Hence, it will not be an understatement to say that the Indian workforce deals with an extreme informality in their employment. The opportunities to find decent wages, formal employment, and job security became restricted for the informal workers, with the increasing unemployment rate at 7.7% in 2019–20 [30].

In the last decade, informal migrant workers have played a pivotal role in ensuring that cities became engines of economic growth. However, their contribution as an informal workforce has remained either underappreciated or unrecognized so far. According to the 2011 census, there are 456 million migrants in India, of which 41.4 million migrated owing to their work [31]. While these numbers seem huge, they are still grossly underestimated, which leaves migrant workers in a blind spot.

The world faced unforeseen challenges when the global pandemic of COVID-19 hit in 2020. In India, the pandemic had the most gruesome impact on the migrants, particularly low-paid workers. The initial restrictions related to COVID-19 impacted the mobility of migrants, impacting their livelihood and wages. Of these migrant workers, circular migrants are more vulnerable than others.

The scale of India's circular migrant labour

There are no specific numbers on the extent of circular migrants[e] in India. However, estimates from [29] suggest that an estimated 120 million Indians travel from one state to another in search of better livelihood opportunities [31]. These numbers include people from both white- and blue-collar jobs. While it is easy to access the data of people with white-collared jobs, a vast number of informal wage workers engaged in jobs in textiles, diamonds, household works, building, and construction. There is no accounting of such workers because there is often no formal contract between the employee and employer, no formal rules and regulations of engagement, and no liability on the employer in the case of any workplace accidents or mishaps.

Circular migration depends upon many socio-economic factors. The vicious cycle of poverty, lack of education, social disadvantages, and risk to life lead people to migrate to cities, often facing occupational hazards, meagre pay, extended working hours, and poor living conditions. Out of 28 states and eight union territories, Haryana, Delhi, Gujarat, Maharashtra, and Kerala are the fastest-growing states

---

[e] Circular migrants are informal workers who migrate mostly from rural areas to urban cities in search of work. They are called circular due to their nature of spending 6–8 months in the destination cities and travelling back to their villages for the rest of the year to perform agriculture and allied activities.

resulting in the highest inflow of people from less-developed states like Uttar Pradesh, Bihar, Odisha, Rajasthan, Madhya Pradesh, Chhattisgarh, Mizoram, and Nagaland [32]. Estimates suggest that the five BIMAROU states[f] (Bihar, Madhya Pradesh, Rajasthan, Odisha, and Uttar Pradesh) contribute to approximately 50% of the out-migration in the country [33].

One of the prominent reasons for underestimating the number of migrants in India is the flexible definition of a migrant. The International Organisation for Migrants (IOM) defines a migrant worker as a person engaged or has been engaged in a remunerated activity in a state of which he/she is not a national [32]. In the Indian context, migrants are defined based on their place of birth and domicile, and a deviation from it categorizes them as migrant workers.

A global pandemic and the genesis of a mass exodus

The global pandemic of COVID-19 prompted massive reverse migration from 'destination'[g] to the 'source'[h] in various parts of the country. A sudden lockdown was imposed on the eve of 24th March 2020 on a short notice of 4h, leaving behind millions of people to manage their lives and livelihoods. The decision was unplanned and made without any consultations with local governments and led to a mass exodus and displacement of migrant workers—primarily the informal workforce. Thousands of migrant workers walked back to their source states without a safety net[i] in destination cities. While millions of them reached their home states only to quarantine in camps, many remained stranded far from home without any money or food.

The mass exodus of migrant workers from urban areas to inter- and intrastate hinterlands exposed the vulnerability of a migrant's urban life. 'I had no money to buy food, returning home was the only option left'—a migrant worker who travelled 1000 miles from Surat (Gujarat) to his hometown in Ganjam (Odisha) on a three-wheeler. Millions of others walked and cycled with their families because of a lack of facilities in their place of work.

The World Bank reported that lockdown, loss of employment, and social distancing prompted a chaotic and painful mass return for internal migrants in India [34]. More importantly, for us to understand are the reasons behind the mass displacement of migrants so that such situations can be dealt with effectively.

Scarcity of resources to survive in the destination

The migrant workers mainly work in blue-collared jobs, are not paid equitably for their work, and are dependent upon their employers. The nexus of recruitment of

---

[f] An Indian economic analyst Ashish Bose coined an acronym BIMAROU, in a paper submitted to then Prime Minister Rajiv Gandhi in the mid-1980s. It is an acronym formed from the first letters of states—Bihar, Madhya Pradesh, Rajasthan, Odisha, and Uttar Pradesh. BIMAROU has a resemblance with the Hindi word 'bimar' meaning 'sick'.

[g] Destination are the states and cities where the workers migrate in search of livelihood.

[h] Source states are considered as their birthplace or place where their family lives.

[i] Safety net can be defined as social protection schemes—a set of programs and policies broadly aimed to protect vulnerable individuals against poverty, marginalization, discrimination, and social exclusion. The social protection system helps to absorb the shocks and minimize the adverse impact on the vulnerable and poor and help them grow.

these workers is predominantly informal, where the worker does not get a contract from the employer and employers save themselves from paying any kind of provident fund. During the imposed lockdown in the first wave of COVID-19, the employers and the contractors (intermediaries engaged in recruiting workers) were the last hope for the informal workers. In the absence of any support from either the employer or contractor, the workers were left stranded in their place of work and had no other option but to start the painful journey back home.

The Stranded Workers Action Network surveyed 11,159 workers stranded in their place of work to document workers' vulnerability during the time of lockdown in the country. The survey findings suggested that 96% of the workers did not receive any food/ration support from the government in those difficult times, of which 72% reported having food left for only the next two days, 78% of the workers had less than Rs. 300 (less than £3) to spend while their wages were either held back by the employer or contractor [35]. Even as the state and employers hail migrants as the invisible backbone of the Indian economy, they were left in the lurch. The employers failed to pay the workers' salaries despite the government's continuous requests. The state, on its part, delayed starting special trains and other modes of transport. Such sociopolitical and economic constraints left the migrants with no choice but to return home by foot, bicycle, or whatever means seemed doable.

Loss of self-esteem and violations of fundamental human rights

Since the migrant workers do not possess voting rights in their place of work, their identity trickles down to a secondary citizen in their own country. Regionalism plays a significant role in keeping this divide intact, and they are left to the charity of the privileged class of the cities. The local governments pay no attention to the condition of migrants in the cities because migrants do not have any say in the framing of the government. Since these workers live in their destination cities most of the time, they also lose all the benefits provided in their hometowns.

In its report, the National Commission for Enterprises in the Unorganized Sector (NCEUS) established a nexus of poverty, vulnerability, type of employment, and social status of migrant workers [36]. Various studies conducted on the internal migration of informal workers show an over-representation of people from Dalit (scheduled caste), Adivasi's (scheduled tribes), women and children, religious minorities, landless people, and people whose primary source of income is agriculture. Due to their vulnerable position in society, these groups are forced to undertake the riskiest, most hazardous, and lowest-paying jobs and are looked down upon due to the nature of their work.

Failure of the socio-political system

As the country started to acknowledge the migrants' pain and suffering, many nongovernmental organizations and individuals came forward to help. The Government of India also announced approximately 350 schemes providing relief measures for the stranded migrant workers [37]. These included arrangements of special trains and buses from destination to source, shelter homes, food, water, and medical facilities while in transit, and instructions to employers for not deducting Rs. 31,000 crores of wages and welfare funds to support the building and construction workers registered with their employers. However, it was delayed by a month, and the delivery of provisions was marred by the failure to identify all migrant workers [37]. Some provisions were made so late that they did not benefit people in need; others did not reach the actual beneficiaries. The lack of trust in government policies was so shallow that most people did not even wait for the government to announce relief packages and start their journey back home.

Future course of action

In February 2021, Niti Ayog came forward with a draft National Migrants Policy. In the context of the ongoing global pandemic, the challenge is for the government to tackle the continuous migration and measures required to avoid another mass exodus.

Rights-based approach

The policy draft suggests a rights-based approach, focusing on cash transfers and reservations for migrants to enhance the capacity of the community to thrive on their own [38].

Institutional mechanism within the government system

The draft migrant policy identifies the Ministry of Labour and Employment as the nodal agency and proposes placing nodal persons of source states in the destination cities to address grievances in their local languages. The policy proposes migration resource centers in high-migration zones, a national labour helpline, financial linkages of workers, and migration management bodies in each state.

Amendments to existing laws

The Interstate Migrant Workmen (Regulation of employment and conditions of services) Act, 1979 is a four-decade-old act which we still follow to provide the safety net to the migrant workers. The draft migrant policy provides a valid argument to amend the Interstate Migrants Workmen Act, 1979 for 'effective utilization to protect migrants' [39]. Of many, the fundamental drawback of the act is that it only covers workers migrating through a contractor and leaves out the majority who migrate individually.

Creating alternative livelihood opportunities in their source states

The primary reason for not getting engaged in any other work within their states is low wages for semiskilled and unskilled workers. States like Kerala, Gujarat, New Delhi, and Maharashtra provide comparatively higher daily wages than other states. The draft policy acknowledges the need for alternative livelihood options and suggests creating more livelihood opportunities by promoting handloom works and alternative agriculture activities and increasing the daily wages for workers to limit migration.

The volume of seasonal migration—A case of migration from Ganjam (Odisha) to Surat (Gujarat)

Ganjam is the most populous district in Odisha on the country's east coast adjacent to the Bay of Bengal, with more than 35 lakh persons, constituting nearly 8% of the state's population. The district holds the distinction of having the highest incidence of interstate migration in Odisha. Ganjam has many of the characteristic push factors involved in migration—small landholdings, low literacy, a large proportion of historically marginalized communities, and frequent incidences of flood in the Bay of Bengal. Estimates suggest that approximately 4–7 lakh migrants from Ganjam have been working in the power loom sector in Surat for more than 30–40 years.

Surat is India's fastest growing business hub with the world's most extensive diamond cutting and polishing industry and India's man-made textile industry engaging more than 11 lakh migrants from all across the country. Surat forms the largest destination for outward migration from the Ganjam district, primarily in power looms and construction work.

Seasonality plays a significant role in circular migration between Surat and Ganjam. Every year these workers migrate to Surat for eight months to work in the power loom industry and move back to Ganjam during the agriculture season. Since Ganjam is a flood-prone area, a lack of livelihood options creates a nexus of migration in the district.

Need for robust data on migrants

To develop effective policies for migrants, the Ministry of Labour and Employment has taken the initiative to develop a National Database for Unorganized Workers to gather all the relevant information on workers engaged in informal/unorganized works across the country. Many proposals were developed for enumeration and robust data collection across the country. However, there is very little progress with the actual implementation of these schemes. In June 2021, the Supreme Court of India, an apex legal body of the country, issued advisories for the Government of India to complete the enumeration of migrant workers across the country by July 2021. This was launched in September 2021 [39].

Need for social protection

Social protection schemes cover public distribution systems of ration, accidental and life insurance, pension schemes, disability allowance, and other government-led benefits that help workers minimize the adverse impact of any mishap. The Supreme Court (SC) of India also acknowledges the need for food security for migrants in its latest judgement. According to SC, 'A government cannot abandon its duties to feed migrant workers, merely because they do not possess ration cards[j], and ordered all the states to implement the One Nation One Ration Card (ONORC)[k] by July 2021 [39]. Recently, the Government of India has launched schemes—such as issuing of E-Shramik Cards, MGNREGA (an employment guarantee scheme) for returnee migrants to source states, Affordable Rental Housing Complexes (ARHCs)—under the umbrella of Atmanirbhar Bharat (self-reliant India), focussing on the urban poor (including migrants) and agriculture and promoting small-scale businesses in villages to provide livelihood options within the district. The draft migrant policy acknowledges the need for portable access to social schemes through the Aadhaar Card for migrants in the country. The Aadhaar

[j] Ration cards are issued by state governments and are considered an official document in India. The households can get subsidized food grains from the Public Distribution System under National Food Security Act (NFSA) through a ration card.

[k] One Nation One Ration Card is a scheme that provides access to subsidized food to any card-holder at any fair price shop with his/her ration card in any part of the country.

Card is a 12-digit unique identification number issued by the Unique Identification Authority of India (UIDAI), for which one needs to present proof of address [40]. The irony is that most migrants do not possess an Aadhaar Card and ration card, essential documents for proof of identity. Even if the benefits of all the government schemes cover the last mile, a huge chunk of the population will remain untouched due to the absence of identity proofs.

As Dr Priya Deshingkar, Professor, Migration and Development, University of Sussex, rightly pointed out in one of her articles, an effective migrant policy must consider where existing labour laws fail. The draft migrant policy is the first step towards addressing the gaps in the existing labour laws which would help reduce the risks and vulnerabilities faced by migrants [40]. Although the policy has its shortfalls, it will be interesting to see how it unfolds and becomes more inclusive to migrants, especially women.

4.5 Lockdown alternatives

More widespread testing access: In late 2020, the government of the Republic of Ireland still did not cover testing for those who believed they had been exposed but were still asymptomatic. From a public health perspective, this was an unfortunate policy akin to only allowing HIV/AIDS testing for those with compromised immune systems (i.e. rather than the asymptomatic or at-risk population), though undoubtedly driven by cost concerns and the risks of people abusing the system. In addition, this lack of access infiltrated healthcare workers in Ireland and worldwide, who were often restricted by initially limited testing resources.

Unfortunately, this limited testing resulted in widespread workforce shortages on the frontline due to frontline workers being advised to isolate themselves if in close contact with a patient or colleague. There was difficulty proving infectivity rates due to asymptomatic cases during the first lockdown alone. With the introduction of low-cost (even if perhaps less accurate) testing assays, the pool of those who transmitted infections unknowingly would unquestionably have been reduced [41].

Strengthened app functionality: The use of tech via the Irish contact tracing app was, to many, hugely disappointing, with data privacy and confidentiality priorities overriding what could have been a beneficial epidemic control tool. In this regard, the lack of public availability of crucial information on the local spread of outbreaks and epidemics unquestionably hampered societal efforts at infection control. Reconsideration of government policies around protected health information (PHI) under these extraordinary circumstances represents a further nonlockdown option for epidemic control that the government failed to utilize.

Port-of-entry testing: Though complicated by the Northern Irish border, the government also appeared to have deliberately chosen not to introduce mandatory port-of-entry testing (along the lines of the Icelandic or United Arab Emirates models, for example) on the grounds of tourism, commerce, convenience, and logistics. Later in 2020, the Republic of Ireland would regret this missed opportunity to build borders, and international health security systems, that are much more effectively sealed against the importation of disease.

A narrative pivot to vaccine preparedness: Though not an alternative to lockdown policies per se, a shift in the governmental and media narrative towards prospective vaccine availability may have been an important one. In late 2020, the Republic of Ireland's vaccine preparedness policy had been conspicuous by its absence: with a pivot in the governmental and media narrative towards this prospect, lockdowns and associated regulations could have been framed in this context. This would also have provided valuable timelines for the public to work and plan around, enabled the preparation of supply chain and logistical systems, provided critical dates for public and policy planning, and initiated the challenging refutation of antivaccine disinformation at societal and individual levels [42].

PICTURE 2.4.3 Hopeful messaging in Ireland—with political overtones. Eimear Bourke.

4.6 Guest essay: Pandemic lessons: A Buddhist perspective

Kathylene Castillo Southard

I remember being on the island of Maui, Hawaii, on January 13, 2018. As a clinical social worker, I was participating in a psychotherapy training in EMDR—Eye Movement Desensitization and Reprocessing—a treatment that helps people work through trauma. On that morning, a chorus of phone alerts went off just as we were starting. And a message came across our phones declaring 'BALLISTIC MISSILE THREAT INBOUND TO HAWAII. SEEK IMMEDIATE SHELTER. THIS IS NOT A DRILL.'

Interestingly, at this moment, we were about to get the down-low on ATIP—acute traumatic incident processing. Our teacher said, before dismissing us, 'This happens every time'. Apparently, many times when our EMDR instructor teaches this acute incident processing, something traumatic happens to disrupt his class in real life, like fire alarms, building crumbling, and other emergencies.

Carl Jung tells us to pay attention to synchronicities. Yes, this was a moment to remember, to be present for. I only started to feel that this was a serious threat when I overheard fellow students calling loved ones in tears saying things like 'If I don't come home, tell everyone I loved them'. It was surreal. I was surrounded by 50+ emotional psychotherapists, after all. Lots of panic. Lots of tears. Weirdly, I felt fine. I thought that I was about to meet my maker or be harmed in some irreparable way and that chaos was definitely coming.

However, I also felt so curious, wondering what would happen next and there was this deep knowing inside, that whatever happens is basically good. I will be basically okay even if death or dismemberment comes to me today. I called a good friend on Maui. We had a very upbeat chat. Yes, we may die today! A few minutes later, a friend in Poland informed me that it was a false missile alert by sharing a Twitter post from Tulsi Gabbard, the former Member of the U.S. House of Representatives from Hawaii's 2nd district. Apparently, Tulsi was the first to alert everyone that it was a false alert. A friend on the other side of the world sent me this. So weird, this modern life of ours.

I mention this experience because it is a concentrated moment in time when the world seemed threatening and we were all going to die, like this ongoing pandemic. It's the end of the world, and I feel fine. How is that? It's from that deep understanding of basic goodness.

> The bad news is you're falling through the air, nothing to hang on to, no parachute. The good news is, there's no ground.
>
> ***Chögyam Trungpa***

Fear, threat, anxiety, death, and illness on a mass scale, destruction, and disaster have made their way into everyday life for all of us for much of 2020, into 2021. People have lost so much: lost loved ones, lost lives. They've also lost normal everyday interactions with other humans, lost jobs, lost homes, or lost the ability to go to their college graduation, the wedding, or vacation that had been planned.

To me, it felt like the veil had lifted: that the usual state we all live in, that we ignore, was being revealed to everyone, that there is no ground and there is no security, despite what you may think. Death is real, and it comes without warning. Everything falls apart. And nothing ever comes together to resolve itself fully as we want. We are always in a state of not knowing, and impermanence, despite tricking ourselves otherwise that the ground is solid underfoot. Those moments of sudden change can kick us right in the gut, but they are a real reminder to bring us back to what is important and what is meaningful.

This can bring tremendous anxiety and fear, yes—but at the same time, understanding this is the nature of existence brings great relief when that fear is fully investigated, felt, and accepted. Then, we can relax with the groundlessness of it all.

My former Buddhist teacher, Sakyong Mipham Rinpoche, has said that the question of human nature is the most important global issue we face today. Is human nature basically good, or is it not? He says if we conclude that humanity is not basically good, that we do not possess inherent wisdom, what hope does the future hold?

I agree. Often it seems like we're all going to kill each other, that humans will be wiped off the planet from climate change or another pandemic. But, actually, what I do believe is that we're going to figure it out and create this good human society where we take care of each other on a global scale, as one massive human family. My opinion is that it's one or the other, in the end, and the outcome depends on what we as a society believe to be true about human nature.

When we don't see that basic goodness in society, the forces of fear and doubt escalate, and we experience tremendous anxiety and upheaval in the face of change and threat. Depression ensues, creating harm to our minds and hearts. It can be so tempting to just want to close off—shut off into ourselves—forget about the rest of the world, and retreat to one's own comfortable cocoon, hoping that the world's miseries don't affect us personally too much. Or perhaps we can just blame it on someone bad: the Chinese, the Mexicans, the immigrants, or that little old Asian lady who lives down the street....

That's when things go really awry. Violence and aggression escalate. This pandemic has taught us just how interconnected we all are. We're forced to confront just how much our actions affect others. If we continue down this path of separateness, thinking our actions have no impact on others, we mindlessly cause more harm. The challenge is convincing people that it's not enough to just protect oneself but to protect everyone else—and that protects ourselves, too. One needs to have empathy to understand this. Everyone has empathy, but it requires connecting with our hearts, a brave act when we're caught in fear. To do that requires trusting that everyone has this same basic goodness, no matter how much it is buried, twisted, or seemingly missing in the other. We have to wake up. All of us. This is the path forward I can see in creating a good human society and a more conscious and enlightened society.

PICTURE 2.4.4 **Empty lockdown landscapes in Ireland.** Aoife Kirk.

4.7 Conclusion: Lockdown trade-offs

Most importantly, it would also have made short-term lockdowns more palatable and thus improved compliance. Instead, the government risked rebellious reactions to fears of *ad infinitum,* suspensions of services, and loss of freedom, which characterized policy decisions earlier this year. Finally, of course, the Christmas meltdown in infection rates was also directly linked to the unnecessary lifting of the lockdown with vaccine availability only weeks away.

All of these strategies unquestionably entailed risks, costs, and trade-offs against other policy objectives and principles. And yet, they still appear to be the only way to avert further lockdowns. In March 2020, no one could have predicted the future, but perhaps if we had appreciated that this is a global and not a national problem [1], we would also understand that the Irish epidemic would not be over until the international pandemic is over. Perhaps by late 2020, Ireland would have been closer to a New Zealand model whereby schools, pubs, hospitals, and sporting events could return to function in relative normality. Should countries, as a last resort, ever have to go through the whole unpleasant process of a lockdown again, the government will have to demonstrate that it has learnt from these events.

PICTURE 2.4.5 A return to nature was a big part of many Irish; people's COVID experiences—if they were lucky enough to have access. Sarah O Dwyer.

4.8 Key points

- In the early stages of pandemic, lockdown policies may have benign effects and may perhaps even be reassuring for the general public to help control the spread of disease.
- However, lockdown policies are also inherently unsustainable and difficult to enforce, as the social costs of limitations on movements counterbalance the health gains.
- A range of options related to testing, contact tracing, and health security are available as an alternative to lockdowns at the point when society can no longer sustain compliance.

References

[1] A. Doyle, W. Hynes, S.M. Purcell, Building resilient, smart communities in a post-COVID Era: insights from Ireland, Int. J. E-Plan. Res. 10 (2) (2021) 18–26.

[2] B. Colfer, Herd-immunity across intangible borders: public policy responses to COVID-19 in Ireland and the UK, Eur. Policy Anal. 6 (2) (2020) 203–225.

[3] Bsg.ox.ac.uk., COVID-19 Government Response Tracker. [online] Available at, 2021. https://www.bsg.ox.ac.uk/research/research-projects/coronavirus-government-response-tracker#data. (Accessed 5 December 2021).
[4] Val Seriana News, https://www.valseriananews.it/2020/03/19/limmagine-che-bergamo-non-dimentichera-mai-bare-portate-via-dallesercito/.
[5] ANSA,https://www.ansa.it/canale_scienza_tecnica/notizie/biotech/2020/03/25/ricerca-il-coronavirus-era-in-lombardia-dal-primo-gennaio-_7c598d66-d4fb-4c2f-95df-519950da17ca.html.
[6] LaRepubblica, https://www.repubblica.it/politica/2020/03/01/news/coronavirus_misure_governo-249980561/.
[7] JHU's Coronavirus Resource Centre, https://coronavirus.jhu.edu/data/mortality.
[8] S. Abramowitz, Epidemics (especially ebola), Ann. Rev. Anthropol. 46 (2017) 421–445.
[9] K. Bardosh, et al., Towards People-Centered Epidemic Preparedness and Response: From Knowledge to Action.SSHAP, 2019.
[10] L. Enria, The ebola crisis in Sierra Leone: mediating containment and engagement in humanitarian emergencies, Dev. Change 50 (6) (2019).
[11] K. Sams, A. Desclaux, From Ebola to Plague and Beyond: How Can Anthropologists Best Engage Past Experience to Prepare for New Epidemics? Cultural Anthropology, 2017.
[12] K.L. Bardosh, et al., Integrating the social sciences in epidemic preparedness and response: A strategic framework to strengthen capacities and improve Global Health security, Glob. Health 16 (1) (2020) 1–18.
[13] S.E. Davies, C. Wenham, Why the COVID-19 response needs international relations, Int. Aff. 96 (5) (2020) 1227–1251.
[14] L. Enria, et al., Trust and transparency in times of crisis: results from an online survey during the first wave (April 2020) of the COVID-19 epidemic in the UK, PLoS One 16 (2) (2021).
[15] T. Hrynick, M. Schmidt-Sane, S. Ripoll, Rapid Review: Vaccine Hesitancy and Building Confidence in COVID-19 Vaccination, 2020.
[16] M. Leach, et al., Covid-19: Key Considerations for a Public Health Response, 2020.
[17] S. Ripoll, et al., SSHAP: 6 Ways to Incorporate Social Context and Trust in Infodemic Management, 2021.
[18] D.B.G. Tai, et al., The disproportionate impact of COVID-19 on racial and ethnic minorities in the United States, Clin. Infect. Dis. 72 (4) (2021) 703–706.
[19] D.A. Widyastari, Understanding community's behavioral responses towards Covid-19: Who comply, who don't and reasons of incompliance, GHMJ 4 (1) (2020) 1–4.
[20] A. Wilkinson, et al., COVID-19 – a social phenomenon requiring diverse expertise, Affairs 114 (454) (2020) 136–148.
[21] La Stampa, https://www.lastampa.it/cronaca/2020/03/08/news/coronavirus-l-assalto-ai-treni-per-il-sud-nel-timore-della-paralisi-1.38566116.
[22] The Guardian, https://www.theguardian.com/us-news/2020/apr/10/new-york-coronavirus-inequality-divide-two-cities?CMP=Share_iOSApp_Other.
[23] ISTAT, https://www.istat.it/it/files/2018/10/Economia-non-osservata_2013-2016_rev.pdf.
[24] V. Daniele, P. Malanima, Il prodotto delle Regioni e il divario Nord-Sud in Italia (1861-2004), Rivista di politica economica 97 (2) (2007) 267–316.
[25] I. Odoardi, F. Muratore, The north–south divergence in Italy during the Great Recession, Manch. Sch. 87 (1) (2019) 1–23.
[26] 4SD, COVID19: Do what is needed to hold the virus at bay | 4SD. [online] Available at, 2021. https://4sd.info/covid-19-narratives/do-what-is-needed-to-hold-the-virus-at-bay/. (Accessed 5 December 2021).

[27] C. Pieh, S. Budimir, J. Delgadillo, M. Barkham, J.R.J. Fontaine, T. Probst, Mental health during COVID-19 lockdown in the United Kingdom, Psychosom. Med. 83 (4) (2021) 328–337.
[28] B. Standard, Urban Population to Contribute 70-75% of India's GDP by 2020-Barclays. Retrieved From Business Standard, 2014. https://www.business-standard.com/article/news-cm/urban-population-to-contribute-70-75-of-india-s-gdp-by-2020-barclays-114032000273_1.html.
[29] Census, Office of the Registrar General & Census Commissioner, Ministry of Home Affairs, Government of India, 2011, Retrieved from D-1: Population classified by place of birth and Sex - 2011: https://censusindia.gov.in/2011census/d-series/d-1.html.
[30] M.Kapoor,TheBestAndWorst-PerformingStatesOfIndia.RetrievedfromBloombergQuint,2021. https://www.bloombergquint.com/economy-finance/the-best-and-worst-performing-states-of-india.
[31] K. Das, S. Saha, Inter-State Migration and Regional Disparities in India. Retrieved from. n.d. https://iussp.org/sites/default/files/event_call_for_papers/Inter-state%20migration_IUSSP13.pdf.
[32] WB, Covid-19 Crisis Through a Migration Lens, World Bank Group, 2020.
[33] ILO, International Convention on the Protection of the Rights of All Migrant Workers and Members of Their Families. (n.d.). Retrieved from United Nations Human Rights- Office of the Commissioner: https://www.ohchr.org/en/professionalinterest/pages/cmw.aspx.
[34] SWAN, 21 Days and Counting: Covid 19 Lockdown, Migrant Workers and the Inadequacy of Welfare Measures in India, Stranded Workers Action Network, India, 2020.
[35] NCEUS, Conditions of Work and Promotion of Livelihoods in the Unorganised Sector, Dolphin Printo, New Delhi, 2007.
[36] G. Order, Covid-19 Government Order Tracker, 2021, Retrieved from Court Orders: www.covid-india.in.
[37] K. Mehrotra, Explained: What is NITI Aayog's Draft National Policy on Migrant Workers? 2021, Retrieved from https://indianexpress.com/article/explained/niti-aayog-migrant-workers-policy-covid-lockdown-7201753/.
[38] SC, In the Supreme Court of India Civil Original Jurisdiction Suo Motu Writ Petition (Civil) NO.6 OF 2020, 2021, Retrieved from IN RE: PROBLEMS AND MISERIES OF MIGRANT LABOURERS: https://main.sci.gov.in/supremecourt/2020/11706/11706_2020_36_1501_28166_Judgement_29-Jun-2021.pdf.
[39] UIDAI, Unique Identification Authority of India, Retrieved from, 2021. About Aadhaar: https://uidai.gov.in/index.php?option=com_content&view=article&id=14.
[40] P. Deshingkar, An Effective Migrant Labour Policy Must Consider Where Existing Labour Laws Fail, Retrieved from The Indian Express, 2021. https://indianexpress.com/article/opinion/columns/national-migration-policy-labour-laws-niti-aayog-7261732/.
[41] M.J. Binnicker, Challenges and controversies to testing for COVID-19, J. Clin. Microbiol. 58 (11) (2020). e01695–20.
[42] K. Krishan, T. Kanchan, Lockdown is an effective'vaccine'against COVID-19: a message from India, J. Infect. Dev. Ctries. 14 (6) (2020) 545–546.

Chapter 5

Creativity and originality: Original responses

Sebastian Kevany and Aoife Kirk

Chapter outline

5.1 A challenging initial response

In the opinion of those who have witnessed and participated in responses to numerous public health emergencies (PHEs) and epidemic outbreaks throughout the world for many years—from Cholera in Zimbabwe, to Ebola in Sierra Leone, to multidrug-resistant tuberculosis in Iraq, to all of the longer-term issues associated with HIV/AIDS in sub-Saharan Africa, or to malaria in the South Pacific—the government of the Republic of Ireland's initial response to Ireland's health security concerns in early 2020 deserves to be complimented, if not yet lauded.

Outbreaks, Epidemics, and Health Security. https://doi.org/10.1016/B978-0-323-91705-6.00002-2

However, one wonders how a less-developed political regime would have dealt with these challenges and the associated need for swift, decisive, and potentially unpopular policies? Such situations demand experienced leadership to navigate the national and international bureaucratic pathways that appropriate choices require [1].

Perhaps timing had been in Ireland's favour—and not merely in the context of the end of winter. Maybe the country had been lucky, as well, to be led at the beginning of March 2020 by a medically trained prime minister (Taoiseach). Yet, this was also a politically fragile time in Ireland—with a recent election and a government in transition [2]. Were national strategies appropriate, timely, and suitable; were there current opportunities to consider both the strengths and the limitations of the initial national response that may better inform policy and individual behaviour? These questions remain to be seen during the initial pandemic response and the long-lasting implications and effects in the future.

Ireland's public health information campaign since the beginning of the pandemic in March 2020 was both well-funded and admirable. However, it would not, for example, have taken a great deal of foresight for the government to have anticipated the domino effects of its announcements, messaging, and media-intensive

PICTURE 2.5.1 Panic buying. Eilis Conneally.

saturation strategy. The risk of hoarding and panic buying was high during the first lockdown [3]. Ensuring resources are fairly distributed by avoiding amoral stockpiling-for-profit, opportunist retail pricing, and excessive hoarding at the individual level—is just as important as personal health safety [3]. Nevertheless, lessons were learned, with future press releases ensuring clear, compelling, and straightforward public health messaging and communication to the general public, for the most part.

Second more frequent updating of public health messaging—as well as inclusion in different languages (in Irish (as Gaeilge)? Polish? Italian? Portuguese?) and on other platforms—was also essential to improving public health messaging in Ireland [4]. An overreliance on the Internet, for example, often resulted in the contamination of objective messages with fake news and conspiracy theories—not just causing further panic and distress, but also further alienating those small but highly susceptible elements of the population who are not always online.

Mediums such as Twitter [5] were used by healthcare professionals, scientists, and epidemiologists to share information fast and often. Will the future for public health messaging rely too much on 'bubbles', where fake news can be easily spread? Should we also hold these companies—which have expanded power and capacity internationally—accountable to ensure accurate reporting and messaging? Academia has long been an area that takes time, often years, to publish, which is, of course, too slow in a health emergency. The focus and medium of the messaging must and no doubt will also evolve.

This public health messaging has had a significant impact, not only in Ireland, but throughout the world. However, in a world that is oversaturated with messaging, there is a risk that messaging of health behaviours gets lost. The risk of oversharing information, along with health anxiety, trust issues, uncertainty, lack of faith in governmental structures, individual risk behaviours, health literacy—which we have experienced on numerous occasions during this pandemic—all contribute to 'alert fatigue' (when too many alerts and changing of health messages result in mixed messages). Ensuring cohesive messaging, ideally from local channels which expand 'bottom-up' towards governmental 'top-down' approach, is yet another lesson public health bodies, scientists, and epidemiologists have learned.

	Level 1	Level 2	Level 3	Level 4	Level 5
Visitors	10 from 3 households	6 from 2 or 3 households	One other household	No visitors permitted	No visitors permitted
Gatherings	50 outside your home or garden	6 indoor 15 outdoor	No social or family gatherings	No indoor gatherings	No Except funerals or weddings
Weddings	100 people can attend	50 people can attend	25 people can attend	6 people can attend	6 people can attend
Indoor	100 Up to 100 can attend indoor gatherings	50 Up to 100 can attend indoor gatherings	No indoor gatherings	No indoor gatherings	No indoor gatherings
Sports fixtures	100 indoor 200 outdoor	50 indoor 100 outdoor	No matches or events	No matches or events	No matches or events
Bars, cafes & restaurants	Open with protective measures in place	Open 6 people from up to 3 households	Open Additional restrictions for indoor dining	Open. Take away or delivery. Outdoor dining only. Max 15 patrons	Open. Take away food or delivery only.
Wet pubs	Open	Open Groups of six from 3 households	Open Additional restrictions	Open outdoor only limited to a max of 15	Only offer takeaway or delivery service

Good guide to new restrictions etc

PICTURE 2.5.2 Covid restrictions. Aoife Kirk.

Third, the malign effects of panic on individual health and immunity should not be underestimated. Without a doubt, Ireland—compared to many of its neighbours—is in recent years a liberal, some might even say relaxed, society. Whilst government responses at times allowed for the casual 'it'll be grand' mentality [6], associated messaging and strategy have to be tailored to counterbalance the country's unique cultural context, as there are also potential negative downstream consequences to this approach.

Therefore, attaining good public health messaging, which can often be fear-inducing, is a tricky balance to strike [7]. Not only do we need to increase public knowledge (what is a pandemic; how does epidemiology measure risk; how can the general public 'flatten the curve'), but also to avoid panic, the messages must try to win over the trust of the general public [8]. In that regard, it is often unclear whether authorities have weighed up related national stress and fear quotients and associated health effects. Without the individual, behavioural, and social responses, the losses are infinitely and immeasurably preferable to a laissez-faire response.

There was virtue in the high-intensity, occasionally Hollywood, occasionally doomsday, messaging that the government employed—and that the media have both benignly advanced and malignly pounced upon. Nonetheless, the sense of panic that this also created weakens national and individual resilience [8], and it remains to be seen how many have suffered the health consequences of excessive paranoia. So a balance needs to be struck here, as well.

5.2 Guest essay: China versus COVID-19: A battle China has won

Sandra Chukwudumebi Obiora[a]

Introduction

The nightmare imposed on humanity with the advent of COVID-19 in Wuhan is an experience that should never be relived. The negative effects brought on by the pandemic have been far-reaching. Millions of lives have been lost, countless economic backbone industries have barely managed to survive, and one of the most dramatic financial market crashes the world has ever seen was witnessed [9]. As the world continues to battle the pandemic, great political, economic, social, and academic minds are working on two major things. First, how can the world return to relative normalcy? And second, how can we ensure that such a calamity does not reoccur?

Such critical questions are ones already being tackled aggressively by China. China's fight against COVID-19 is something of a spectacle that has left most of the world greatly astounded. Beyond achieving a state of zero new infections before the summer of 2020, China managed to pick its economy back up despite the negative global trend even as the rest of the world experienced a severe economic deficit between 2020 and 2021 [10]. From the discovery of the virus, to the rapid rise in the number of infected cases, to effective control of the spread, and finally to the attainment of apparent normalcy, China has indeed managed to set a working template on how to conquer a virus spread.

This reflective essay on China and its fight against the pandemic will present the following: first, China's experience and response during each critical stage of the fight against COVID-19; second, an outline of the template for success China set in place, as well as the plans for the future; and lastly, an assessment of the lessons the world and future generations can learn from China's fight against COVID-19.

The beginning of a global health crisis

In December 2019, the first cases of human infection as a result of a novel virus, now known as COVID-19, were reported in Wuhan, in the Hubei province of

[a]School of Management and Economics, University of Electronic Science and Technology of China, Chengdu, Sichuan, People's Republic of China. Email: sandra_ora2000@hotmail.com (Corresponding author).

China [11]. Soon, the rapid increase in daily case counts reached 81,000 cases within mainland China and 148,000 global cases. During this time, China had already begun its fierce fight against COVID-19 as the nature of the virus was confirmed. Several definitive approaches were taken in attempts to ultimately stop the rapid spread and bring infection counts down to zero. Armed with wisdom and know-how after suffering the SARS outbreak between 2002 and 2003, the Chinese government promptly prioritized standard infection control measures beginning from Wuhan. A nationwide mandate to wear face masks, wash hands, keep a 1-m distance, amongst other such regulations was promptly enforced. Rigid restrictions were enforced first on Wuhan from January to June of 2020.

This complete lockdown of the city ensured that it was effectively sealed off from the rest of mainland China. By the start of February, China stunned the rest of the world with its construction and completion of the Huoshenshan hospital holding 10,000 beds created to isolate the infected in just 10 days and with the effort of 7500 labourers [12]. This was in effect China's promise that the fight against the virus would be won. China is one of the only countries that after April 2020 did not appear to experience a second or third wave of infections.

Taking the hard but effective route on a national scale

One key thing that cannot be neglected in the analysis of China's fight against the virus is, first, its tenacity to overcome the virus, and more especially its strict enforcement of pandemic control regulations despite economic cost [13]. From the highest in position to the lowest, from the strongest to the weakest, from the young to the old, most Chinese people, to a large extent, consciously obeyed the regulations created. Beyond the lockdown on Wuhan and then Hubei province at large, this also included an almost nationwide lockdown involving every province, city, district, street, and housing unit where infection cases were noted or rising. In Chengdu Sichuan province, for instance, where I resided, the international and local students living on the university campuses were sealed off from the students living outside the campuses. COVID-19-related regulations were enforced by every housing, schooling, and business unit.

For instance, whilst students were not allowed to leave the campuses unless in exceptional cases, students and residents outside campuses were not allowed to leave their compounds more than once a day. Even then, each person in each housing unit had to have some specific documents showing the residence area they belonged to, and was not allowed to cross into other districts even in the same city during that time. Strict 2-week long quarantine rules were set in place for those who were returning to China or even travelling from or to a different city. In addition, its borders were soon closed off at the end of March 2020 to avoid a second wave of the virus due to imported cases.

Another thing that China did was to deploy 450 military medics to the epicentre of the virus outbreak [12]. Beyond this effort to assist the fight against the virus spread, as many as 1700 doctors and other medics travelled to Wuhan of their own volition to assist in the fight against the virus [14]. Almost all of them took this step at their own risk whilst leaving their families behind. This combined effort of both government and citizens led to the results seen today. There was undoubtedly a single mindedness about the fight. Whilst many could not open their businesses, and countless had to work from home, overwhelming support was still shown daily for those in Wuhan, and those working in the medical and service industries. These were, in effect, popularly tagged as the 'soldiers and heroes on the front lines'.

Additionally, the Chinese Confucius virtues of integrity, harmony, and benevolence were perhaps the foundational mindset most evident during the toughest of times, and undoubtedly played a big role in bringing the average Chinese person through those months of lockdown and heavy mobility restrictions.

The utilization of technology and the return to 'relative normalcy'

A major tool that assisted China in maintaining accurate infection records and returning to normalcy was its utilization of technology. Additionally, it utilized already widely used means to make infection spread control possible. Present-day China is broadly cashless. Especially for those living in urban areas, there is almost no time when the use of cash is needed. Technology giants like WeChat and Alipay have made it possible to pay for goods and services, no matter how small, by simply scanning QR codes and punching in payment passwords. From using public transport to calling a taxi, to even buying snacks on the street, these payment methods have made life quite simple for the average Chinese person in possession of a phone and Internet.

In China's fight against COVID-19, the first critical step that was taken to ensure that relative normalcy could be achieved and greater mobility could be made possible was the invention of health codes. Health codes were created to reduce cross-border transmission of the virus from both within and outside China. They are in effect the implementation of contact-tracing meant to detect when an individual has been at risk of exposure. Health codes have three colours: green meaning 'no exposure', yellow meaning 'possibility of exposure', and red meaning 'high chance of exposure' within the last 14 days.

Depending on the colour of the individual's health code, they may or may not be allowed to move around freely or travel. Widely used applications like Alipay and WeChat have embedded these health-code apps as mini-programmes, making access easy for over 1 billion users in China [15,16]. Registration of national ID/passport details and addresses was mandatory to receive a health code. Thus, soon after the lockdown ended, individuals could not enter public spaces such as public transport, housing units, business buildings, or schools without first presenting a valid green-coloured health code at the door, wearing a face mask, and getting their temperature checked. Universities also implemented private applications which kept track of all the students inside and outside China, within and outside the campuses. Students are expected to fill out a daily health report before they can be allowed inside or outside the campus, and movement is still generally restricted to 12 h daily. All these measures have ensured that on a national scale, infection numbers remain low or even zero in some cities and provinces despite having a population of over 1.4 billion people.

China today, vaccinations, and the sustained hope for the future

Whilst China has, in effect, won the current battle against COVID-19 with the enforcement of many strategies on national, provincial, and local scales keeping infection rates low, it is now on its way to solving the second presented issue. How can it be ensured that such a calamity does not reoccur? Indeed, infection counts have been kept low, and possibilities of second and third waves are being kept at bay. For instance, places with new cases are closely monitored, and entire districts and, in some cases, cities are promptly sealed off. Recent situations like Beijing in early 2021 and Guangzhou in June 2021 experienced lockdowns as new cases were found.

Earlier in 2021, Chengdu saw about 12 cases, and this number of infections led to rapid COVID-19 testing of millions of people. For instance, in a particular district in Chengdu, as many as 2 million residents were tested within 2 days to ensure quick tracing of new infections. Additionally, these tests were administered free of charge.

The current phase (at time of writing, July 2021) in the fight against COVID-19 is the fight to vaccinate the majority of the population. China was amongst the first to create a vaccine, and its type of vaccine differs fundamentally from most others currently being used in the world. In particular, when comparing the Chinese vaccines such as Sinovac or Sinopharm with Moderna or Pfizer vaccines, for instance, Sinovac works by utilizing killed viral particles which expose the body's immune system to the virus. This does not risk any serious disease response. In essence, the Chinese vaccines have been developed using the traditional methods of vaccine development, unlike the new mRNA vaccine types. Moderna and Pfizer for instance are mRNA vaccines whereby a part of the virus' genetic code is injected into the body to trigger the production of viral proteins that then train the immune system to attack. Sinovac's key advantage is that it can be stored in standard refrigerators at 2–8 degrees. Comparatively, vaccines like Moderna and Pfizer require temperatures down to −20°C and −70°C, respectively, for the components not to break down. This simple advantage, regardless of effectiveness percentage, makes the Chinese vaccines easily exportable to the Global South where many countries may be unable to properly afford storage of vaccines that require extremely cold temperatures [14].

As of June 2021, China has already administered over 1.23 billion COVID-19 vaccine doses accounting for close to half of the global total vaccine doses administered. China's goal is to vaccinate over 70% of its population before 2021 ends [17]. It became the first country to have administered over a billion doses in 2021 superseding its half-year goal [18]. This vaccination rate is thanks to several incentives gradually being set in place for those who are vaccinated. Besides the fact that the doses are covered by a variety of affordable and commercial insurance packages, incentives like food packages, cash prizes, and vouchers are being offered to those who get vaccinated in areas like Taiwan.

In mainland China, local districts, residential administration boards, schools, and universities are persuading residents to take the vaccine [19]. As the vaccination rate increases by millions daily, news of restrictions to mobility for those who are not vaccinated have already begun to spread. A future where a vaccine certificate is necessary when using public spaces does not at all seem far off. As China continues to rapidly increase its rate of inoculation, nationwide immunity to COVID-19 is just around the corner. Despite China's large population, it will soon join countries like Canada, Israel, the United Kingdom, and the United States who currently hold the highest percentage of the population vaccinated, further pushing the 25.6% count of global vaccinations higher [20].

Conclusion

It is indubitable that some of China's success in fighting COVID-19 stems from the very attributes that set China apart from most of the world. Its unique governance style, thriving global economic presence, cultural values, and Confucian ideologies create a foundation upon which its 1.4 billion residents can depend. All that China stands for, and its long-understood ability to maximize its resources to achieve outstanding results, has come into play in its fight against COVID-19. Strict and very rigid pandemic control regulations were actively enforced all over the country starting from the total lockdown of Wuhan and Hubei provinces. Information was made readily available to raise awareness, and technology was utilized and made simple to understand, and used towards widespread contact-tracing. Infrastructural prowess and brilliant utilization of human resources were showcased in the building of its 10,000-bed hospital within 10 days. The rising death toll was promptly cut short with all these measures. Vaccines were created in record time, and inoculation rates are already superseding local and global expectations. China's ability to quickly respond to new situations, enforce regulations where necessary, make and work avidly towards the achievement of short- to long-term goals, and its capacity to mobilize and effectively utilize resources are lessons the world can indeed observe and learn from. Without a doubt, China has set down a template on how to fight and overcome a pandemic that the rest of the world may count as a success story.

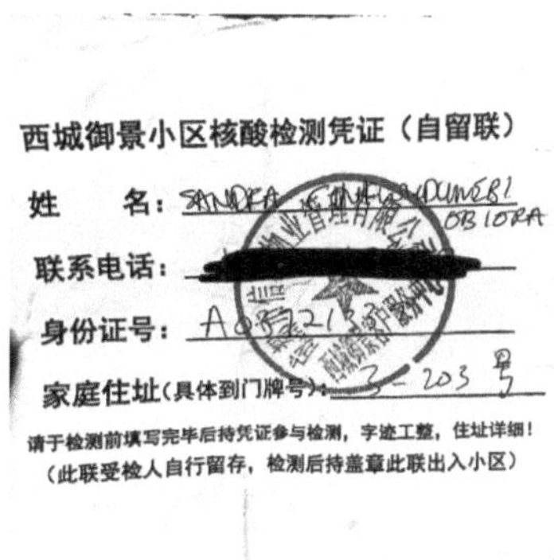

PICTURES 2.5.3, 2.5.4, AND 2.5.5 Images from China's Covid experience. Sandra Obiora.

5.3 The forgotten concept of health security

The role—or lack thereof—of the defence forces was also moot. Whilst their deployment would no doubt risk increasing individual and collective anxiety even further, in Ireland—as in many other countries—this valuable resource currently initially stood idle, at a time when essential services could therein be provided. The historical stigma associated with such mobilizations is combined, in Ireland as elsewhere, with a passé deficit in military public health education training [21].

The use of defence forces eventually occurred in Ireland to aid the public health workforce [22], underfunded for years prior. The efficiency and speed of cooperation and achieving tasks such as contact tracing were notable. Are we underutilizing a service that could be given training in pandemic preparedness and also willing to engage for the national 'greater good' [22]?

And yet—on the occasions when armed forces have been involved in humanitarian efforts in the United States and elsewhere—a documented spin-off has been the strong sense of accomplishment amongst (and goodwill towards) soldier-caregivers: valuable downstream benefits.

Should this be an essential part of future Irish military training—if it is not already, is there a plan in place to convert barracks into bedsteads, battalions and brigades into benevolent enforcers and providers of public health decrees and directives? Likewise—in similar future scenarios—all individuals, industries, professions, and services may benefit from voluntary or enforced consideration of their potential alternate roles in public health emergency (PHE) situations. These ethically, morally, and even spiritually vital contingency plans not only raise social awareness but also stand to provide essential support to a strained public sector.

A further set of considerations relate to the newfound national consciousness towards health security—in terms of budget allocated to public health, determination of resource distribution within the sector, and the conversion of day-to-day resources into health security assets [23]. Both allocative and technical efficiency in the early stage of the 2020 pandemic response should be closely reviewed, in Ireland and elsewhere, in due course: with all possible respect to the urgent considerations of climate change, housing, and other concerns, it appears that health (and, more specifically, health security) considerations ultimately trump any of these issues in the public consciousness. Perhaps, rightly so—without human health, no other initiative matters—or can hope to succeed.

Not only that, but politically, as with Brexit—Ireland was once again a country in limbo at the early stages of the epidemic. Also, as with Brexit, Ireland was subject to its powerful neighbour's decisions and response strategies. During periods when political and economic choices differed significantly from European Union approaches, given the current state of geopolitical limbo on the island, the country's efforts were at times in vain. Moreover, UK policy often diverged significantly from that of the Republic of Ireland. As a result, without adopting equally progressive, precautionary, and sensible approaches, Ireland's response was, on occasion, severely curtailed. Therefore, from a public health perspective, it is essential that further efforts to both cooperate and resolve territorial integrity be made [24].

In this regard, Independent SAGE was a group set up to approach the 'all-island' consideration [25]. Whenever the United Kingdom went for a different approach, during both the first lockdowns and the vaccine roll-out, it raised the question of borders, equity, nationalism, and cooperation. With the complications of Brexit, which required both ensuring peace was maintained on the island whilst also forging a new identity, ongoing collaboration throughout sectors and borders to ensure messaging and health security is consistent and clear was (and remains) essential.

Yet, much remained within Irish society's power—and there still remained in the early stages of epidemic responsiveness many avenues left for policymakers to explore. At many times, both locally and internationally, there was an uneasy sense that current policy is a step behind the epidemic; would a related acceleration to higher levels of escalation help to contain exponential growth whilst also curtailing duration? Would a call for decreased economic productivity—combined with associated social support and an associated focus on providing essential services only—have ensured that changes in the production-consumption cycle left none disadvantaged? Would it mean that social and individual energy could have been preserved without penalty?

5.4 Guest essay: Indian–American diaspora boosts India's fight against COVID-19[b,c]

Rahul M. Jindal[d], Deena Mehjabeen[e], and Lyndsay S. Baines[f]

Introduction

There is a socially active and politically savvy diaspora from the Global South in several countries of the Global North. The Indian–American diaspora, in particular, has achieved spectacular economic and political achievements as seen by the election of Vice President Kamala Harris and appointment of key cabinet members in the United Kingdom [26]. In addition, members of the diaspora hold important senate-confirmed positions in the current Biden administration, such as Dr. Vivek Murthy, the Surgeon General.

---

[b]**Acknowledgement:** The Fulbright Commission & the United States—India Educational Foundation (USIEF) https://www.usief.org.in/ for supporting RMJ's work.

[c]**Disclaimer:** The views and opinions expressed in this article are solely those of the authors and do not reflect the official position or policy of the Department of the Army, the Department of the Navy, the Department of Defense, or the US Government. No financial conflict of interest exists.

[d]Department of Surgery & Division of Global Health, Uniformed Services University, Bethesda, MD, United States.

Adjunct Professor, SUNY-Downstate University Medical Center, Brooklyn & the Indian Institute of Public Health, Gandhinagar, India.

Fulbright-Nehru Distinguished Chair, India (2016 & 2022), Commissioner, Office of Human Rights, Montgomery County, MD, United States.

Email: jindalr@msn.com.

[e]Translational Health Research Institute, Western Sydney University, Australia. Email: mehjabeen.deena@gmail.com.

[f]Head of School (Health & Social Care), Anglia Ruskin University, London.

Lead Global Health Diplomacy Working Group, University of Oxford.

Email: lyndsay.baines@conted.ox.ac.uk.

The COVID-19 pandemic has seen an increase in the humanitarian response of diaspora around the world. These highly organized and effective groups are usually motivated by common cultural, ideological, political, and religious values with their country of origin. India has a track record of engaging with the diaspora to develop global collaborations to promote 'Brand India'. Increase in infection rates has overwhelmed the Indian healthcare system which has been attributed to both the emergence of the delta variant and a sluggish and poor strategic response from the Indian government. However, it was really thanks to Indian–Americans and the political pressure they leveraged with the US government that led to health-related cooperation between the two governments during the pandemic [27].

In this essay, we discuss the engagement of the Indian–American diaspora in the acute pandemic crisis in India [28]. We make a case for the diaspora to learn the art of global health diplomacy by engaging with health attachés or the official aid agencies such as USAID to ensure transparency and accountability in their humanitarian response.

Humanitarian activity of the Indian–American diaspora

The Indian–American diaspora has been active in several humanitarian activities in India even in times of peace [29]. These activities are in building schools, colleges, transfer of knowledge, or providing specialized technical assistance individually or via professional organizations. Noteworthy are the Association of American Physicians of Indian Origin (https://www.aapiusa.org/) and Wheels Global Foundation formed by the alumni of Indian Institute of Technology, seen as the leaders in innovation and technology globally (https://www.wheelsglobal.org/). In the education field, EKAL Global (https://www.ekal.org/), a nonprofit organization, has made significant impact in bringing basic education to rural India, whilst the SEVAK project (www.sevakproject.org), initiated in 2010, empowers local people to look after the health of the people in their own village. The acronym SEVAK stands for 'Sanitation and Health, Education in Village communities through improved Awareness and Knowledge of Prevention/Management of Diseases and Health Promotion'. These organizations swiftly moved their focus to the COVID-19 crisis in India. Their assistance included supplying oxygen cylinders, masks, and vital medications, and providing medical consultations [30]. Some of these consultations were to family and friends, whilst other efforts were formal via tele-health platforms organized by the American Association of Physicians of Indian Origin to connect volunteer doctors to patients with mild-to-moderate coronavirus symptoms in India [31].

In most cases, there was a genuine desire to share lessons learnt from the US viral crisis to their Indian counterparts [32]. Several of us in the diaspora, who are not infectious disease or pulmonary physicians, were making video calls to friends and relatives to provide moral support and check on their breathing and the care they were receiving. Despite these noteworthy efforts, there were numerous reports in the social media and in the Indian press that their efforts were misdirected or oftentimes failed to reach their target population [33]. Initially, the Indian government was not willing to forego taxes on imports by NGOs that were helping in pandemic relief as traditionally exemption was only applicable for contributions made to the Indian government through the Indian Red Cross Society [32].

There was utter confusion with massive supplies landing unannounced at the airports without proper documentation. Most of the information to Indians and the diaspora was through newspapers and social media, which was unverified, precisely why humanitarian assistance needs to be regulated for quality.

Technological companies provide virus assistance to India

India's Ministry of Health and Family Welfare partnered with Google to launch a search tool to locate COVID-19 testing centres, hospital resources, and vaccination sites in India. Their engineers are also trialling a new feature that allows people to share information on available hospital beds and medical oxygen in real time. Facebook is also collaborating with the Indian government to launch a vaccine locator tool. However, the underlying problems are a long-standing shortage of healthcare infrastructure and professionals—there is only one attending physician for every 1445 patients, 0.7 beds for every 1000 people, and 40,000 ventilators for the entire population of India. Unfortunately, temporary assistance by individuals, high-tech, and pharmaceutical companies makes a minimal dent in the structural health inequities, which can only be resolved by increased allocation of funds in the national budget with careful management of scarce resources.

Humanism and vaccine equity

Western medicine has a long tradition of humanitarian service in low-resource countries and in crisis and disaster situations. However, advances in social network technology have dramatically changed the manner in which global health services are delivered. A new generation of professionals within the diaspora, modelled as social entrepreneurs who use collaborative and nonprofit models, has established relationships with healthcare professionals with their Indian counterparts, to facilitate early disease detection, and scale up services and research. Several influential Indian–Americans, such as Dr Ashish Jha, Dean, Brown University School of Public Health, and 4 elected Indian–American members of the US Congress, Raja Krishnamoorthi, Ami Bera, Pramila Jayapal, and Ro Khanna, have vigorously advocated donating vaccines and other critical supplies to India and countries in the Global South. They have effectively used Twitter and Facebook to promote their case for vaccine equity. News reports of harrowing scenes of death and funeral pyres lead to a call between US President Joe Biden and Indian Prime Minister Narendra Modi, in which the US promised to share its 60 million doses of the Oxford/AstraZeneca jab.

Oftentimes, technology allows healthcare professionals to contribute to these efforts remotely, without detracting from their routine clinical work. Technology also provides more flexible pathways for global health training in postgraduate education. Social network technology has changed the landscape of global healthcare delivery across low-resource settings by giving people who live in isolated rural areas the ability to connect with healthcare systems across the globe. A phone call can compress the time it takes healthcare professionals to assess risks and make decisions for patients. Social network technology has also changed the way in which healthcare professionals view global health in terms of the ability to impact populations, collect data in real time, and develop intervention strategies that were previously inconceivable. However, challenges go beyond simply determining ways to reach more people using this technology.

The diaspora has connected with their counterparts in their countries of origin to use virtual technology using social networks. However, there is no hard evidence beyond scattered anecdotes and photographs on social media to confirm positive outcomes [34]. More research is needed to determine how human capacity and resources can keep pace with technology, how social network technology should be used ethically, and how these advances could be integrated into the training curricula of healthcare professionals.

Advances in tele-radiology and tele-psychiatry

In the modern era, the term globalization is not limited to establishing a franchise or business centre in other parts of the world. Information technology has expanded and redefined the meaning of globalization. Tele-radiology is a healthcare product of information technology and networking and is rapidly becoming popular worldwide. The use of radiological imaging is constantly increasing, whilst the number of radiologists has not kept pace with it. Using broadband Internet, high-speed telephone lines, and the latest cloud services, a radiologist sitting in one country can evaluate scans from other countries.

Tele-psychiatry is also maturing. With the help of communication equipment, software, and high-speed Internet, psychiatric consultation is accessible through real-time videoconferencing. The results of the consultation are saved in the patient's medical record electronically. Thus, information technology and networking have transformed the world into a compact and connected domain.

The diaspora has used these newer modalities to counsel and give a second opinion on advanced and complicated cases of COVID-19 individually or via formalized tele-health consultations. A good example is eGlobalDoctors (https://www.eglobaldoctors.com/home), set up by physicians of Indian origin in the United States and the United Kingdom as a source of unbiased healthcare at an affordable cost. This would potentially eliminate people having to travel long distances to get expert medical treatment. A common platform could reduce duplication of effort and standardize advice given from the Global North [35]. However, the potential for litigation should be investigated if tele-health consultations become widespread across national boundaries.

Political implications of humanitarian activity by diaspora

Numerous Indian media outlets have reported that donated goods are often caught in bureaucracy at airports and are not reaching the intended targets [36]. In addition, the government of India has denied that there is a shortage of critical supplies such as oxygen or trained personnel, and claimed that COVID-19 crisis is overblown by the Western media [37]. Human Rights Watch [38] has alleged that the current government seems to care more about its image than the crisis itself. It was reported that when the New Delhi state lawyer complained about supply shortages, India's solicitor general told him to stop being 'a cry baby'. It was also reported that the chief minister of Uttar Pradesh, the most populous state in northern India, talked about using the country's draconian National Security Act against people who complain about lack of oxygen supplies on social media. Several NGOs have urged that the government of India should ensure equitable distribution, particularly to the vast hinterland, of vital life-saving supplies that are pouring in from around the world.

We are particularly worried about supplies reaching rural areas in India, which has poor infrastructure and low penetration of the Internet. It is also alleged that the PM-CARES Fund has not been transparent about how tax-payer money is being deployed. It is thought that there may be cases of denying or hiding bad news instead of setting up systems and organizing consultations with experts to build a prompt and efficient response [38].

It is clear to the authors who have visited India on numerous occasions that some of the shortages could be due to supply chain problems rather than actual lack of supplies. However, it is impossible for the diaspora in the Global North to micro-manage distribution of vital life-saving supplies, and ensure transparency. In this situation, the diaspora is advised to work with reliable and ethical local partners to ensure that supplies reach targets without favouritism or nepotism. Local partners have to be vetted in advance so that the diaspora has collegial working relationships with them, which cannot be fostered in an acute crisis. In addition, diaspora should develop working relationships with the Indian government and US health attachés (or equivalent officers), and with various US or UK aid agencies, such as the United States Agency for International Aid (USAID).

Conclusions

We make a case for the diaspora to train themselves in the art of global health diplomacy, which will enable them to target vital supplies and professional expertise to their intended targets. There should be transparency and accountability in the collection of funds, and a mandatory provision of outcome measurement by independent monitors. Currently, most of the information about distribution of vital supplies to India is gleaned from social media, which may well be inaccurate. Ideally, there should be a common tele-health platform to standardize advice given from the Global North to their counterparts residing in areas of critical need, as protocols change rapidly in acute pandemics. We advise that diaspora should ensure that their medical recommendations are consistent with those of the WHO, the CDC, or the NIH, as the case may be. Steps should be taken to ensure that members of the diaspora adhere to medical licensing and legal requirements of the host country. They should also focus on long-term infrastructure needs of the country in crisis and not merely short-term assistance.

The lessons learnt from various diaspora efforts should be independently evaluated and written as 'best practices' for future pandemics and humanitarian crises.

5.5 The value of originality and creativity

There are many other questions. Would disclosure of further geographic or demographic details of the epidemic have helped to establish cordon sanitaires—or merely have induced panic travel or mass evacuation (as occurred in India during their first lockdown) whilst also compromising individual health information privacy? [39]

In Ireland, the government's decision to reveal specific clusters of infections throughout Ireland felt relatively unhelpful in itself; the disclosure of county-by-county infection levels was likewise of limited use and potentially dangerous unless appropriate population denominators are provided and applied. Publicly identifying those subcounty geographical areas most affected may, conversely, have helped the general public to avoid those areas—whilst also both focusing resources and heightening awareness at the local level. There is no stigma in that.

Similarly, whilst the local demographic information is more informative, why not also in future reveal which professions, or other stratifications, are most at risk—so that appropriate contingency and safety plans can be made? For example, COVID-19 has drastically highlighted social inequalities amongst the front line workers of bus drivers, shopkeepers, and emergency care staff. Would it have been more beneficial to have specific breakdown data? In this regard, a cohort of COVID-19 positive cases was from meat factory workers—who are often migrants, with little to no rights. These workers often have accommodation in cramped, unfavourable living conditions. However, would revealing this to the public merely enforce further stigmatization to a cohort who are already vulnerable?

On more abstract levels, there remains the question of whether society, nationally and internationally, has now paid the price for a too-rapid embrace of globalization. For example, should (after repatriation deadlines) airports have been closed sooner, should borders have been sealed earlier [40].

In conclusion, we have learned that success in public health emergencies and infection control relies not just on these big-picture considerations. It was as much, if not more, thanks to the citizens as it was to the government. To take Ireland's example, the modern, cosmopolitan, and educated society, a society that has (to date) rejected xenophobia and the more malign elements of isolationism, that triumph in a time of adversity will have been achieved. Many achievements of Irish society over the last decade have been due to the power of citizens, with the Citizen Assembly used as a form of political policy-making, along with numerous appeals to question the power of various institutions. For each individual involved, finding a way to contribute to the national and international response is thus both rewarding and essential.

The further development of platforms for innovative approaches and shared challenges at the community level was thus of great value in this regard. This also facilitated capacity to draw on available yet unrecognized or overlooked local resources; transcended turf battles; empowered; enabled the sharing of original, timely, and creative macro- and micro-level response concepts, systems, and ideas; and informed and—above all—facilitated consensus and cooperation.

5.6 Key points

- Though the COVID-19 pandemic was a truly global issue, it was also very much a local one, requiring different policy responses in different places.
- Pandemic responses were also very much trial-and-error in the early stages, with different countries taking highly individualistic and experimental approaches.
- Each country, over time, contributed something to the global response through their own paths—successful strategies were swiftly adopted worldwide.

References

[1] A. Doyle, W. Hynes, S.M. Purcell, Building resilient, smart communities in a post-COVID Era: insights from Ireland, Int. J. E-Plan. Res. 10 (2) (2021) 18–26.

[2] P. Cunningham, Significant progress towards new government—now what? RTE.ie (2021). Available at: https://www.rte.ie/news/politics/2020/0311/1121563-politics/>. (Accessed 5 December 2021).

[3] C.M. Hall, P. Fieger, G. Prayag, D. Dyason, Panic buying and consumption displacement during COVID-19: evidence from New Zealand, Economies 9 (2021) 46, https://doi.org/10.3390/economies9020046.

[4] B. Nezafat Maldonado, J. Collins, H. Blundell, L. Singh, Engaging the vulnerable: a rapid review of public health communication aimed at migrants during the COVID-19 pandemic in Europe, J. Migr. Health 1–2 (2020) 100004.

[5] J.M. Banda, R. Tekumalla, G. Wang, Y. Jingyuan, T. Liu, Y. Ding, E. Artemova, E. Tutubalina, G. Chowell, A large-scale COVID-19 Twitter chatter dataset for open scientific research—an international collaboration, Epidemiologia 2 (3) (2021) 315–324.

[6] C. O'Shea, Dr Holohan Says One Very Irish Habit Will Be Thing of the Past Post Coronavirus, 2020, June 22, Retrieved December 5, 2021, from https://www.irishmirror.ie/news/irish-news/dr-tony-holohan-says-one-22235866.

[7] S. Barari, S. Caria, A. Davola, P. Falco, T. Fetzer, S. Fiorin, L. Hensel, et al., Evaluating COVID-19 public health messaging in Italy: self-reported compliance and growing mental health concerns, MedRxiv (2020).

[8] K.H. Ng, R. Kemp, Understanding and reducing the fear of COVID-19, J Zhejiang Univ Sci B 21 (9) (2020) 752–754, https://doi.org/10.1631/jzus.B2000228.

[9] J.J. Szczygielski, A. Charteris, P.R. Bwanya, J. Brzeszczyński, The impact and role of COVID-19 uncertainty: a global industry analysis, Int. Rev. Financ. Anal. (2021) 101837, https://doi.org/10.1016/j.irfa.2021.101837.

[10] K. Vaswani, Covid-19: China's economy picks up, bucking global trend, BBC News (2021). https://www.livescience.com/first-case-coronavirus-found.html. (Accessed 16 July 2021).

[11] J. Bryner, 1st known case of coronavirus traced back to November in China, Live Science (2020). https://www.livescience.com/first-case-coronavirus-found.html. (Accessed 14 July 2021).

[12] BBC News, Time-lapse shows Wuhan hospital 'built in 10 days, BBC (2020). https://www.bbc.com/news/av/world-asia-china-51348299. (Accessed 15 July 2021).

[13] G. Emma, K. Lily, China's coronavirus lockdown strategy: brutal but effective, The Guardian (2020). https://www.theguardian.com/world/2020/mar/19/chinas-coronavirus-lockdown-strategy-brutal-but-effective. (Accessed 16 July 2021).

[14] C. Buckley, W. Sui-Lee, Q. Amy, China's doctors fighting coronavirus, beg for masks, New York Times (2020). https://www.nytimes.com/2020/02/14/world/asia/china-coronavirus-doctors.html. (Accessed 14 July 2021).

[15] M. Iqbal, WeChat revenue and usage statistics (2021), Business of Apps (2021). https://www.businessofapps.com/data/wechat-statistics/. (Accessed 16 July 2021).

[16] F. Liang, COVID-19 and health code: how digital platforms tackle the pandemic in China, Soc. Media Soc. (2020), https://doi.org/10.1177/2056305120947657.

[17] L. Shumei, China completes half-year vaccination goal early, will cover as many people as possible in H2, Global Times (2021). https://www.globaltimes.cn/page/202106/1227529.shtml. (Accessed 16 July 2021).

[18] GT Staff Reporters, China passes 1b shots mark; eyes herd immunity by Dec, Global Times (2021). https://www.globaltimes.cn/page/202106/1226579.shtml. (Accessed 16 July 2021).

[19] K. Lily, L. Lyric, China's covid vaccine drive is lagging. Free food could help turn things around, Wash. Post (2021). https://www.washingtonpost.com/world/asia_pacific/covid-china-vaccine-rollout/2021/04/05/58b5d376-95cf-11eb-8f0a-3384cf4fb399_story.html. (Accessed 15 July 2021).

[20] F. Erika, R. Nicole, 25.6% of the world has received a COVID vaccine. See how your country is doing, Fortune (2021). https://fortune.com/2021/07/14/covid-vaccine-tracker-update-worldwide-us-countries-pfizer-moderna-johnson-johnson-numbers-data/. (Accessed 16 July 2021).

[21] J.P. Kalkman, Military crisis responses to COVID-19, J. Conting. Crisis Manag. 29 (1) (2021) 99–103.

[22] Houses of the Oireachtas, Dáil Éireann Debate—Wednesday, 3 Jun 2020, 2020, June 3, Retrieved December 5, 2021, from https://www.oireachtas.ie/en/debates/debate/dail/2020-06-03/4/.

[23] N. Kandel, S. Chungong, A. Omaar, J. Xing, Health security capacities in the context of COVID-19 outbreak: an analysis of International Health Regulations annual report data from 182 countries, Lancet 395 (10229) (2020) 1047–1053.

[24] Institute of Medicine (US) Committee on Assuring the Health of the Public in the 21st Century, The Future of the Public's Health in the 21st Century, National Academies Press (US), Washington, DC, 2002. 3, The Governmental Public Health Infrastructure. Available from: https://www.ncbi.nlm.nih.gov/books/NBK221231/.

[25] L. Freedman, Scientific advice at a time of emergency. SAGE and Covid-19, Polit. Q. 91 (3) (2020) 514–522.

[26] Social Realities of Indian Americans: Results From the 2020 Indian American Attitudes Survey. https://carnegieendowment.org/2021/06/09/social-realities-of-indian-americans-results-from-2020-indian-american-attitudes-survey-pub-84667. (Accessed 29 July 2021).

[27] D. Kapur, The power of Indians abroad, 2021, May 13 https://foreignpolicy.com/2021/05/13/the-power-of-indians-abroad/. (Accessed 29 July 2021).

[28] Every time I'm calling, someone has died': the anguish of India's diaspora, The New York Times (2021). May 18 https://www.nytimes.com/2021/05/01/world/europe/uk-india-diaspora-Covid.html. (Accessed 29 July 2021).

[29] Migrants Play Key Role in Disaster Response, IOM Explores Diaspora's Engagement in Humanitarian Assistance, International Organization on Migration, 2021. https://www.iom.int/news/migrants-play-key-role-disaster-response-iom-explores-diasporas-engagement-humanitarian. (Accessed 29 July 2021).

[30] Patients in India are desperate for care. This family turned to a doctor 7,700 miles away, Wash. Post (2021). https://www.washingtonpost.com/health/2021/05/23/indian-doctors-us-helping-covid-patients-india/.

[31] Doctors Provide Free Telehealth Services for COVID-19 Patients in India. https://www.healthcareitnews.com/news/apac/doctors-provide-free-telehealth-services-covid-19-patients-india.

[32] On the Ground and Afar, Diaspora Boosts India's Virus Fight, The Associated Press, May 6, 2021. https://accesswdun.com/article/2021/5/1003205. (Accessed 29 July 2021).

[33] A long wait: Indian diaspora sends Covid aid, but it gets stuck in the process, India Today (2021). May 4 https://www.indiatoday.in/coronavirus-outbreak/story/a-long-wait-indian-diaspora-sends-covid-aid-but-it-gets-stuck-in-the-process-1798879-2021-05-04. (Accessed 29 July 2021).

[34] How social media is saving lives during India's COVID crisis, PBS News Hour (2021). https://www.pbs.org/newshour/world/how-social-media-is-saving-lives-during-indias-covid-crisis. (Accessed 29 July 2021).

[35] Heeding Lessons From India's Second Wave, Bill & Melinda Gates Foundation, 2021. https://www.gatesfoundation.org/ideas/articles/covid-india-response. (Accessed 29 July 2021).

[36] The world sent India millions in Covid aid. Why is it not reaching those who need it most? CNN (2021). https://www.cnn.com/2021/05/05/india/india-covid-foreign-aid-distribution-intl-hnk-dst/index.html. (Accessed 29 July 2021).

[37] The lurid orientalism of Western media, Project Syndicate (2021). https://www.project-syndicate.org/commentary/western-media-double-standard-for-crises-in-foreign-countries-by-brahma-chellaney-2021-05. (Accessed 29 July 2021).

[38] The human rights crisis behind India's COVID 19 surge, Human Rights Watch (2021). April https://www.hrw.org/news/2021/05/10/interview-human-rights-crisis-behind-indias-covid-19-surge#. (Accessed 29 July 2021).

[39] A. Zwitter, O.J. Gstrein, Big data, privacy and COVID-19–learning from humanitarian expertise in data protection, J. Int. Humanit. Action 5 (1) (2020) 1–7.

[40] S. Mas-Coma, M.K. Jones, A.M. Marty, COVID-19 and globalization, One Health 9 (2020).

Chapter 6

Lessons from—and for—Developing countries

Sebastian Kevany and Aoife Kirk

Chapter outline

6.1 Limited learning opportunities?

Throughout 2021, Dr. Michael (Mike) Ryan, leader of the World Health Organization's pandemic response, would often observe, 'Nobody on this planet is safe until everyone is safe' [1]. This highlighted that the COVID-19 pandemic, as with many other infectious disease outbreaks and epidemics, can affect populations and nations unequally and deepen divides. In 2020, the Organization for Economic Co-operation and Development (OECD) highlighted Irish Aid as a leader for sustainable development [2], ensuring that policy dialogue is driven from needs on the ground up. At the time of writing, less than 1% of the African continent has been vaccinated and an Omicron variant recently announced by South African/Botswana scientists is a significant threat. This vaccine inequity

Outbreaks, Epidemics, and Health Security. https://doi.org/10.1016/B978-0-323-91705-6.00006-X

may result in years of setbacks for regions that have historically struggled with lack of access to health services, climate change impacts, political instability, and conflict [3].

There are only a few ways that data or examples from developed countries are likely to help the developing world, and sub-Saharan Africa in particular, in their preparations for the global pandemic. There are too many variables in play to compare notes—not least geographical differences; possibilities of diminution of virulence over time and space; preparation times; available resources; culture; religion; and a host of other possible confounders to enable a retrospective evidence-based comparison between countries.

Moreover, the political nature of border controls and other response mechanisms interferes with significant comparisons, along with broader social and economic factors [1]. Response systems which have shown encouraging quantitative results in one nation may thus be ineffective, or even counterproductive, on other levels and in other places.

Timing, chance, and circumstance (such as in the case of Italy) have played critical roles in country-by-country impact, meaning that ostensibly positive data cannot always be meaningfully linked to policy, either. Moreover, even within Europe, unhelpful comparisons on the interaction between policy and results abound: Sweden, for example, has greater health system capacity than Ireland—where, ultimately, the only point of any social or economic intervention has been to avoid health system overload. Within policy-making in public health, considerable thought is needed to avoid unintentional consequences, along with ensuring that interventions attain stated goals to provide evidence-informed decision-making [4].

Questions remain regarding the various approaches taken by different countries, which may result in future inquiries into governmental bodies of their handling of data, lockdowns, and implications. Nonetheless, there were some benefits developing countries accrued from attention to initial health security experiences and response strategies in Southeast Asia, Europe, and elsewhere.

6.2 Lead time

Much of the developing world was uniquely placed in terms of the global pandemic. However, for many geographical and socio-economic reasons, such countries may have had more significant time to prepare, and thus—at least theoretically—more opportunities for fine-tuning and tailoring their interventions to local contexts. For example, Rwanda—a country that has invested heavily in healthcare infrastructure over the last two decades—had the foresight to close borders early in March 2020 to prevent spread [5].

Along with innovative channels of disseminating information through the use of drones and official communications through social media channels, these measures were all vital to the national response. A comprehensive, cohesive 6-month plan was also drawn up, establishing a national incident management system and phases for a national response through the Coronavirus National Preparedness and Response Plan [6].

The African continent has many infectious disease experts and epidemiologists, given the higher day-to-day threat of infectious disease in comparison with the Global North, which mostly consists of chronic and cardiovascular disease [7]. Two organizations, in particular, stood out during the initial response—the Centre for Disease Control & Prevention (CDC) Africa led by Cameroonian epidemiologist Dr. John Nkengasong and the African regional headquarters of the World Health Organization led by Dr. Matshidiso Moeti and team.

During this initial outbreak time in early 2020, the coordination of various countries was, perhaps, a result of recent and previous fear of the Ebola virus outbreaks. Along with a complete lack of equipment, ICU beds, health infrastructure, and oxygen access, this perhaps led to some countries leaning heavily on basic public health measures.

PICTURE 2.6.1 A roadside makeshift illegal pharmacy store in a market in Port-au-Prince, Haiti, when it was oblivious to the pandemic. Kiran Kamble.

Initial measures included ensuring public health messaging; contact tracing, triage and protocols; and rapid and (where possible) inexpensive testing.

Ironically, the developing world was perhaps better equipped for the situation than Europe or the United States in many ways. For example, most populations in sub-Saharan Africa are highly experienced in living with the threat of infectious diseases and associated life expectancy issues, unlike those of more affluent countries [8]. As such, the day-to-day living habits of those in the Global South may not have been subject to such dramatic change as those in more affluent countries. Global South populations may also be more adaptable, based on their past experiences.

Nonetheless, specific basic policies, without high levels of investment and with an increased focus on health security, were often advanced. These included health education in many countries that both informed and avoided the hype, paranoia, and associated social unrest and economic disruption; it related to the adaptability of epidemic control efforts to different cultural contexts, including religion and political climate. Should this have been internationally led, with a national focus? What about localized, targeted responses? Answers to these questions remain to be seen, though comparisons between countries may help us interpret the lessons required to prepare for future pandemics.

A further consideration was avoiding information overload—how can developing countries, any more than anywhere else, navigate the increasingly complex global information that flows on the updates of the outbreak? [9] And how can interventions be done in a way that does not disrupt essential services when implemented? Finally, the flux of fake news in this regard was a key consideration: developing country governments should not be encouraged to leverage public health efforts for other political agendas, as has already happened in many parts of the developed world.

Disease control is no excuse for authoritarianism or the limitation of human rights, as was seen with the controversies around internment camps in the 2014 Sierra Leone Ebola outbreak. For example, in the case of South Africa, the conflation of public health and political agendas was open to a particular question: the ban on alcohol seemed extreme, though logical [10]. Meanwhile, the provision of financial assistance only to Black Economic Empowerment companies had less inherent sense and is suggestive of, rightly or wrongly, leveraging public health for the political advancement of other agendas.

Thus, whilst COVID-19 required a health response from international governments, it also led to further impositions on human rights, often in vulnerable populations with questionable track records. Refugee access to health services; minorities being excluded from vaccination programmes; quick lockdowns with no access to social services support or employment benefits; limitations on press freedom; increasing social inequities—all this has left academics, political scientists, and economists questioning the definition of public health in both democratic and authoritarian states.

6.3 Guest essay: Lessons from South Africa

Maryam Rumaney

> In time, we shall be in a position to bestow on South Africa the greatest possible gift—a more human face.
>
> ***Steven Biko***

The grim picture and the state of South African affairs

I am writing this piece from Cape Town, South Africa. As I am writing this, my country is experiencing the third wave of COVID-19, which happens to be the deadliest wave we have witnessed to date. We are currently in a level four lockdown state, which tries to curb the spread of the virus whilst maintaining as much of the economy to be operational as possible. Today I shall tell the story for those in the low-income groups, who are fighting to stay alive every single day, even in the absence of COVID-19. South Africa is a country with an approximate population of 58.5 million, where 55.5 percent is living in poverty at the national upper poverty line, whilst 25 percent are experiencing food poverty [11].

The Minister of Health warned that between 60% and 70% of South Africans are expected to become infected with COVID-19, with 10% to 20% of those developing into severe cases, likely requiring hospitalization [11].

COVID-19 plagues my country, but so does poverty, hunger, and crime. COVID-19 has exacerbated the effects of poverty, hunger, and crime. These 'ailments' then form one large spiraling cycle, one fuelling the other. In addition, we are fighting corruption at a grand scale and we have an ailing public health system. We, as South Africans, are fighting a bigger battle than merely the pandemic. If you have ever visited my city, you would have observed the obvious income inequality (South Africa's Gini index is at 63) [11]. The effects of apartheid still plague us, and now the battle really begins, between the 'haves' and 'have-nots'.

Those who 'have' experience access to private medical care, spacious homes, and the ability to work from home. The 'have-nots' are the true people at the frontline of the pandemic, where they are faced with the possibility of infection on a daily basis. This portion of the population are those stemming from low-income groups who are reliant on minimum wage, high-density public schools, overcrowded public transport options, and the public healthcare system.

The South African public healthcare system has some of the best medically trained professionals in the world, but they are stretched beyond what they can handle, and operate in immensely resource-limited settings. These medical professionals are at the frontline, but so are those that work at grocery stores and those who run public transport facilities—the low-income earners who are most at risk of mortality. Yet those who work in grocery stores are not considered a high-priority risk group eligible for vaccination, as healthcare and related workers were.

Access to vaccines appears to be the one factor that may ease the burden of disease in South Africa, but efforts to vaccinate are slow. We have had a phased approach beginning with healthcare workers and personnel, followed by anyone that advances work in primary healthcare, vaccinated via the Sisonke trial, where the Johnson & Johnson vaccine was administered. Thereafter, the actual vaccine roll-out began, with the Pfizer brand being administered to those that are 60 years and older. This was followed by teachers and individuals 50 years and older. Currently, 39-year-olds and upwards qualify for vaccination. From September, anyone 18 years and older will be able to get vaccinated against COVID-19. To date, we have 4.4% of the population fully vaccinated [12]. When compared to the Global North, Africa is years away from reaching herd immunity, whereas countries like the United States and the United Kingdom are half away already [12]. This inequity leaves us with a great loss of economic normality.

Dangers of lockdowns

As time lapses, the number of variants increase, with higher mortality rates. But, data from the United States and United Kingdom has proven that whilst the infection rate is still fairly moderate to high, the mortality rate associated with COVID-19 has greatly decreased as a result of vaccination efforts [13,14]. As there are further delays in vaccine access in South Africa, new waves may bring about more frequent lockdowns, which impact quality of life. These impacts also affect the economy negatively and affect women and children to a much larger extent. With the closing of schools comes the closure of a haven of safety for many children. Some children get their only meals from school. Thus, with these closures also comes an elevated degree of malnutrition. Immunity requires good nutrition, which further places these children at risk [15].

Lockdowns have also resulted in the rise of gender-based violence (GBV) globally, and the South African context is no different [16,17]. The home is not a place of safety for everyone. Whilst people may be safe from potentially contracting COVID-19, for some, the home turns into a battle ground where physical abuse prevails, because lockdowns trap victims indoors with their abusers. South Africa is notoriously known as the *'destination of femicide'*, where women and children are most susceptible to abuse.

The government's call centre, known as the GBV and Femicide Command Centre, recorded more than 120,000 victims in the first three weeks of the lockdown [18]. Weeks later in Pretoria, a similar call centre was receiving up to 1,000 calls a day from women and children who were confined to abusive homes and thus seeking urgent help [18]. Prior to the pandemic, femicide in South Africa was already five times higher than the global average [17]. Growing GBV rates, less access to civil organizations providing help, associated stigma, and the dismissal of cases by the police are contributing to the underreporting of actual numbers. GBV may thus be defined as a *'secondary pandemic'*. Restricted movement, social isolation, and economic insecurity are increasing female vulnerability to domestic abuse.

> Those who invest in South Africa should not think they are doing us a favor; they are here for what they get out of our cheap and abundant labor, and they should know that they are buttressing one of the most vicious systems.
>
> ***Desmond Tutu***

As I write this, South Africa is also plagued by the biggest acts of looting and arson ever witnessed in this country. Shops, homes, and medical facilities were looted, buildings burnt down, and many innocent civilians have been harmed in two provinces (Gauteng and KwaZulu Natal). This, amid a rising global pandemic. Law enforcement and the army have sat still; citizens have been forced to stand up and protect their lives and livelihoods. Life in these provinces is currently at a standstill. Our government has failed us. Poor management of every sector of society, and a pressing sense of inequality, is ripping this 'free' country apart. This has further negatively impacted vaccination efforts because vaccine administration sites are closed, and all efforts have been halted in these regions. We are on the brink of what feels like a civil war, and it could shred our country into pieces.

Perhaps there may be a silver lining—Solutions for painting a brighter picture
Corruption continues to flourish and is the root cause of all mismanagement related problems in the country. There is no accountability and no consequences. This is the first course needed for change. The government needs to be taken to task on all issues pertaining to how state funds are spent. There needs to be a local accountability task-team that is composed of members of the community at various tiers that hold some degree of power. This is a tall order, but only governance in line with what people need is going to effect change.

The harsh COVID-19 preventative effort lockdowns that persist for weeks to months should also perhaps be changed. In future, the minute one case comes to light the country should be placed into a short, hard lock down for 4–5 days, with complete contact tracing. This will enable a far less negative impact on the economy and particularly on small businesses.

Since the onset of the pandemic and a lockdown level five that lasted for 35 days, many citizens also lost their jobs [19]. This resulted in poverty to the extent where many people were unable to afford food. A solidarity fund was set up, but there were corruption issues here too [20]. The solidarity fund was meant to help any citizen in need; the government needs to create a feeding scheme that allows for basic nutrition, because malnourished individuals cannot fight off illness as well as individuals with adequate nutrition [15]. Otherwise, South Africa may enter into another pandemic soon—malnutrition.

Activism for GBV, where action for the protection of the vulnerable is taken into serious account, is also required. In particular, there must be a drive from the government that enforces better support for GBV despite the lockdown. GBV support services must be marked as *'essential'*. Law enforcement must be trained to eliminate conscious and unconscious bias when GBV cases are reported. Clearer pathways of support must also be established so that those seeking help are not turned away.

The scale-up of vaccination efforts will help to manage the disease burden. An increase in vaccinated citizens should result in less hospitalizations, and less GBV, which would in turn mean a decrease in the need for state resources.

Health literacy is another huge piece in the COVID-19 solution puzzle. Ongoing efforts to educate the country, irrespective of educational level, on social distancing, mask-wearing, hand hygiene, and vaccination are required. In the absence of accurate information to the public in a manner that is easy to understand, this remains one of the biggest efforts to manage this and future pandemics.

The uncertain reality

> The greatest glory in living lies not in never failing, but in rising every time we fall.
>
> *Nelson Mandela*

As we look on into the 'horizon' amid the pandemic, the number of positive COVID-19 cases are continuing to rise in South Africa [21] and we also have no idea of the true extent of the economic impacts on this country due to the looting. Evidence from previous crises, of a far smaller scale, suggest that the negative effects of the pandemic and the looting could be experienced for many generations to come, and thus further exacerbate South Africa's high inequality ratio. As our president ends off after every address, *'May God bless South Africa and protect her people'*.

PICTURE 2.6.2 **Beaches during lockdowns.** Sebastian Kevany.

6.4 The primacy of health security

A further consideration was the necessity of protecting developing country populations not from the virus but from international arrivals. Worldwide, the partial freeze in international travel had been essential, even though (in hindsight) it was perhaps often implemented far too late. In the same way, there was often no need to change the day-to-day practice of life in developing country townships, as long as residents could be protected from interactions with international or regional arrivals [22].

PICTURE 2.6.3 **Empty lockdown landscapes.** Ruchika Gallani.

Epidemic progression in developing countries raised many health security questions regarding ethnicity, culture, and other risk factors that often differed between regions. In this regard, prior epidemics have also shown us, if nothing else, that adaptable interventions that are suitable to local socio-economic conditions are essential [23]. This includes focusing on the utilization of local resources in epidemic response design, careful consideration of branding in health messaging, making feasible requests that do not deter service utilization and paying close attention to a range of other possible sensitivities [23].

Similarly, ethical considerations related to the provision of services for other diseases need to be carefully contemplated. The importance of cross-sectoral responses going beyond ministries of health and utilizing resources from a broader range of government departments and public service providers has to be considered. Should any pandemic shift to low-level 'acceptable risk' endemic status in developing countries, responses also need to be designed with long-term surveillance and knowledge (as well as emergency response) considerations in mind [24].

6.5 What worked?

The COVID-19 pandemic has been, in terms of recent history, a 'novel' crisis. However, it has resulted in far greater significant social, political, and economic consequences. The main difference between this pandemic and the other coronaviruses (such as SARS) is that this pandemic has been transboundary in geography [25].

Closure of borders: By far, the most influential international public health measure, from a health security perspective, has been the closure of borders and limitations of population movements, both internally and between countries. This has worked due to cooperation between transport providers, legislators, and public health experts and the decisive intervention of politicians.

As such, curtailments of travel between countries, districts, and regions (in other words, the constriction of relatively loose *cordon sanitaires* or reverse *cordon sanitaires*) were always most likely to be highly effective. Moreover, in many developing countries, partially enforced borders between counties, districts, and regions were already in place, so continuing suspension of long-distance travel was practical, effective, and enforceable in this context [26].

Similarly, the curtailment of the movement of international populations arriving into and living in developing countries was crucial. The development of tagging systems for new arrivals to wait out quarantine periods was also an enforceable initiative in this regard; the further development of effective and efficient 'bush telegraph'[a] surveillance information sharing was also of crucial importance in curtailing local outbreaks.

6.6 Guest essay: Strengthening pandemic preparedness in Indonesia: Reflections from COVID-19 pandemic response

Yodi Mahendradhata

Introduction

As of 5 September 2021, Indonesia has reported 4,116,890 confirmed cases of COVID-19 with 134,930 deaths, placing Indonesia amongst the worst-hit countries in the Southeast Asian region in terms of total cumulative cases and deaths [27]. The Indonesian Government, by time of writing in August 2021, has already started to prepare a strategy for living with COVID-19 in the longer term. This policy shift has been based on the expectation that the disease will hopefully become nationally endemic by 2022, moving away from the current aim of Zero-SARS-CoV-2 [28]. Endemic SARS-CoV-2 is expected to present a less significant threat to Indonesians, as it is hoped that most people should be protected

[a] An informal and rapid means of communication, usually through word-of-mouth.

against severe cases, and that resurgences will be contained via regular testing and outbreak tracing. Becoming endemic has been the natural progression of many infections in humans, arguably including HIV, tuberculosis, or other newly emerging/re-emerging diseases [29]. Thus, arguably, endemic COVID-19 is where Indonesia is heading as well, sooner or later. However, it is difficult to accurately estimate the timeline for how and when COVID-19 will move into endemic territory in the country.

The only sure thing is that when this pandemic is under control, another will likely eventually take its place. Pandemic threats are inevitable [30]. Therefore, being prepared nationally and internationally is the key. This opinion essay aims to distil key lessons from the COVID-19 response in Indonesia to strengthen future pandemic preparedness efforts in the country and similar settings.

Improving healthcare surge capacity

Hospitals and community health centres were overrun entirely during the mid-2021 surge [31]. By August 7, more than 3,013 people died in isolation at home, unable to find medical help [32]. Those admitted to hospitals often struggled to get care due to staff shortages and limited medical supplies. To make matters worse, an alarming number of health workers were getting infected [33]. Thousands of doctors and nurses were often unable to work as were in isolation. At least 1977 healthcare workers have been recorded as having died by 5 September 2021 [34]. The number of qualified medical staff was far from sufficient to deal with the patient surge. The pandemic highlighted the human resource challenges that Indonesia's healthcare system has struggled with for decades [35]. The pandemic has further exposed the weakness of medical supply chains in the country as surges of patients requiring hospitalization led to a shortage of medical supplies. The COVID-19 pandemic also revealed the fragility of the patient referral system and the limited capacity of the healthcare system to maintain essential health services during a pandemic [35]. The Indonesian Government thus needs to ramp up the country's healthcare capacity to better prepare for the next pandemic. However, the challenges of increasing healthcare capacity during the COVID-19 pandemic highlight that such efforts could represent only a part of the pandemic preparedness equation. Effective pandemic preparedness ultimately requires governments' commitment to increase healthcare capacity and improve capacity to flatten the curve through proven public health interventions.

Strengthening capacity for public health interventions

The surge of COVID-19 cases which overwhelmed healthcare facilities in mid-2021 [31] highlighted the failure to flattening the curve through public health interventions. Indonesian public health advocates throughout the pandemic have pleaded with the Government to enforce measures to curb COVID-19, mostly in vain. The Government notably views the pandemic primarily as a threat to economic stability rather than a public health crisis [36]. At the beginning of the pandemic, rather than implementing aggressive testing and tracing; social distancing policies; and isolation, quarantine, and other public health interventions to dampen the rate of new infections, the Government was subjugating public health imperatives to economic interests. From the beginning, the Government ruled out lockdowns to contain the virus [36]. The Government initially implemented large-scale social restrictions that set capacity caps on transport and businesses in affected regions instead of closing them down. The surge in mid-2021 eventually forced the Government to impose emergency restrictions on travel and nonessential businesses. Nations can only go into a lockdown when there is strong governance, a population largely willing and able to follow lockdown orders, police and defence personnel equipped to enforce such orders, and resources to ensure people are sheltered, fed, and provided with healthcare [37]. Indonesia evidently did not have these preconditions well in place. Most notably, many poor households cannot follow public health recommendations that compromise their ability to provide food and shelter for themselves and their families. Many people faced a dilemma between staying at home and enduring conditions of deprivation and slipping into poverty or risking infection by leaving home to earn a living [38]. Such a tough decision disproportionately affected Indonesian women, many of whom make a living in the informal sector. Thus, during the pandemic, vast inequality has led to disparities in disease, particularly amongst people working in informal sectors, and those who are paid low wages and are working or living in crowded environments. Consequently, Indonesia's Government is afraid of the consequences of stopping workers in the vast informal sector from earning an income or doing anything that would cause mass redundancies in the formal sector [39]. Addressing inequity, therefore, should be a critical part of efforts to prepare for future pandemics [40].

Enhancing capacity to manage infodemic

Notably, there are competing understandings about COVID-19 within Indonesia at the national, institutional, and individual levels [38]. Without clear and unified messaging from the Government and national health bodies, people receive mixed messages about COVID-19 and measures to prevent its spread. These competing narratives surrounding COVID-19 created uncertainty for those seeking reliable information about the disease and how to avoid infections. Low digital literacy rates across socio-economic classes compounded the lack of a unified message [38]. A nationwide survey conducted in 2020 by the Communications and Information Ministry and Katadata Insight Center (KIC) found that between 64 and 79 percent of respondents could not recognize misinformation online [41]. An overwhelming majority said they primarily seek information through social media. Health messages are further diluted by the presence of multiple and often contradictory claims proliferating on social media platforms. Social media notoriously provides a platform for countering expert knowledge and spreading misinformation and disinformation. Claims that the COVID-19 vaccines contain microchips and wearing masks inhibits breathing have been amplified on social media, undermining prevention and vaccination efforts [41]. Immediately after the vaccination campaign began on 13 January 2021, the amount of fake news against vaccines also accelerated [42]. The disinformation strategy also propagated misleading narratives from institutions, including the Government [38]. In addition, the COVID-19 denial narrative played a role in how health professions in Indonesia were perceived. Rumours spread that health professionals were profiteering by falsely diagnosing patients and requiring unnecessary and pricey COVID-19 tests. In mass and social media, doctors were accused of extracting economic benefits by asking every patient to get a COVID-19 test, regardless of the presence or absence of symptoms [38]. Given such misinformation, people were scared to visit hospitals, fearing expensive misdiagnoses or unnecessary tests. Consequently, Indonesians are paying the price for littering social media with hoaxes, misinformation, and rumours, which has claimed real lives [43]. That will only worsen in the next pandemic unless the Government strengthened its capacity to establish unified evidence-based messages, infodemic management, and increasing digital literacy.

Conclusion

COVID-19 in Indonesia is far from disappearing, but important lessons are already evident to prepare the country for future pandemics. First, the Government must ramp up healthcare capacity and access to ensure all cases can be managed adequately, particularly during surges. Second, the Government should prioritize measures to correct inequities, minimizing disparities that hindered the poor from following public health recommendations. Third, the Government must increase capacity for evidence-based policy-making and managing infodemics, whilst increasing digital literacy. Most importantly, the Government's strategy to fight future pandemics must be centred on the human security concept in navigating through political, social, and economic pressures. Indonesia should fight future pandemics through a whole-of-government approach, engaging key stakeholders across multiple levels, incorporating all sectors, including local authorities and community, and building on clear and solid coordination [44]. Future pandemics are inevitable, but the missteps that have accompanied the spread of COVID-19 in the country were not. Indonesia must ensure that the nation is better prepared to avoid the critical systemic shortcomings that have led to excess suffering and deaths when the next pandemic occurs. The Indonesian Government must commit considerable political, capital, and economic resources to reduce the systemic vulnerabilities that have jeopardized the nations' health security.

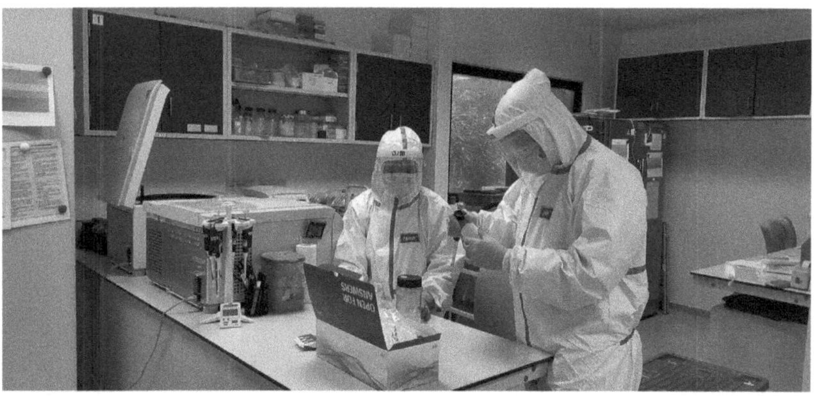

PICTURE 2.6.4 A high-tech Covid lab. Yodi Mahendradhata.

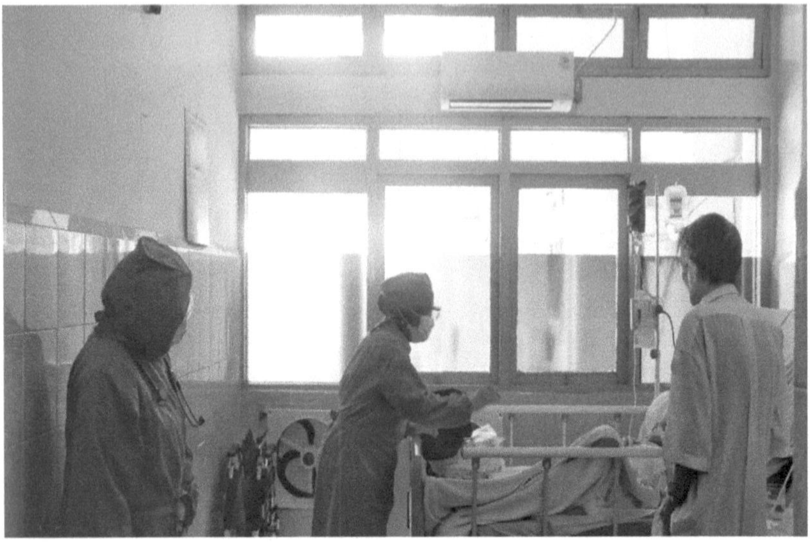

PICTURE 2.6.5 **Covid treatment in action.** Yodi Mahendradhata.

PICTURE 2.6.6 **Innovative self-screening efforts were a global phenomenon.** Yodi Mahendradhata.

6.7 A checklist of health security concepts for developing countries

Border control support: During the pandemic, this included the provision of fact sheets or other health checks at air and seaports, and donor support of efforts to maintain border security within partner countries.

Social distancing: Social distancing policies were also successful, mainly because of their humanistic nature—in many cases, the previous status quo of large crowds concentrated in small spaces has been revealed as optional; from overcrowded commuter trains to other examples of current congestion, few are essential to human progress and survival. The ascendancy of remote working was also critically important in this regard to the Global North, though the latter was, and unfortunately still is not, an option open to most developing country populations.

Shielding elderly and vulnerable populations: The compliance of vulnerable and elderly populations with national directives has been critical. Without these measures, those at the highest risk would likely rapidly fill many hospital systems to capacity—as was the case in Italy.

This has highlighted the limited facilities and institutions for elderly patients in many countries: the UK struggled to contain nursing home outbreaks, with hospitals sending patients directly back without testing, resulting in many mini-outbreak scenarios. Shielding these at-risk patients from visitors was implemented as a measure, which resulted in many difficult situations related to end of life care, and in ensuring adequate socialization and care during often the last years of life [45].

Optimal measures in developing countries focused on these lessons learned—with significant adaptation to local conditions. Social distancing efforts were, by contrast, almost impossible to enforce, for example, because there is traditionally less mobility amongst older and vulnerable populations in the developing world compared to developed countries.

Health security preparedness: A Nascent Health Security Checklist for the Developing World Health security [46] is too often considered a purely developed country concern: protecting borders and affluent populations from the arrival of Ebola from West Africa during outbreak periods, for instance. This is, perhaps, a misplaced assumption: health security, today, is of greater importance than ever for developing countries, who are now equipped with justifiable rationales for lamenting ingress.

In conflict ridden settings such as Sudan, entry by nonnationals into Khartoum, the capital city, is relatively easy compared to the stipulations required for further progress into other areas of the country. Once regarded as inherently malign, such policies now likely will have a crucial emergency response rationale—provided they are not abused. The following may thus help to serve as a draft framework for future donor health security assistance in this regard.

Utilization of local networks: This occurred via many existing foreign assistance initiatives and partners; these networks played a pivotal role in health education messaging and grassroots surveillance and knowledge transfer. Donor collaboration with local religious leaders, for example, played a significant role, along with health education messaging which was effectively integrated into religious ceremonies and teaching [47].

Health education and improving health literacy: Providing specific, locally adapted health education messaging was critical for developing countries. These campaigns not only relied on internet promotion but instead on text messaging, poster campaigns, advertising, and community awareness efforts; radio and popular drama messaging were also met with success in this regard. Wherever possible, such initiatives were tightly integrated with existing health education programmes rather than supplanting them.

Positive messaging: Combining health protection measures with education on the benefits of limited mobility in the climate change and environmental context was also of assistance to developing countries—this was similarly a significant incentive for the developed world to comply with emergency regulations.

Xenophobia avoidance: It was necessary, in developing country health security messaging, to avoid conflation of lockdown efforts with tribalism, regionalism, or other forms of domestic or international xenophobia. At the pandemic's beginning, the terms 'Chinese flu' and 'Asian flu' were often carelessly used by politicians. Conversely, such policies were on other occasions also leveraged to promote the opposite: the necessary integration of migrant communities during periods of limited mobility.

Convertibility: It was rightly emphasized in donor-partner dialogues that, strategically, all emergency services should also be convertible to day-to-day use under nonemergency conditions, generalizable into treatment of other health conditions, but also ready for reactivation. This helped to ensure that white elephant investments that remained unused after the emergency period were avoided. As a result, the building of 'field hospitals', for example, in Huoshenshan Hospital in China, began with impressive speed in January–February 2020 to assist with increasing capacity for patients. Organizations such as Medecins Sans Frontieres also have a long history of ensuring that similar 'field hospitals' are quickly constructed in areas of most pressing need.

Home and community prevention and care: In many ways, developed country responses were based on health system capacity: in that sense, their policies were as much to protect health systems as populations. In the developing world however, due to minimal health infrastructure that is often already overwhelmed, the immediate focus was thus on community-led care and home-care strategies.

Prevention and treatment campaigns: Information promotion is the basis of prevention of infectious disease spread. Some examples of prevention campaigns included the essential measures required in these regards. In particular, this also had water and sanitation health (WASH) guidelines, risk factor recommendations such as smoking, and locally relevant advice such as avoiding indoor fires and other pulmonary and respiratory health initiatives.

Industrial change preparations: Industries likely to be highly affected in many developing countries were also advised on possible consequences at early stages: recommendations on diversifying away from sporting events, tourism, alcohol sales, restaurants, and other services into more resilient industries helped many to prepare for, or avoid, financial hardship or economic loss.

Local ownership: In many cases, government policies were often conflated with other agendas. Thus, in emergency situations, there is a solid case to be made for civil obedience, but also for social authority. Trusting the local populations to make informed decisions based on local circumstances, rather than exclusive reliance on (often unenforceable) overly paternalistic or controlling efforts that might be construed as didactic or authoritarian is preferable.

Reconsidering protected health information: Developing countries were also required to consider the risks and shortcomings of protected health information in the emergency context. Greater clarity, granularity, and resolution in the local control of outbreaks will help in the future with *cordons sanitaires*. In addition, they will likely reduce, rather than increase, damaging stigma and paranoia considerations.

Avoiding overreliance on testing: Since the outset of the pandemic, testing strategies were unfeasible for the developing world. Associated costs and logistics continued to be prohibitive—though point-of-entry strategies have on occasion been considered. In the HIV/AIDS realm, testing was of limited practical use in controlling epidemics unless partnered with community referral systems; likewise, contact tracing strategies, mainly when epidemics take place on large scales, were often largely ineffective in the developing world.

A focus on geography: As with many other infectious disease outbreaks, the nature of the pandemic was inherently urban, with transmission, incidence, and prevalence rates all significantly higher in high-density population areas. Developing countries often focused efforts, *cordon sanitaires*, and other border control measures on protecting rural areas (and associated essential food production efforts).

A focus on demographics: The COVID-19 pandemic has been revealed to have affected the elderly and other vulnerable populations with much greater severity than different demographics. Developing countries subsequently made this an inherent part of their response strategies—by emphasizing the importance of protecting those population groups without limiting their fundamental personal freedoms. Where possible, such efforts should be undertaken from a humanitarian, rather than a health system, perspective.

Based on a combination of geographical and demographic considerations, there was a particular need for developing countries to focus on protecting vulnerable or elderly populations in urban areas. Though less common in low-resource settings, hospices, care homes, and retirement centres nonetheless remain high-risk areas: developing countries may reflect and learn from some of the possible oversights of developed countries in this regard.

Looking beyond treatment or prevention paradigms: Epidemic responses are often simplistically divided between treatment and prevention. In 2020–21, the treatment options were limited—particularly for severe cases in the developing world. Until the vaccine's arrival, it was uncertain when the pandemic would recede—and unfortunately for many countries, access to preventative vaccine supplies remains a significant issue. Even with less severe cases, limited curative or treatment options existed beyond basic recovery measures.

Similarly, whilst effective, social distancing measures (as noted above) were doubtful to be effective in many parts of the world due to educational, literacy, and population density considerations. A more creative approach, using local triage, ensuring contact tracing systems, and looking to countries who showed leadership, such as Rwanda (see above), is a lesson for developing and developed countries to begin learning from the Global South.

Prevention and containment: A key consideration for developing countries was the difference between prevention and containment. Containment strategies, in geographical terms, are essentially large-scale infection prevention efforts—even at the micro-level, they may theoretically reduce risk within restricted areas. Therefore, geo-containment strategies represent one of the few feasible alternatives to micro-level prevention efforts in developing countries with such a community-level focus.

Feasible local responses: The use of local resources (including recognizing local expertise) is key to epidemic containment in any setting, but particularly in low-resource areas. Ebola contact tracing is an example of this, through the success of using locals to contact trace via the innovative use of GPS systems and tracking. In addition, in the 2014 response to Ebola, local measures such as house-to-house information campaigns and messages on local radio stations in regional dialects were used to ease public fears through solid leadership and quick action.

Donor funding towards global health is also likely to be severely curtailed as affluent countries count the cost of their epidemics. Thus, the advancement of practical and culturally acceptable measures such as sneezing into elbow crooks, recommendations on time limits spent near others, or the use of homemade face masks may be far more appropriate and effective for the developing world than more socially or resource-demanding recommendations.

A need for bipartisan health security approaches: It was, in all cases, critically crucial that developing countries avoided overreacting or conflating policies with other agendas: to do so was to undermine public health efforts and risk the erosion of public trust in authority. Instead, it was vital to steer a line between hype, paranoia, and despair (as characterized by many liberal democratic responses) versus political or social recklessness (associated with more conservative, libertarian, or authoritarian regimes).

Moderate bipartisanism may be the best overarching advice for any developing country. Considerations of quality of life are essential; above all, extreme health security measures should not be allowed to excessively or unnecessarily affect other areas of public health, human rights, or economic productivity beyond emergency circumstances [48].

6.8 Key points

- Lead time is a crucial factor in pandemic response, and such responses should be tailored to local economic capacity and conditions to ensure feasibility.

- Pandemic responses are inherently iterative, with different countries making containment breakthroughs throughout the containment period.
- There is currently no widely accepted, consistent methodology for estimating impacts of pandemics.
- In many cases, the most effective infectious disease control responses are those that have already been established in cost-effective ways in the developing world.

References

[1] P. Carmody, G. McCann, C. Colleran, C. O Halloran, COVID-19 in the Global South, Bristol University Press, 2020, https://doi.org/10.2307/j.ctv18gfz7c.

[2] OECD, OECD Development Co-Operation Peer Reviews: Ireland 2020, OECD Publishing, Paris, 2020, https://doi.org/10.1787/c20f6995-en.

[3] A.Z. Chowdhury, K.S. Jomo, Responding to the COVID-19 pandemic in developing countries: lessons from selected countries of the global south, Development 63 (2) (2020) 162–171.

[4] K. Oliver, T. Lorenc, J. Tinkler, et al., Understanding the unintended consequences of public health policies: the views of policymakers and evaluators, BMC Public Health 19 (2019) 1057, https://doi.org/10.1186/s12889-019-7389-6.

[5] S. Djankov, U. Panizza, COVID-19 in Developing Economies, Centre for Economic Policy Research, 2020.

[6] N. Karim, L. Jing, J.A. Lee, R. Kharel, D. Lubetkin, C.M. Clancy, D. Uwamahoro, E. Nahayo, J. Biramahire, A.R. Aluisio, V. Ndebwanimana, Lessons learned from Rwanda: innovative strategies for prevention and containment of COVID-19, Ann. Glob. Health 87 (1) (2021) 23, https://doi.org/10.5334/aogh.3172.

[7] C.M. Michaud, Global burden of infectious diseases, in: Encyclopedia of Microbiology, 2009, pp. 444–454, https://doi.org/10.1016/B978-012373944-5.00185-1.

[8] A.K. Singh, A. Misra, Impact of COVID-19 and comorbidities on health and economics: focus on developing countries and India, Diabetes Metab. Syndr. Clin. Res. Rev. 14 (6) (2020) 1625–1630.

[9] S.L. Vériter, C. Bjola, J.A. Koops, Tackling COVID-19 disinformation: internal and external challenges for the European Union, Hague J. Dipl. 15 (4) (2020) 569–582.

[10] Y. Ngqangashe, M. Heenan, M. Pescud, Regulating alcohol: strategies used by actors to influence COVID-19 related alcohol bans in South Africa, Int. J. Environ. Res. Public Health 18 (21) (2021) 11494, https://doi.org/10.3390/ijerph182111494.

[11] World Bank, Poverty & Equity Brief. povertydata.worldbank.org. April 2020. Available from: https://databank.worldbank.org/data/download/poverty/33EF03BB-9722-4AE2-ABC7-AA2972D68AFE/Global_POVEQ_ZAF.pdf.

[12] Global Change Data Lab, Coronavirus (COVID-19) vaccinations, in: Our World in Data, Global Change Data Lab, 2020. [Cited: July 10, 2021]. Available from: https://ourworldindata.org/covid-vaccinations?country=OWID_WRL.

[13] A. Christie, et al., Decreases in COVID-19 cases, emergency department visits, hospital admissions, and deaths among older adults following the introduction of COVID-19 vaccine—United States, September 6, 2020–May 1, 2021, Morb. Mortal. Wkly Rep. 70 (2021), https://doi.org/10.15585/mmwr.mm7023e2. 23, s.l.

[14] J. Wise, Covid-19: Is vaccination roll out reducing cases and deaths in the UK? The Brit. Med. J. 372 (2021) n506, https://doi.org/10.1136/bmj.n506.

[15] P.C. Calder, Nutrition and immunity: lessons for COVID-19, Eur. J. Clin. Nutr. (2021), https://doi.org/10.1038/s41430-021-00949-8.

[16] P. Mlambo-Ngcuka, Violence Against Women and Girls: The Shadow Pandemic, *UN Women*, 2020. April 6 [Cited: July 15, 2021]. Available from: https://www.unwomen.org/en/news/stories/2020/4/statement-ed-phumzile-violence-against-women-during-pandemic?gclid=CjwKCAjwlrqHBhByEiwAnLmYULtJTySwNQyqVCmjNzrLNdnpiH_AHMNChSSkunJykaFrsbwXO2JBSBoClucQAvD_BwE.

[17] V. Minisini, South Africa's secondary pandemic: a crisis of gender based violence, Global Risk Insights (2021). March 28 [Cited: July 15, 2021]. Available from: https://globalriskinsights.com/2021/03/south-africas-secondary-pandemic-a-crisis-of-gender-based-violence/.

[18] T. Farber, Shocking stats on gender-based violence during lockdown revealed, Times Live (2020). September 1 [Cited: July 28, 2021]. Available from: https://www.timeslive.co.za/news/south-africa/2020-09-01-shocking-stats-on-gender-based-violence-during-lockdown-revealed/.

[19] Department of Health, Republic of South Africa. COVID-19 Risk Adjusted Strategy, COVID-19 South African Online Portal, 2020, March [Cited: July 24, 2021]. Available from: https://sacoronavirus.co.za/covid-19-risk-adjusted-strategy/.

[20] Solidarity Fund, Who We Are, *Solidarity Fund*, 2020. March [Cited: July 24, 2021]. Available from: https://solidarityfund.co.za/about/.

[21] John Hopkins University, COVID-19 Data Repository by the Center for Systems Science and Engineering (CSSE) at Johns Hopkins University, GitHub, 2021. [Cited: July 10, 2021]. Available from: https://github.com/CSSEGISandData/COVID-19.

[22] S. Devi, Travel restrictions hampering COVID-19 response, Lancet 395 (10233) (2020) 1331–1332.

[23] N. Madhav, B. Oppenheim, M. Gallivan, et al., Pandemics: risks, impacts, and mitigation, in: D.T. Jamison, H. Gelband, S. Horton, et al. (Eds.), Disease Control Priorities: Improving Health and Reducing Poverty, third ed., The International Bank for Reconstruction and Development/The World Bank, Washington, DC, 2017 November 27. (Chapter 17). Available from: https://www.ncbi.nlm.nih.gov/books/NBK525302/doi. https://doi.org/10.1596/978-1-4648-0527-1_ch17.

[24] P. Hunter, The spread of the COVID-19 coronavirus: health agencies worldwide prepare for the seemingly inevitability of the COVID-19 coronavirus becoming endemic, EMBO Rep. 21 (4) (2020) e50334.

[25] H.A. Aboubakr, T.A. Sharafeldin, S.M. Goyal, Stability of SARS-CoV-2 and other coronaviruses in the environment and on common touch surfaces and the influence of climatic conditions: a review, Transbound. Emerg. Dis. 68 (2) (2020) 296–312, https://doi.org/10.1111/tbed.13707.

[26] M.Y. Ni, L. Yang, C.M. Leung, N. Li, X.I. Yao, Y. Wang, Q. Liao, Mental health, risk factors, and social media use during the COVID-19 epidemic and cordon sanitaire among the community and health professionals in Wuhan, China: cross-sectional survey, JMIR Ment. Health 7 (5) (2020) e19009.

[27] WHO, WHO Coronavirus (COVID-19) Dashboard, 2021. https://covid19.who.int/table. (Accessed 9 May 2021).

[28] D. Septiari, Government Outlines Endemic Strategy for COVID-19, But Experts Urge Caution, 2021. https://www.thejakartapost.com/news/2021/08/27/government-outlines-endemic-strategy-for-covid-19-but-experts-urge-caution.html. (Accessed 9 May 2021).

[29] A. Martinez, Covid-19: Hopes for 'Herd Immunity' Fade as Virus Hurtles Toward Becoming Endemic, 2021. https://www.nytimes.com/live/2021/05/09/world/covid-vaccine-coronavirus-cases. (Accessed 9 May 2021).

[30] G20 High Level Independent Panel on Financing the Global Commons for Pandemic Preparedness and Response (2021). A Global Deal for Our Pandemic Age. https://pandemic-financing.org/ (Accessed 05.09.2021).

[31] G.H. Cahya, Indonesia's Hospitals in Covid Crisis as Car Parks Turned Into Emergency Rooms, 2021. https://www.theguardian.com/global-development/2021/jul/02/broken-indonesias-hospitals-in-crisis-as-doctors-treat-covid-patients-in-streets. (Accessed 9 May 2021).

[32] Lapor Covid-19a, Total Deaths of Self-Isolation and Non-Hospitalized Cases (in Indonesian), 2021. https://laporcovid19.org/data/kematian-isoman. (Accessed 9 May 2021).

[33] T. Allard, K. Lamb, COVID Infections Imperil Indonesia's Vaccinated Health Workers, and Hospitals, 2021. https://www.reuters.com/world/asia-pacific/covid-infections-imperil-indonesias-vaccinated-health-workers-hospitals-2021-07-07/. (Accessed 9 May 2021).

[34] Lapor Covid-19b, Healthcare Workers Digital Memorial (in Indonesian), 2021. https://nakes.laporcovid19.org/statistik. (Accessed 9 May 2021).

[35] Y. Mahendradhata, N.L.P.E. Andayani, E.T. Hasri, M.D. Arifi, R.G.M. Siahaan, D.A. Solikha, P.B. Ali, The capacity of the Indonesian healthcare system to respond to COVID-19, Front. Public Health 9 (2021) 649819, https://doi.org/10.3389/fpubh.2021.649819.

[36] S. Jaffrey, How the Global Vaccine Divide Is Fueling Indonesia's Coronavirus Catastrophe, 2021. https://carnegieendowment.org/2021/08/05/how-global-vaccine-divide-is-fueling-indonesia-s-coronavirus-catastrophe-pub-85107. (Accessed 9 May 2021).

[37] S. Davies, Lockdown Might Seem Harsh, But It's a Luxury Many Nations Don't Have, 2021. https://lens.monash.edu/@coronavirus-articles/2021/08/24/1383683/lockdown-might-seem-harsh-but-its-a-luxury-many-nations-dont-have. (Accessed 9 May 2021).

[38] K. Najmah, S. Andajani, S.G. Davies, T.G. Davies, COVID-19 Denial in Indonesia, 2021. https://www.insideindonesia.org/covid-19-denial-in-indonesia. (Accessed 9 May 2021).

[39] L. Gammon, Indonesia Stuck in the Middle With Its COVID-19 Crisis, 2021. https://www.eastasiaforum.org/2021/08/22/indonesia-stuck-in-the-middle-with-its-covid-19-crisis/. (Accessed 9 May 2021).

[40] A. Maxmen, Has COVID taught us anything about pandemic preparedness? Nature 596 (7872) (2021) 332–335, https://doi.org/10.1038/d41586-021-02217-y.

[41] J. Kruglinski, Countering an 'Infodemic' Amid a Pandemic, 2021. https://www.unicef.org/indonesia/coronavirus/stories/countering-infodemic-amid-pandemic. (Accessed 9 May 2021).

[42] I. Ningtyas, Indonesia Battles Spread of Vaccine Misinformation, 2021. https://www.aa.com.tr/en/asia-pacific/indonesia-battles-spread-of-vaccine-misinformation/2124319. (Accessed 9 May 2021).

[43] D. Suhenda, Deadly 'Infodemic' Looms Large Over Indonesia's Fraught Pandemic Response, 2021. https://www.thejakartapost.com/news/2021/07/23/deadly-infodemic-looms-large-over-indonesias-fraught-pandemic-response.html. (Accessed 9 May 2021).

[44] V. Haldane, C. De Foo, S.M. Abdalla, A.S. Jung, M. Tan, S. Wu, A. Chua, M. Verma, P. Shrestha, S. Singh, T. Perez, S.M. Tan, M. Bartos, S. Mabuchi, M. Bonk, C. McNab, G.K. Werner, R. Panjabi, A. Nordström, H. Legido-Quigley, Health systems resilience in managing the COVID-19 pandemic: lessons from 28 countries, Nat. Med. 27 (6) (2021) 964–980, https://doi.org/10.1038/s41591-021-01381-y (Epub 2021 May 17).

[45] P.G. Lloyd-Sherlock, A. Kalache, M. McKee, J. Derbyshire, L. Geffen, F.G. Casas, et al., WHO must prioritise the needs of older people in its response to the covid-19 pandemic, BMJ 368 (2020) m1164.

[46] D. Meyer, D. Bishai, S.J. Ravi, H. Rashid, S.S. Mahmood, E. Toner, J.B. Nuzzo, A checklist to improve health system resilience to infectious disease outbreaks and natural hazards, BMJ Glob. Health 5 (8) (2020), https://doi.org/10.1136/bmjgh-2020-002429.

[47] World Health Organisation, Practical considerations and recommendations for religious leaders and faith-based communities in the context of COVID-19, 2020, Retrieved December 6, 2021, from https://apps.who.int/iris/bitstream/handle/10665/331707/WHO-2019-nCoV-Religious_Leaders-2020.1-eng.pdf.

[48] S.L. Greer, E.J. King, E.M. da Fonseca, A. Peralta-Santos, The comparative politics of COVID-19: the need to understand government responses, Global Public Health 15 (9) (2020) 1413–1416.

Part III

Vaccine issues

Chapter 7

Vaccine preparedness

Sebastian Kevany and Aoife Kirk

Chapter outline

7.1 Were we ready?

There is no way that Ireland, like so many other small countries, could have been expected to lead the way in vaccine research. Irish innovations in contemporary epidemic responses were often limited to social and behavioural interventions, many of which functioned well inside the Irish setting and perhaps, inspired others—even if they were sometimes challenging to export to other countries without adaptation.

Yet Ireland has also often been the triumphant underdog in the medical field device world, with examples ranging from William Campbell, the Irish scientist who won the Nobel Prize (along with his Japanese scientist colleague Satoshi Omura) for the invention of Ivermectin [1]. This drug coincidentally came under the spotlight on numerous occasions during the COVID-19 pandemic. Ireland also claims the inventor Robert Collis, who invented a cheap and affordable incubator for neonates [2], Arthur Leared who invented the stethoscope as we know it today [2], and many others. As the race for COVID-19 vaccines began, the Oxford team, co-led by Dublin-born Professor Adrian Hill, Co-Director of the Oxford Martin Programme on Vaccines, received an honorary knighthood in the UK.

Outbreaks, Epidemics, and Health Security. https://doi.org/10.1016/B978-0-323-91705-6.00013-7

Meanwhile, Professor Teresa Lambe, also from Ireland, was appointed an honorary OBE (Order of the British Empire) for science and public health services.

Thus, despite the lack of research resources, Ireland's responsibility, as with every country around the world, is to be fully prepared for vaccine roll-out when the time comes. Yet in late 2020, there was little evidence of the Irish government's preparedness in this respect, even though the country had clearly decided to adopt a 'hold-your-breath' approach and wait for vaccines—rather than attempting normal societal functioning under the new status quo, Sweden-style.

Ireland had instead, perhaps, been forced into a vaccine orientation as a price for two critical limitations: preexisting public health demands, including those imposed by the high prevalence of alcohol, smoking, and sedentary lifestyles [3] and the failure of successive governments in Ireland to strengthen and resource the public health system. Neither of these issues could have been addressed in the short term. This was further compounded by the lack of ICU beds in Ireland—calculated by the OECD in 2020 as 5 per 100,000 population [4]—one of the lowest numbers in Europe. These shortages, along with medical staff shortages and migration, resulted in a 'collision' of events that caused a health system crisis during the pandemic in Ireland in 2020.

Yet, Ireland undertook a range of preparatory activities to make sure the country was prepared for a successful vaccine roll-out. However, it was perhaps far better to have these in place—and the subject of public debate—in advance of any vaccine announcements rather than choosing to wait for the medicine development and approval before beginning this challenging, complex, urgent, and large-scale process of roll-out.

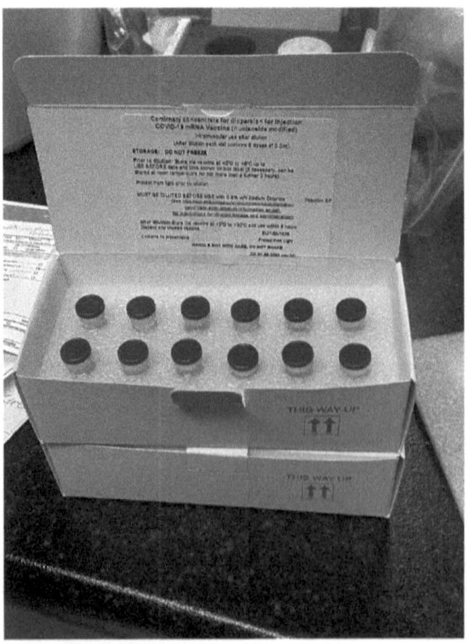

PICTURE 3.7.1 A delivery of early Covid vaccines to a clinic in County Mayo, Ireland, 2021. Eilis Conneally.

7.2 Guest essay: Vaccine equity—A lesson or a harbinger for future global crises?

Eamonn Faller

The Irish satirist, artist, and podcaster Blindboy Boatclub once stated that a good starting point in the analysis of any issue is to imagine the situation in question unfolding before you from the perspective of an alien visiting Earth. The human race currently lives under a system often called 'capitalism'. Individual humans are all extremely similar biologically, but are divided into groups called countries according to where they live on Earth.

Humans have, for the last year, been gripped by a virus that has killed approximately one out of every thousand. By my own estimation, an average of ten to fifteen times more would have possibly died had they not made efforts, with varying degrees of success, to physically distance themselves from each other as advised by experts, and decided upon by political leaders. A huge effort, sponsored largely by public money, was undertaken across this planet to come up with a vaccine to protect these humans from this virus. And, lo and behold, in this endeavour they had success—within a year of the emergence of this virus, a number of safe and effective vaccinations had been invented.

However this, from the alien's theoretical perspective, is where things start to become confusing. The risk of death from this virus varies markedly depending on the characteristics of the host. A human over the age of 75 is about 230 times more likely to die than a human aged 25. Yet, countries with a lot of resources are vaccinating all their 25 year olds before the 75 year olds living in countries with fewer resources. Furthermore, these 75 year olds do not have access to the expensive health care that the rich 25 year olds have. The alien might ask the question, 'Why not use these vaccines in the people who are at least 200 times more likely to die first? Are the rich young people's lives 230 times more valuable than those of the poor people? Perhaps, they must be'.

At the time of writing this essay, over a quarter of the world's population have received one vaccine dose, but only one out of every 100 people in those countries classified as low income has received one. This state of affairs is wildly unfair but, in this world where ten individuals own more wealth than the poorest 3.8 billion, it is about as wildly unsurprising. The wealthy people of the world have been allowed to drink indoors with lower risk of catching this virus, even if the cost is tens of thousands of deaths of people who were unlucky enough to be born outside Europe or North America.

The organization nominally tasked with overseeing the health of all the world's citizens (the World Health Organisation), for some of its flaws, have repeatedly pointed out this amoral approach. And don't get me wrong: vaccinating young, healthy people who do not work in health care is important, just nowhere near as important as vaccinating vulnerable people and healthcare workers in places with extremely limited health infrastructure.

Not only this, but a proposal by the less wealthy countries to allow an intellectual property rights waiver on technologies to combat COVID-19 such as the vaccines has, to this point of writing, been denied by the wealthy and now well-vaccinated nations. These technologies were largely funded by public money and charitable donations—but the patents, and ultimately the profits, belong to multibillion dollar pharmaceutical and biotech companies who have thrown their considerable lobbying power into ensuring that this waiver does not happen. Moderna, for example, received almost $2.5 billion in public money to develop their mRNA vaccine: its stock price has quintupled since November and they are expected to earn $18.4 billion in revenue this year. But even in the direst emergency the world has faced in 70 years, the interests of the pharmaceutical and biotechnology industries may not match those of the majority of humanity, according to western governments.

Token words from the wealthy world around an equitable worldwide vaccination rollout were matched with a token programme (COVAX) to source vaccinations for low-income countries. Or, more importantly, to be seen to be doing so: at the time of writing, in the last five days, my own wealthy country of origin, Ireland, with a population of 5 million has administered over 300,000 vaccines, the vast majority of these to people at extremely low risk of severe disease. These are more than have been administered in Yemen, a poor country of about 30 million people, reliant on COVAX for supply.

Voices in our media calling for equitable worldwide vaccine rollout were comfortably drowned out by loud complaints that other rich countries were managing to vaccinate their populations quicker than we were. The extreme tactics of some countries in not exporting so much as a dose of a vaccine in the early days of rollout were ignored by our media, and praise was heaped upon them and their industriousness, in moving on from vaccinating their vulnerable populations to their healthy ones. This coverage was a depressing manifestation of a capitalist and individualist societal outlook.

This has all already happened, and we seem to have decided that it is okay. Tens of thousands of people are dead in India and South America for want of a vaccine and most western countries are happily eating, drinking, and partying indoors. The WHO has had its say, and possibly history will not look kindly on this, but we have built a society where the rich and powerful can be mind-bogglingly selfish and get rewarded for being so. Come 2023, the world may actually be leaving this nightmare behind—even with our lopsided rollout—so, in our eyes, this extraordinary selfishness will have been worth it.

Unfortunately, this pandemic, or the next one, is unlikely to be the biggest challenge humanity will face this century. When the next pandemic arrives, as it surely will, or as climate breakdown accelerates, are there any potential parallels in the nature of how we will be required to respond? Will those in the West bear to be discommoded in order to blunt the effects of this catastrophe on the parts of the world that will be most severely affected in the early stages? Imagine large swathes of the tropics threatened with drought and desertification, Pacific Islands being submerged under rising sea levels. Will we recognize this problem and our role in it? Will we give up our cars; drastically reduce the amount of meat we eat; shrink our economy; and make do with less? Or will we act in the same self-interested manner that we have on this occasion?

How about the massively polluting—but likewise massively wealthy—industries and billionaires lobbying our governments? As the planet heats up, will our governments side with humanity in a desperate attempt to save the planet? Or, as they did for this emergency, side with industry, money, and lobbyists? Unfortunately, the crucial difference is that climate breakdown will not be something we can muddle through—whilst ignoring the consequences of our unfair actions and policy decisions—and expect to come out the other side with something salvageable.

In the early days of the pandemic, belief in our ability to act collectively on a larger scale for the common good was reignited. I do not think anyone realized that what happened in March 2020 was possible before it actually happened—a drastic disruption of everyday life at a substantial economic cost in order to protect our vulnerable. Our alien visitors would, perhaps, have been impressed. But it took the disaster coming to our doorstep in a dramatic fashion for us to do so.

Climate breakdown is more insidious and will affect other, mostly poorer, parts of the world in an existential manner before it will affect us—a situation much more akin to that of our vaccination equity predicament. Will it be with the spirit of March 2020 or that of March 2021 with which we meet this challenge? Can we recognize this worldwide injustice of vaccine apartheid for what it is, so as to learn from our actions? Or will we just sleepwalk towards oblivion, with tokenism and cosying up to powerful interests? This pandemic, like all traumatic experiences, is an opportunity to learn, to reflect on how our world acts in the face of an existential threat. On this occasion, it was decidedly not in favour of the majority of its citizens. It has to be on the next.

7.3 Simple initial paradigms

A straightforward place to begin, in any situation, was with target populations. Most would agree that vaccine roll-out should, ethically and otherwise, first focus on those populations most at risk of severe illness. These populations included those in older age brackets, those with underlying conditions or immunocompromised, and other vulnerable groups. Either in concert with those target demographics or possibly preceding it, the needs and risks faced by healthcare workers were, naturally, a priority from the very beginning [5].

A second straightforward consideration was in terms of spatial dynamics: Would it have been fair to suggest that certain regions and areas within countries that had been most affected, should have been vaccinated as a priority? Similarly, those countries and intercountry regions that had been least affected might be considered better protected until later in the process—without compromising ethical or equity standards—if required.

PICTURE 3.7.2 **Irish pop-up vaccine clinics.** Aoife Kirk.

Licensing, supply chains, and production capacity were all considerations that were necessarily negotiated by the government and the cost of the effort. These areas required logistical planning, regulation, and possibly even legislation (Should vaccine uptake be considered, even enforced, in the legal context?). These issues could add months to the roll-out effort without intense logistical planning if left to the last minute. Similarly, it is never by no means too soon to consider the production and use of vaccination certificates (e.g. for use when attending sporting or other events). Adverse event reporting also plays a key part in the early stages of any such roll-out, even after such processes have been completed at the clinical trial level.

The potential networks of vaccine centres during future pandemics should also be carefully considered: it may be fair to assume that Ireland will again, as with the influenza vaccine, focus its supply efforts on general practitioners, and selected chemists, for general availability. Yet was this, one wonders, enough? There may be a need, under future extraordinary circumstances, for the government to consider other options: extending availability to all chemists and pharmacists, door-to-door services, and pop-up vaccination centres (using the same model as for existing testing centres).

7.4 Guest essay: Journey from misery to hope—Experience from state of Bihar, India (COVID vaccination—Strategies)

Sourav Neogi

Introduction

The COVID-19 pandemic, which started in 2019 in Wuhan, China, spread across the globe in just a few months. Soon, it will be 2 years since the pandemic ravaged the entire world, including India. The Government of India confirmed India's first case of Coronavirus on 30 January 2020 [6]. By the third week of March, the total cases in the country increased to 500 [7]. Meanwhile, in Bihar (one of the lowest ranked states in India, based on developmental indicators), the first case was confirmed on 22 March 2020 [8].

Since March 2020, two further waves of COVID-19 cases have devastated the country. The health system has literally crumbled under intense pressure, especially rural health care. However, even amidst all the madness, vaccines came as a ray of hope for everyone.

COVID-19 vaccine—A hope

The COVID-19 vaccine, which provided hope to the people and the government, has also posed numerous challenges to the health system. The nature of vaccine management differs significantly from one institution to another, especially in terms of maintaining the temperature of the vaccines. Significantly, the world's largest vaccine manufacturing company, Serum Institute of India Pvt. Ltd., is located in India.

Vaccination in India started in January 2021, and in the last 7 months, millions of people got vaccinated, and the number is still growing. The Indian vaccination drive could be divided into four phases (stated below); this population-wise targeted approach was adapted based on the vulnerability of the population:

1. Vaccination for healthcare workers including doctors, paramedical staff, and other associated workers in the healthcare sector (16 January 2021) [9].
2. Vaccination for the elderly population (60-plus population) and 45-plus people with comorbidities (1 March 2021) [10].
3. Vaccination for the 45-plus population (1 April 2021) [11].
4. The 18-plus population (1 May 2021) [12].

These target approaches could be considered as strategies to reach out to different segments of the population; however, other than these broader strategies, there are various specific strategies that have been adapted to reach out to the masses.

COVID-19 vaccine strategies—Bihar state

India is a vast country, comprising 29 states and seven union territories. Although the Government of India time and again developed several guidelines and shared advisories, the strategies adopted by various states varied significantly. Different strategies have also been adapted by different state governments (provisional governments) to encourage and vaccinate large numbers of their populations.

Bihar is one of the union states in the eastern part of India. The state has a population of 130 million, with almost 40% of its population under 18 years of age. Vaccinating large numbers of people in the state was a huge challenge for the health sector in Bihar. However, since the possibility emerged regarding the availability of COVID-19 vaccines, the Government of India and the state of Bihar geared up to strengthen the vaccine distribution system and devised several strategies.

One such move by the government of Bihar set a stiff target of covering 60 million (total state population 130 million) of its population in 6 months under vaccination. In order to strengthen the health system, a range of decisions and actions have been taken up by the immunization cell within the health department. In this section, such strategies are highlighted chronologically.

Prioritizing healthcare workforce (along with frontline workers)—Vaccines were not available in sufficient quantity in the initial phase as the supply was very limited; hence, it was decided that the COVID-19 vaccine would be made available initially for healthcare workers (HCWs) only. After that, it would be available for people above the age of 60 years and people who are immunocompromised. Updating the database of healthcare workers was also completed to ensure everyone eligible got vaccinated.

Other than healthcare workers working in the hospitals, frontline workers of the health department were also prioritized and vaccinated. Integrated Child Development Services (ICDS) workers (Anganwadi workers)[a] were also enlisted as frontline workers. Other than government health workers, private healthcare workers were also vaccinated on a priority basis. Nodal private agencies and associations were also engaged in the vaccination listing process.

Vaccination for all the healthcare workers has thus been a priority for the state of Bihar. Vaccination has always been a voluntary process; however, state machinery encouraged healthcare workers to come forward to get their vaccines. In order to improve access, the decision was taken to increase the number of sites and sessions to expedite the vaccination process for all the healthcare workers.

After initial progress and enthusiasm, the number of healthcare workers receiving vaccines dipped considerably. Initially, the need to motivate and mobilize them was not felt; however, a dip in coverage amongst healthcare workers prompted a focused mobilization campaign for them. Special sessions were also organized to vaccinate the healthcare workers that were left out. In the subsequent months, it was decided that frontline workers other than healthcare workers will get vaccination on fixed days to make it convenient for them.

Prioritizing other essential workforce—Defence personnel, including security personnel of the Department of Home Affairs, were also prioritized. A decision was taken to vaccinate defence personnel on specific days of the week. A decision was also taken to start the vaccination session early in the day, so that defence personnel did not have to take leave to get the vaccine. Other than defence personnel, cleaning workers, including sweepers working in various municipalities, were also included in the list of frontline workers and were prioritized to get vaccinations. Special drives were organized to encourage and motivate other frontline workers of the municipality and defence personnel of the Home Affairs Department to get vaccinated.

Extending the vaccination campaign to the general population (45 to 59 with comorbidities and 60 plus) was the next logical step. It was evident that extending vaccine coverage to the general population will increase workload exponentially. The number of facilities for vaccination was increased manifold: private health facilities and lower level government. Facilities like primary

[a] The Anganwadi worker is a functionary of the Integrated Child Development Scheme (ICDS), an early childhood development programme aimed at addressing the health care and development needs of young children and pregnant and nursing mothers. Anganwadi is a type of child and mother care centre that was established as a part of the ICDS.

health centres (PHCs) which match selection criteria were included. For all, registration through an online portal was made compulsory to avail vaccination.

It was also estimated that roughly 10 million people in the state are over the age of 60 years. Envisaging online registration may pose a challenge in rural areas; instructions were given to all districts to increase slots for self-registration. For engaging and mobilizing the 60-plus population especially in rural areas, other states were involved. Scheduling for vaccination session sites in all areas was performed 20 days in advance, to avoid crowding in the session sites. Specific strategies were adopted; medical colleges were informed to target outpatient department (OPD) patients with matching selection criteria to be included in the vaccination. Vernacular media was also wisely used to mobilize local communities.

Prioritizing women beneficiaries (over 60 years old) on national women's day (8 March 2021)—The strategy was also adapted to mobilize women beneficiaries on women's day; this bore positive results and a significant number of elderly women got vaccinated on the 8th of March 2021. The strategy was named 'facilitated cohort mobilization' by state strategists. Keeping in view the success of the women's day vaccination drive, state institutions were also informed to observe a similar event on the 12th of March, 2021. Using a similar approach, a list of pensioners was shared with the district authorities and they were mobilized to the session sites. Finally, due to the increasing number of people coming for vaccination, session site timing was increased from 9 am to 5 pm.

Prioritizing vaccination in urban areas—Usually, urban areas have better access to healthcare services and may not need special provision; however, evidence from the past suggests that even urban areas need equal attention to improve vaccination coverage, especially in Southeast Asia, where a large number of urban populations live in slums. There are usually pockets in urban areas which need greater attention; a separate strategy for motivating all urban citizens through use of line listing of all eligible candidates and approaching them through ward members (elected representatives) was devised.

Accredited journalists (print, electronic, and web media) by the Information and Public Relations Department were enlisted as frontline workers. This decision by the health department on the behest of the government paved the way for the media persons to get free vaccines from government run vaccination centres, irrespective of their age.

Adult men and women working as *private security guards and in firefighting-related services* were also considered and listed as frontline workers and got priority vaccines. People engaged in private security jobs are highly exposed, and even during lockdown, these men and women were always at work; hence, placing them on a priority list was considered.

Vaccination for people without identity proof—In order to get vaccinations, online registration (onsite or offsite) was made compulsory along with a photo ID proof. However, even those without identification have the right to have vaccination; hence, a special initiative was undertaken by the government. In accordance with GOI guidelines to enrol undocumented individuals in a separate vaccination effort, support from other departments (minority affairs, social justice and social welfare) was also sought to create a list of such beneficiaries.

Vaccination for Nepali citizens—Based on the recommendation of the National Expert Group on Vaccine Administration for COVID-19 (NEGVAC), support was also extended to all Nepali Citizens working in the state for COVID-19 vaccination. It has been realized that many Nepali Citizens working in various parts of India including Bihar needed support to get the COVID-19 vaccine. However, they did not have proof of identification; hence, a special provision was recommended to enlist and vaccinate Nepali citizens working in the state of Bihar.

Vaccination for pregnant women—Initially, there was not enough evidence to understand whether pregnant women could be vaccinated or not; however, gradually, global evidence was generated and the state started forming policy around vaccination amongst pregnant women. Based on the national government's instruction in July 2021, vaccination of pregnant women was initiated in the state of Bihar. Special emphasis was placed on motivating pregnant women using community-based platforms run by HCWs.

*24*7 COVID vaccination centre*—To ensure people had access to COVID vaccination round the clock, the Government of Bihar proposed 24/7 COVID vaccination centres in all the district hospitals across the state. The 24/7 vaccination centres are currently functional in a selection of facilities and provide round the clock vaccination service to the citizens of Bihar. These centres are also manned by a dedicated doctor.

District communication task force to manage vaccine hesitancy—Even though more and more people are getting vaccinated in various centres across the state. Vaccine hesitancy still exists amongst the people. Along with it, rumours and false information are generating dilemmas amongst people about getting vaccinated. In order to manage these situations in the state, a District Communication Task Force was formed under the guidance of the District Magistrate. Panchayati Raj[b] Institution members, religious leaders, reputed medical practitioners, members of private school associations, and press and media persons are all part of the Task Force. The District Communication Task Force discusses issues around rumours and vaccine hesitancy amongst residents and then devises a plan by engaging local media to counter those false claims.

The Bihar government's continued effort to cover the maximum number of people under COVID vaccination to ensure the state is better prepared to combat subsequent waves of the COVID-19 pandemic resulted in setting up an ambitious target of covering almost 50% of its population within the next 6 months.

Learnings

Appropriate strategies and willingness at the highest level of leadership are key for the success of any programme. Efforts to reach the unvaccinated are key for the success of any vaccination programme. Even though strategies have adapted to improve COVID vaccination, they are still in a nascent stage; however, the majority of the strategies to date have been well thought out. The majority of the strategies adopted are also based on consultative processes with development partners and experts; this further highlights the importance and significance of collaborative efforts bearing positive outcomes.

Periodic review meetings to gauge progress further accelerated the process of mass vaccination. Meetings conducted by the highest leaders of state have also created a positive atmosphere amongst the healthcare workers and citizens of the state. Leadership and governance are one of the six pillars of the health system; proactive and concerted efforts from the top level make a huge difference.

[b] Panchayati Raj is a three-tiered system of local self-governance in rural India.

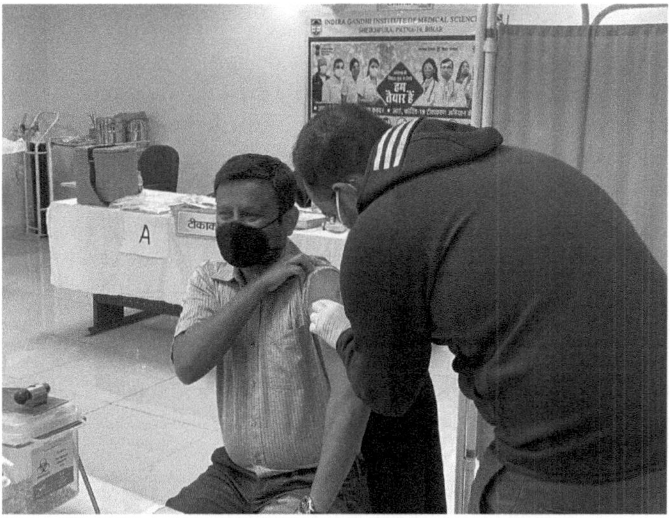

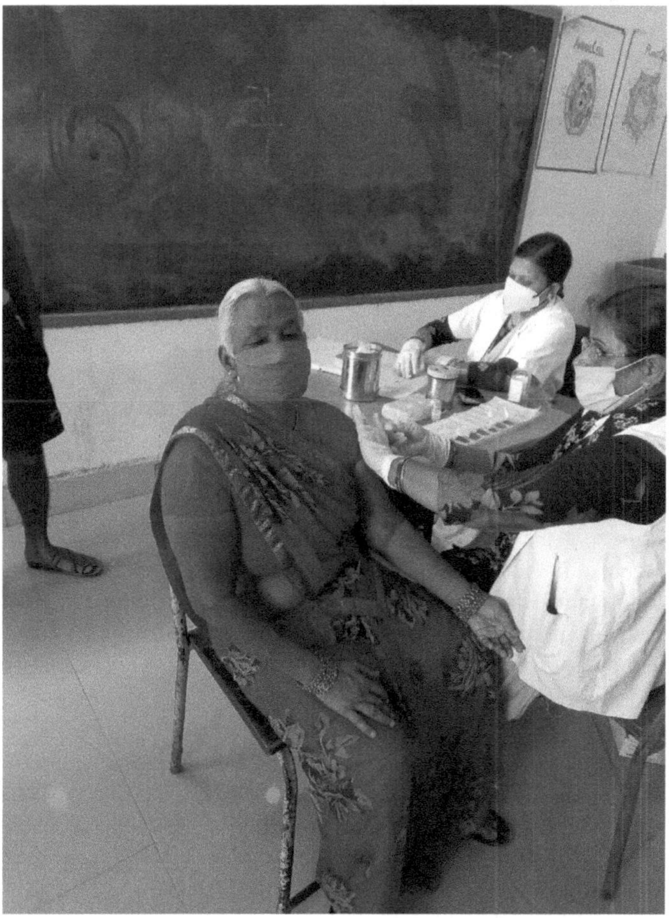

PICTURES 3.7.3 AND 3.7.4 Vaccination efforts underway in India. Sourav Neogi.

In 2021, the need to go one step further by the Irish government was eventually met—with mass vaccination days held in large hall venues; door-to-door vaccinations for vulnerable populations; and the international football stadium of Croke Park taken over for weeks, along with other local and national venues, in an effort for mass vaccination events. Cost-effectiveness, of course, is also a key consideration.

7.5 The European Union and 'setting the tone'

In Ireland, as with many other countries worldwide, there was also an urgent and contemporary need to combat fear, disinformation, and suspicion surrounding vaccines. In other countries, inappropriate and misguided suspicion that vaccine development had been manipulated for political or electoral gains both threatened to delay the process and eroded public trust [13]. This parallel fake news and the media-driven epidemic of fear and suspicion undoubtedly affect any vaccine uptake. Through general information campaigns and other measures, the government addressed these issues far in advance of the actual medical roll-out [14].

Perhaps, one could argue that examining what data was already available internationally on vaccine hesitancy would have been helpful before implementing a national public vaccination [14]? One example of such a project which has looked at data over the last 10 years is the Vaccine Confidence Project (VCP), founded in 2010 to explore the response at the time of suspicion, hesitancy, and misinformation on vaccine programmes such as polio. It has since developed an early-warning system that looks to identify the geographical location and evaluate public confidence in vaccines. Whilst it remains challenging to analyse certain psychological aspects and create programmes targeting these individuals, the VCP shows that such challenges often have little to do with vaccines themselves—and instead are a 'window into the broader issues of our polarized, mistrustful society' [15].

We have already experienced the adverse effects of too much guardedness and pessimism in the face of the virus: many experts, and at times the World Health Organization itself, failed to recommend using face masks to prevent infection. In the same way, there is the risk that related negativity and guardedness around vaccines inhibit or delay such processes unnecessarily. Of course, there were going to be risks with vaccines, just as there are with face masks—but society had to be willing to accept some form of risk for the common good, as well.

Vaccine preparedness policies also represented a unique opportunity for Ireland to be at the fore in the European Union context. Despite the reassurance of EU presence and related directives, there remains a high degree of autonomy in national-level decision-making, one that should, rightly, have been maintained. As a result, Ireland could in future help to lead other related processes at both the EU and the national level.

In this context, it was unfortunate to see countries resort to the blunt (and worryingly authoritarian) employment of the police force, once again, for epidemic control rather than adopt other more innovative, civilized, and creative measures.

By contrast, a proper future preparedness plan will mean that the nation is leading the way in creative, just, and effective responses. In that way, something also would undoubtedly be given back to other EU actors and international partners. Ireland must not forget that the EU is a union rather than a governing body, and it is the responsibility of each member state to contribute to the well-being of the whole and benefit from it.

7.6 Guest essay: Has the right to health been sidelined by a right to profit?

Ciara Conlan and James Larkin

The COVID-19 pandemic has had a massive impact on the lives and livelihoods of humans around the globe. It has highlighted and exacerbated existing deep structural inequalities between rich, powerful countries and the rest of the world. We find ourselves living in an unfortunate reality that was predicted by many—a massive skewing of diagnostics, treatments, and vaccines for COVID-19 away from the Global South, where the pandemic is escalating relentlessly. At the time of writing, six billion doses have been administered globally, but only 2.2% of people in low-income countries have received at least one dose [16].

Early in the course of the COVID-19 pandemic, leaders of powerful countries such as France, Germany, and China declared that any future vaccine against COVID-19 should be considered a global public good [17,18]. That is to say, nonexcludable and nonrivalrous. So why then was this shortfall predictable? Why are wealthy countries currently seeking vaccine derived herd immunity and booster shots whilst the Global South can't get its foot in the door of the vaccine market? This essay will explore how this situation occurred and highlight what we can learn, both now and for future global health crises.

At the root of this issue are the monopolies on the production of COVID-19 vaccines. Due to intellectual property (IP), the Global South generally have not been allowed to manufacture vaccines for their domestic markets. The vaccines produced so far have been sold mainly to wealthy, powerful countries—countries who have had the purchasing power to make bilateral agreements in advance with pharmaceutical companies to ensure their own national supply. For example, Pfizer/BioNTech and Moderna have sold over 90% of their vaccines so far to higher-income countries [19]. Due to dose hoarding by these countries, lower- and middle-income countries have had difficulty competing in the market for vaccines. Sadly, this is not the first time a pharmaceutical company's economic monopoly has decided who lives and who dies across the world, by proxy of address.

> COVID-19, like HIV and AIDS, is laying bare the underlying inequalities in our world and in our societies.
>
> ***Winnie Bwanyima, executive director of Joint United Nations Programme on HIV and AIDS (UNAIDS) [20]***

In order to provide historical context on the dissonance between the human right to health and the intellectual property (IP) rights enshrined by the 1994 Trade-Related Aspects of Intellectual Property Rights (TRIPS) agreement at the World Trade Organisation, we can look to the HIV/AIDS pandemic, which peaked in the 1990s. The HIV/AIDS pandemic raged on in the Global South long after being controlled by new life-saving medications in the United States and Europe. The pharmaceutical company monopoly on these HIV medications led to extremely high prices—unaffordable for most of the world. According to UNAIDS, nine million lives were needlessly lost in the Global South, whilst awaiting access to HIV medications [21]. In the face of opposition from the pharmaceutical industry and high-income countries, a massive mobilization by civil society and political actors was needed, which cleared the way for the production of generic versions of these patented medicines. This lowered the price dramatically and allowed for much wider access to treatment.

Like the HIV/AIDS pandemic, solutions to the COVID-19 pandemic are being blocked by pharmaceutical companies and high-income countries. The proposals to ensure ethical and rational distribution of COVID-19 vaccines include the COVAX facility, voluntary or mandatory IP sharing, and technology transfer.

COVAX was created as a risk sharing mechanism so that countries could source and equitably distribute vaccines globally. Originally, it was hoped that COVAX would be the mechanism for all countries to procure their vaccines [22], but many high-income countries decided to pursue their own procurement and thus undermine COVAX. COVAX is now, at the time of writing, focused on vaccinating low- and middle-income countries with the modest aim of vaccinating 20% of the population in 92 low- and middle-income countries in 2021. However, this goal has been hampered by a lack of funding, vaccine scarcity, and obligations to donate vaccines to high-income countries [23]. As of April 2021, only one-fifth of the doses projected to have been delivered at this point had been shipped [24].

Donation-based models to deliver vaccines to the Global South have been supported by powerful elites, such as the G7 country leaders and philanthropists, but have fallen vastly short of their projected targets. Not only are donations unreliable, they also risk removing autonomy from health departments as to when and which vaccines they offer to their populations. The arrival of the vaccines may not be optimally timed in line with local health priorities. As stated by Lioba A. Hirsch in the Lancet in relation to decolonizing global health: 'Patient communities across the world do not need global health leaders to determine access to which medicines should be prioritized or what disease eradication deserves priority' [25].

Some countries are not even fully engaging with a hamstrung COVAX initiative, but instead they are further undermining it by engaging in vaccine diplomacy by making bilateral donations. Dr Tedros Adhanom Ghebreyesus, WHO director-general, has stated that *'Bilateral [donation] arrangements run the risk of fanning the flames of vaccine inequity'* [26]. Furthermore, donated vaccines do not address the root causes of vaccine inequity and they do nothing to increase the overall supply of vaccines. The holders of the rights to make vaccines have so far blocked efforts to increase production capacity around the world. Unfortunately, it now appears likely that vaccinations may only become more widely available in the Global South when the rich North decides that ongoing transmission elsewhere in the world could put them at risk of variants which could overcome their vaccine-derived immunity.

The solutions which would deliver outside of traditional aid models have been slow to gain support. Leadership from the Global South has been strong in proposing and campaigning for these more progressive approaches. For example, the Government of Costa Rica proposed the COVID-19 technology access pool (C-TAP) in early 2020. The C-TAP was designed as a platform to pool IP and know-how for COVID-19 technologies. This would allow developers of COVID-19 medicines, diagnostics, and vaccines to enter voluntary arrangements to scale up production of these technologies. Unfortunately, no COVID-19 vaccine developers have so far shared IP-protected vaccine knowledge with the C-TAP. There is a huge financial incentive for these companies not to do so, and this failure has increased the pressure for mandatory measures, such as the TRIPS waiver.

The TRIPS waiver is a proposal by the governments of South Africa and India to the World Trade Organization to temporarily waive intellectual property rights protection for technologies needed to prevent, contain, or treat COVID-19. This waiver would remain in place 'until widespread vaccination is in place globally, and the majority of the world's population has developed immunity'. The TRIPS waiver for COVID-19 vaccines is supported by over 100 countries, including the United States, who view it as essential to boost vaccine supply and improve access in the Global South. This measure should open up vaccine manufacturing capacity all around the world. For example, in Canada, India, Bangladesh, Indonesia, South Africa, Ghana, Turkey, Brazil, Argentina, the United Arab Emirates, China, and Russia, there are pharmaceutical companies with the capacity to produce many more vaccines [27].

Proponents of the waiver argue that massive profits have already been made from COVID-19 vaccines despite the fact that their development was largely publicly funded. In addition, and as described above, other initiatives such as COVAX and C-TAP are not working. In the 'Nature' editorial in March 2021, it was stated that arguably, the strongest argument for a temporary waiver is that patents were never designed for use during global emergencies such as wars or pandemics [28]. Opponents of the waiver are concerned that this would undermine industry's incentive for future scientific innovation. However, the success of COVID-19 vaccine research and development (R&D) is a reflection, in large part, of public funding, which accounted for 97% of the R&D costs of AstraZeneca and most of the funding of Moderna [29,30].

The manner in which we approach this pandemic from here on can be an opportunity to decolonize global health and create the infrastructure we need to tackle other infectious diseases. Leaders in wealthy countries with a plentiful supply of vaccines must ask themselves if they are truly committed to ending this pandemic or if they are more focused on protecting the profits of the pharmaceutical industry. The countries proposing and supporting the TRIPS waiver and the C-TAP are not asking for charity, but for the right to manufacture and import their own vaccines. The scaling up of vaccine manufacturing capacity in the Global South will be an important step for future global health security.

As we look to the future, it is clear we need new rules in terms of how we view intellectual property rights during a global health emergency. Whether the next emergency is another pandemic or a climate disaster, the default should be to immediately waive patent rights for technologies needed to protect public health during a crisis. Given such urgent technologies are largely paid for upfront by the public system, it should be a given that they be available and affordable to whoever needs them in crisis moments around the globe.

7.7 Optimism through structure

But this is not all about supranational politics: in terms of social, societal, and community-level morale—and other dimensions of health—as the winter of 2020 approached, it was important for Irish society to have clear potential targets mapped out in terms of vaccine efforts. Even if these efforts needed to be pushed back, in the final reckoning. Without such a temporal-chronological-psychological framework, the risk of low morale associated with potentially unlimited lockdowns was a high one [31].

Yet, by the time of writing of this book, Ireland has come to the forefront regarding vaccine uptake, with over 6.9 million doses administered to the general population (>90% compliance rate), the top amongst European countries on uptake by September 2021 [32]. Has this uptake been due to hospitality incentives, with the necessary vaccine passport proof required to sit indoors to enjoy a meal or drink? Or is it due to the more authoritarian lockdown approach? Perhaps it may be weariness or the fear of another winter of sitting indoors with no live music or concerts? Whatever the answer is, a behavioural study on vaccine uptake should be undertaken in each country to guide future responses. It remains to be seen if additional boosters will be required beyond 2021 if certain vaccines protect from hospitalizations more than others, yet Ireland outshone many others regarding general public uptake.

Without understanding behind the behavioural motivation, Ireland also faced, like many other countries, the risk of Stockholm syndrome—as both the government and some aspects of society, risk becoming overly comfortable with the social control that came with lockdowns. Even if vaccines were still months away in early 2021, people needed hope in the form of targets, timelines, and

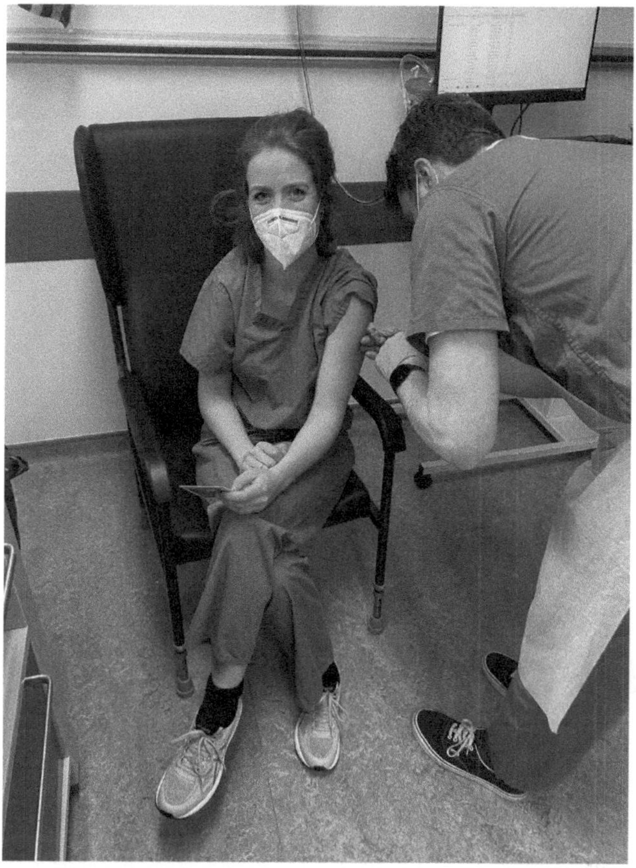

PICTURE 3.7.5 Getting the first-dose vaccine—sigh of relief. Aoife Kirk.

structures. Would Sir Ernest Shackleton's men, at the South Pole, have survived their winter had they have not had a rescue firmly set in their consciousness: would they have managed, had they not been ready—every day—for rescue, and expected it to happen?

Of course not: in this way, future adherence and compliance to prevention and containment measures will undoubtedly only improve if they become framed as short- to medium-term emergency efforts, rather than long-term impositions. Likewise, in the absence of lobbying, marches, singing, and other forms of protest, the Irish civil society voice was often silent, with only journalists, politicians, and academics in control of the narrative. In the context of both future planning and containment measures, it is a lesson for countries to ensure public preferences and needs will be both heard and responded to.

PICTURE 3.7.6 **Even vaccinated surfers kept their distance!** Sebastian Kevany.

7.8 Key points

- Vaccine preparedness cannot start too early, as it serves as a key psychological marker that helps with adherence to lockdown policies.
- Vaccine preparedness measures, in 2021, assumed a widespread acceptance of vaccines and did not adequately consider the antivaccine movement (to be examined in the next chapter).
- A strengthened public health system will help ensure that all the necessary infrastructure is in place for the effective and efficient roll-out of vaccines.

References

[1] E.A. Ottesen, W. Campbell, Ivermectin in human medicine, J. Antimicrob. Chemother. 34 (2) (1994) 195–203.

[2] Irish Scientists and Inventors, 2021. Retrieved December 6, 2021, from https://www.ipoi.gov.ie/en/understanding-ip/student-zone/irish-scientists-inventors/.

[3] J. Meyer, C. McDowell, J. Lansing, C. Brower, L. Smith, M. Tully, M. Herring, Changes in Physical Activity and Sedentary Behaviour Due to the COVID-19 Outbreak and Associations With Mental Health in 3,052 US Adults, 2020.

[4] Intensive Care Beds Capacity, 2021. Retrieved December 6, 2021, from https://www.oecd.org/coronavirus/en/data-insights/intensive-care-beds-capacity.

[5] A. Doyle, W. Hynes, S.M. Purcell, Building resilient, smart communities in a post-COVID Era: insights from Ireland, Int. J. E-Plan. Res. 10 (2) (2021) 18–26.

[6] M. Kumar, A.K. Patel, A.V. Shah, J. Raval, N. Rajpara, M. Joshi, C.G. Joshi, First proof of the capability of wastewater surveillance for COVID-19 in India through detection of genetic material of SARS-CoV-2, Sci. Total Environ. 746 (2020 December 1) 141326.

[7] P. Kumari, D. Toshniwal, Impact of lockdown measures during COVID-19 on air quality—a case study of India, Int. J. Environ. Health Res. 14 (June) (2020) 1–8.

[8] A. KumaR, M. Kumar, COVID-19 and the public health system in Bihar, Econ. Polit. Wkly 55 (16) (2020) 16–19.

[9] V.K. Niyas, R. Arjun, Breakthrough COVID-19 infections among health care workers after two doses of ChAdOx1 nCoV-19 vaccine, QJM Int. J. Med. 1 (2021) 2.

[10] J. Jain, S. Saurabh, A.D. Goel, M.K. Gupta, P. Bhardwaj, P.R. Raghav, COVID-19 vaccine hesitancy among undergraduate medical students: results from a nationwide survey in India, medRxiv (2021). January 1.

[11] https://www.mohfw.gov.in/pdf/GuidelinesforCOVID19VaccinationatWorkplace.pdf.

[12] A. Varghese, S.S. Surachna, Improving last-mile delivery of COVID-19 vaccines: the cluster strategy, Nepal J. Epidemiol. 11 (3) (2021) 1079.

[13] Enhancing Public Trust in COVID-19 Vaccination: The Role of Governments, 2021, May 10. Retrieved December 6, 2021, from https://www.oecd.org/coronavirus/policy-responses/enhancing-public-trust-in-covid-19-vaccination-the-role-of-governments-eae0ec5a/.

[14] J. Murphy, F. Vallières, R.P. Bentall, M. Shevlin, O. McBride, T.K. Hartman, R. McKay, et al., Psychological characteristics associated with COVID-19 vaccine hesitancy and resistance in Ireland and the United Kingdom, Nat. Commun. 12 (1) (2021) 1–15.

[15] The Vaccine Confidence Project, 2021. Retrieved December 6, 2021, from https://www.vaccineconfidence.org/.

[16] H. Ritchie, E. Mathieu, L. Rodés-Guirao, C. Appel, C. Giattino, E. Ortiz-Ospina, et al., Coronavirus pandemic (COVID-19), Our World in Data (2021). [cited 24 September 2021]. Available from: https://ourworldindata.org/coronavirus.

[17] M. Botenga, A Covid-19 Vaccine Must Have a 'Global Public Good' Guarantee, 2020. www.euractiv.com. Available at: https://www.euractiv.com/section/health-consumers/opinion/a-covid-19-vaccine-must-have-a-global-public-good-guarantee/.

[18] S. Wheaton, Chinese vaccine would be 'global public good,' Xi says, Politico (2020). [cited 3 September 2021]. Available from: https://www.politico.com/news/2020/05/18/chinese-vaccine-would-be-global-public-good-xi-says-265039.

[19] Oxfam International, Vaccine Monopolies Make Cost of Vaccinating the World Against COVID at Least 5 Times More Expensive Than It Could Be, 2021, Available from: https://www.oxfam.org/en/press-releases/vaccine-monopolies-make-cost-vaccinating-world-against-covid-least-5-times-more.

[20] C. Devine, WHO's Dr Mike Ryan: 'We Have Been Brought to Our Knees by Covid-19, Not Because the Virus Is Smart, But Because We Are Stupid', 2021, Trócaire. Available from: https://www.trocaire.org/news/whos-dr-mike-ryan-we-have-been-brought-to-our-knees-by-covid-19.

[21] W. Byanyima, UNAIDS Executive Director Report, UNAIDS, Geneva, Switzerland, 2021. p. 10. Available from: https://www.unaids.org/sites/default/files/media_asset/20210701_48_PCB_Speech_En.pdf.

[22] S. Berkley, COVAX explained, 2021, Gavi.org. [cited 13 September 2021]. Available from: https://www.gavi.org/vaccineswork/covax-explained.

[23] A. Usher, A beautiful idea: how COVAX has fallen short, Lancet 397 (10292) (2021) 2322–2325.

[24] P. Erfani, A. Binagwaho, M.J. Jalloh, M. Yunus, P. Farmer, V. Kerry, et al., Intellectual property waiver for covid-19 vaccines will advance global health equity, BMJ 374 (2021) n1837, https://doi.org/10.1136/bmj.n1837.

[25] L. Hirsch, Is it possible to decolonise global health institutions? The Lancet 397 (10270) (2021) 189–190 (cited 2 September 2021).

[26] Coronavirus: WHO chief criticises 'shocking' global vaccine divide, BBC News (2021). [cited 25 August 2021]. Available from: https://www.bbc.com/news/world-56698854.

[27] A. Prabhala, C. Clinton, Biden has the power to vaccinate the world, The Atlantic (2021). Available from: https://www.theatlantic.com/ideas/archive/2021/05/biden-has-power-vaccinate-world/618802/.

[28] It's time to consider a patent reprieve for COVID vaccines, Nature 592 (7852) (2021) 7.

[29] S. Cross, Y. Rho, H. Reddy, T. Pepperrell, F. Rodgers, R. Osborne, et al., Who funded the research behind the Oxford-AstraZeneca COVID-19 vaccine? Approximating the funding to the University of Oxford for the research and development of the ChAdOx vaccine technology, medrxiv (2021). [cited 14 September 2021]. Available from: https://www.medrxiv.org/content/10.1101/2021.04.08.21255103v1.

[30] J. Stone, The people's vaccine—Moderna's coronavirus vaccine was largely funded by taxpayer dollars, Forbes (2020). Available from: https://www.forbes.com/sites/judystone/2020/12/03/the-peoples-vaccine-modernas-coronavirus-vaccine-was-largely-funded-by-taxpayer-dollars/?sh=2de493446303.

[31] P. Hyland, M. Shevlin, O. McBride, J. Murphy, T. Karatzias, R.P. Bentall, A. Martinez, F. Vallières, Anxiety and depression in the Republic of Ireland during the COVID-19 pandemic, Acta Psychiatr. Scand. 142 (3) (2020) 249–256.

[32] R. News, Demand for testing to rise as 1,180 new cases confirmed, 2021, September 05, Retrieved December 6, 2021, from https://www.rte.ie/news/2021/0905/1244790-covid-figures-ireland/.

Chapter 8

Antivaccine campaigns and the far left and right

Sebastian Kevany and Aoife Kirk

Chapter outline

8.1 Pandemics and Edmund Burke

The Irish philosopher Edmund Burke suggested that all it takes for evil to triumph is for good men to do nothing [1]. How can we interpret this statement in the contemporary public health context? Perhaps the best place to start is with the premise that proven, safe, and effective curative treatments, approved at the highest national and international levels, are a force for good in our society. If this premise is accepted, and we are dependent on successful vaccination campaigns to return to normal societal functioning—thereby ending a miserable period for many people, particularly the elderly and vulnerable populations—then, one might well ask, wherefore the evil?

Outbreaks, Epidemics, and Health Security. https://doi.org/10.1016/B978-0-323-91705-6.00008-3

Yet, in Ireland, as elsewhere, there remains the risk that far-right political and social movements may threaten this force for societal good. This may often occur through high-profile antivaccination campaigns that often use alarming rhetoric and inaccurate statements about vaccine safety and effectiveness, leaving the national effort to roll out vaccines safely and efficiently under threat. Misinformation can be spread rapidly through social networks, and unfortunately, fixed false beliefs can rarely be corrected. COVID-19 has introduced pandemic deniers and those who oppose wearing face masks or quarantine measures, which sadly may detriment the more vulnerable population in their vicinity [2].

As we are all aware, we live in an era of fake news [3], not least thanks to the proliferation of Internet platforms; and the promulgation and presentation of opinions as facts is sadly rife. It is almost impossible for many to distinguish between the two. The situation is compounded when scientific evidence is presented as false. The end result, to say the least, is confusing.

Yet, to return to Burke, what can the individual or community actually do in the face of such threats? Such misinformation organizations thrive on the publicity of any kind, and therefore, to many, even challenging them merely feeds their energy and draws attention to inaccurate public health statements. Many people, with this concept in mind, will choose to look the other way.

Further encouragement to these organizations may be provided by the culture of fear, and the threat of being subject to public ridicule or censure. Antilockdown protests, conspiracy theorists, and the uncertainty due to the original playing down of the threat worldwide have had a significant impact. The World Health Organisation recognized, even before the pandemic, that antivaccine campaigns and the spread of false information are major global health threats for the forthcoming decade [4].

8.2 Guest essay: Pandemic reflection—An Irish Surfer's perspective

Conor Burns

I'm sitting here in Ireland, on one of the hottest days of the year in 2021, and find myself staring at a blank sheet of paper, having been tasked with trying to write a reflective essay on securing future global health security and stopping future pandemics before they start.

Wow—I'm just a middle-aged surfer who likes to talk radio and enjoys reading the news—and to that end it makes this task even more daunting, as, even as I type this, I'm currently reflecting on news from the United States which is stating that only 12 people or accounts are responsible for the majority of vaccine misinformation on Facebook [5]. The shocking thing isn't that only 12 accounts are responsible for misleading so many millions of people worldwide, and it is that on reflection on the great political divide, the talk isn't about how did we get here, how do we stop this, or even how do we hold these people accountable. Instead, bizarrely, the media portrayal is about government crackdown on free speech and alleging snooping on private citizens on social media platforms [6]. Social media are searching for likes, or meaning—or are they geopolitical operators, trying to push a global agenda?

I think Timothy Snyder sums the situation up well with his theory of Sada Populism: manipulating society through fear, hate, and fake news—as it has been working for many political parties in recent years. For example, some global leaders pushed their followers to protest or overthrow states' leaders that imposed lockdowns, or supported public health restrictions [7]. Many politicians, worldwide, have pivoted to fully embrace the antivaccine movement at the state level—even in Ireland. But, even more worryingly.... According to a related article in Medical News Today, "The study provides statistical evidence about the danger of political polarization" [8]. The world feels like it's morphing into George Orwell's *1984*, and a corrupted version of H.G. Wells' Shape *of Things to Come*.

To me, these questions about the 12 Facebook accounts perhaps hold one of the keys to solving the global pandemic—as well as preventing future ones.

Science has delivered the answers, but we are left with an ongoing political/cultural division which may be stopping the world from embracing the solutions.

8.3 Manipulating the facts

The far right, in the public health context, is also (rightly) protected by the law [9]. As long as they stay within the realms of jurisprudence, they are, rightfully, allowed to express themselves in any way they choose. This key feature of liberal democratic societies is, of course, unimpeachable and beyond question. The aspects like having a space for public demonstration, allowing issues to be aired and expressed, and questioning political decisions all are considered a necessary part of a democratic state [10].

Yet, when different narratives arise about the virus itself—safety concerns, use of alternative medicines, or when stories of vaccines emerge from unfounded rumours—issues occur. Unfortunately for many, these issues often result in challenging conversations between doctors and patients at the ICU or emergency department bedside related to the truths about the vaccines which had been scientifically accepted by the medical community. So, why decline the vaccine?

Perhaps here, we can return to Burke, with his further observation in the 1780s of "Nobody made a greater mistake than he who did nothing, because he could do only a little" and contemplate how individually we may have an impact. Despite all of the challenges confronting public health misinformation, there is a lot that we as communities, individuals, and society as a whole can do to voice alternate opinions. Just as Ireland essentially ignored the rise of local fascism in the 1930s, there is power in avoiding, disdaining, and isolating these forces of the far right. Freedom of movement, speech, and expression must therefore be maintained.

But, a show is nothing without an audience: What if the disinformation presented to Irish society by extremist groups, via the morally abhorrent tactic of threatening the public's health for political or personal gain, is exposed for what it is? At best, the selective use of facts (which in itself is little more than dishonesty) and the manipulation of statistics are used. Perhaps, interest will quickly fade away as such groups are exposed as, at best, fringe elements of society.

PICTURE 3.8.1 Irish humour. Eilis Conneally.

Not only that but it is estimated that a proverbial antivaccine empire runs out of hundreds of websites and has an estimated 50 million followers on social media [11]. Conspiracy theories proliferate where there is deep epistemic anxiety, and when people feel uncertain about what is true or false [12]. So, not only is it easy to escape down a rabbit hole of antivaccine information, but perhaps we too need to explore how our health agencies and structures may underestimate or deny the far reach of antiscience. How can these structures—through public health campaigns, bedside conversations in hospitals, and general practice information campaigns—prevent the rise of this in the future? [13].

8.4 Guest essay: India's COVID-19 crisis: A lesson for the world

Gladson Vaghela

"Those who cannot remember the past are condemned to repeat it"—George Santayana. This is unquestionably true as we battle one of the world's worst zoonotic threats of late. COVID-19 has had a profound effect on every aspect of our life. As the world navigates through this nightmare, COVID-19 has shown that we live in a big global village, a world in which economics, industrialization, education, environment, and, most crucially, health are all delicately interwoven. However, the pandemic's effect and repercussions are perceived differently depending on our individual and societal status. Whilst some people attempt to adapt by working online, home-schooling their children, and ordering home-delivering food, others are forced to step out of their homes to keep society safe and running.

India, with a billion population, surpassed the global record for the highest single-day spike in COVID-19 case count and fatalities, positioning itself as the global epicentre of the COVID-19 pandemic. What India experienced during the first wave in 2020 was just a glimpse, heralding the possible tragedy to follow. In just 24 days, from 4 April 2021 to 30 April 2021, the daily reported COVID-19 case tally increased from 100,000 to more than 400,000 reported cases; the second wave in India is a lesson for countries around the world. With an alarming shortage of medical oxygen and beds in its hospitals, the second wave outpaced the rate of growth seen in the first wave. India was experiencing fewer cases of COVID-19 during its first wave, causing the Indian government to lift restrictions that allowed for the resumption of superspreading events (SSEs) like large social gatherings, weddings, political rallies, and religious celebrations—acting as fuel to the fire.

Lloyd-Smith et al. [14] coined the term *'superspreading event'*, meaning a single infected individual infects an 'unusually high' number of secondary cases. SSEs act as a catalyst for spreading infectious diseases, resulting in an explosive growth of infections in the community. In the past, SSEs have facilitated the MERS-CoV outbreak in South Korea and the Ebola crisis in Western Africa, highlighting the possible adverse outcomes of SSEs during the COVID-19 pandemic. In India, millions of devotees gathered to take a holy dip in the water at pilgrim river sites during *Maha (or Grand) Kumbh Mela*—cited as the world's largest religious gathering, with most people ignoring COVID-19 guidelines. SSEs of such scale are associated with high public health risks and implications like communicable disease transmission due to high mobility and intermixing of healthy and diseased populations, causing an increased burden on healthcare. Williams et al. [15] identified acute respiratory infections (ARIs) and influenza-like infections (ILIs) as the leading cause of morbidity amongst *Prayagraj Kumbh Mela* visitors in 2019, calling for urgent action for better health surveillance to curb respiratory pathogens during SSEs. These warnings should not have been ignored, especially during an out-of-control situation like the COVID-19 pandemic.

One of the ritualistic practices followed in *Kumbh Mela* is *Snan* (*bathing*) by immersing in the river, a practice which is believed to absolve one's sins and attain *moksha* (*salvation*). It attracts a large number of people to gather at the river banks. Millions of devotees take a dip in the river water in just a span of a few hours. Such mass bathing practice has caused increased bacterial load in the river bodies, exposing millions of devotees to increased risks of infections and diseases. Sridhar et al. [16] found an increased occurrence of respiratory infections 2 days after the mass bathing ritual at *Kumbh Mela*, concluding that *Snan* was a public health risk event. Meanwhile, pilgrims often belong to senior adult age groups having comorbid conditions. A study of hypertension screening at the *Grand Nashik Kumbh Mela* by Balsari et al. [17] found that more than 27% (of the total screened individuals) were previously unaware of their diagnosis and only 2.7% with previously known hypertension were compliant with their prescribed medications, highlighting the increased risk of COVID-19 transmission amongst patients with comorbid conditions during SSEs.

However, is it justified to blame only religious celebration as the chief cause? SSEs were not solely limited to religious festivals. With the ongoing state elections, political campaigners continued to hold mass rallies, often overcrowded with mask-less supporters and very little social distancing. Though rallies were usually held in open spaces, proximity to others for long periods and not socially distanced without personal protective equipment significantly increased chances of transmission. States across the country faced a rapid surge in the COVID-19 cases as the aftermath of SSEs, as visitors of various SSEs travel back to their native place, further spilling the virus in their local community. According to COVID-19 data in India, the increase in cases peaked around the beginning of April, just when the *Kumbh Mela* and political rallies were in full swing.

During the second wave, COVID-19 patients having comorbid conditions like diabetes and patients recovering from the illness reported several cases of mucormycosis. This opportunistic fungal infection was caused by excessive use of steroids and immunosuppressants in the treatment of COVID-19. Although the first wave has reported few cases of black fungus, it has become more widespread in several cities in India in the second wave, causing states to proclaim this an epidemic. These resulted in increased fatalities. With a single-day death count of 4233, as of 7 May 2021, India surpassed the global records of the highest single-day COVID-19-related deaths. The COVID-19 spread was not limited to big cities; rather, it has penetrated rural communities where people are often devoid of basic healthcare services. With the increase in COVID-19 deaths daily, several rural communities faced an increasing shortage of wood needed for cremation as per Hindu rituals. Often, dead bodies were piling up, and the waiting period at the crematories had left people helpless.

With the discovery of a new double-mutant COVID-19 variant in India, known as the Delta variant, the SSEs were the worst potential ground for superspreaders. The situation worsened due to the shortage of medical oxygen, resulting in horrifying incidents of patients being left to die. Several hospitals around the country had to deny new admissions due to limited oxygen stock, forcing the Indian government to import 50,000 metric tonnes (MT) of medical oxygen from abroad to meet the increasing demand. There was also a shortage of vital medications like Remdesivir, resulting in a long queue of patients' relatives waiting to acquire essential drugs for their loved ones. This has encouraged black market sales of medical oxygen and critical drugs, leaving the disenfranchised to die. With most healthcare practitioners being overloaded, civilians have been left without a choice but to act as a proxy for healthcare practitioners, resulting in the healthcare system's responsibility given to the 'public'. Some family members or relatives had to actively participate in caring for their COVID-19-infected family members, even within

COVID-19 wards, often without any protective gear. Although public participation is essential for public health systems, the pandemic has seen an inverse trend in India, proving dangerous for both the citizens and the government.

With the ongoing chaos, something else acted as a silent killer, sabotaging worldwide efforts to eradicate COVID-19. This silent killer is 'misinformation'. Countries worldwide have witnessed the worst consequences of the infodemic, exacerbated by misinformation. Not a single nation, developed or developing, was spared! After India reported its first case in January, key communication channels were flooded with an infodemic of false news covering every aspect of the pandemic, from its genesis to its treatment. Social media channels were flooded with misleading pictures, showcasing doctored material on various topics, including false diagnosis and treatment, fabricated quotations from prominent people, customized instructions, and promising home cures. In India, where the vast majority of the population practices Hinduism (more than 80% of the total population), followed by Islam (more than 13% of the total population), fake news concerning particular religion was also prevalent. Members of the Hindu community encountered false news stating that cows' flesh is utilized to manufacture vaccines. Similarly, members of the Muslim community encountered false news alleging that the vaccines consisted of elements of pork. This fake news is in addition to the growing, small, yet significant antivaxxers groups thriving across the nation.

Faith-based misinformation was not limited to vaccine hesitancy. In March 2020, following a large Muslim gathering in Delhi, communal allegations accusing Muslims of spreading the COVID-19 virus were circulated. Sensational news reports aggravated the existing obstacles, which further instils fear within the community. Aside from all of this turmoil, COVID-19 has a significant psychological effect on healthcare employees. During the COVID-19 pandemic in India, there was a substantial increase in violence against healthcare workers, often violent and fatal. Violence against healthcare professionals is a worldwide issue, posing significant difficulties and dangers, particularly in emerging and less-developed nations. What seems to be spontaneous outbursts of violence by patients or their attendants against doctors are manifestations of India's blatant disregard for public health.

Therefore, healthcare violence, which reinforces and reintroduces deeper and broader forms of dissatisfaction and mistrust amongst patients, patient-families, medical providers, and health institutions, needs a context-specific critical examination. Physicians have long had a high social and professional standing in India; the loss of this status owing to individual medical malpractice and political deception is affecting doctors and health professionals as a whole. In this environment, it is regrettable that the social and economic standing of nurses, paramedics, hospital janitors, or even healthcare workers like Accredited Social Health Activist (ASHA), Anganwadi Worker (AWW), and Auxiliary Nurse Midwife (ANM) who tremendously support public health systems, has not improved. These workers who often work for the hard-to-reach population at the lowest governance rung are indeed the unsung heroes of our time! Combating violence against healthcare professionals involves first acknowledging it as a major public health issue.

One should also not forget the sacrifice and efforts given by healthcare students. Healthcare students formed a significant proportion amongst frontline workers, along with other healthcare professionals. Healthcare students who are often devoid of required exposure and not even fully trained had to work shoulder-to-shoulder supporting the crumbling healthcare institution, often without due pay. Medical education was greatly hampered in the light of the COVID-19 outbreak. It is crucial to train our future healthcare professionals with the required skills to tackle and combat the future outbreaks.

With the increasing incidence of natural calamities due to climate change, the world will witness an exponential increase in tropical and infectious diseases, coupled with noncommunicable diseases. The ongoing pandemic has exposed the loopholes in the existing healthcare systems. The impact of the COVID-19 pandemic was accentuated by decade-long underfunding in public healthcare infrastructure, lack of stringent law enforcement of COVID-19 appropriate behaviour, a new variant of concern, and the burst of misinformation, accounting for this tragedy. Learning from the past and formulating better-informed policies for the future is critical. Future generations should never have to go through what their ancestors did!

8.5 Trusting science

Despite the promise of certain companies providing a cheap and readily available vaccine—aiming for a not-for-profit vaccine until the pandemic was over—developments and setbacks experienced by the Astra-Zeneca and Sputnik vaccines, even without any solid evidence of causative links and affected conditions, may have contributed to fuel further future vaccine hesitancy. These setbacks only provided further coverage for antivaccinators' messaging towards the general public. Furthermore, reports of abnormal bleeding cast doubt on this particular vaccine's safety and resulted in several European countries in early 2021 postponing its use.

PICTURE 3.8.2 Vaccine clinic social distance. Sebastian Kevany.

Returning to Burke's statement, we may therefore be echoed by another: Alexander Pope's sentiment that "a bit of knowledge, in the wrong hands, is a dangerous thing" [18]. No doubt Ireland's extremist groups may wish to draw attention to the occasional failure in vaccine effectiveness: in point of fact, the current success rates, whilst magnificently high, will inevitably have, as with any treatment, a proportion of patients for whom it is ineffective, or requires further booster doses due to individual responses.

But, perhaps, by heightening the voices of the far right and antivaccinators through amplifying their voices on social media and news channels, we are adding to the problem—and not, in turn, addressing the solution [11]? Indeed, to present these isolated case reports as the status quo through selective reporting and sensationalist coverage, with all the associated viral transmission of such news by the susceptible, is simply wrong.

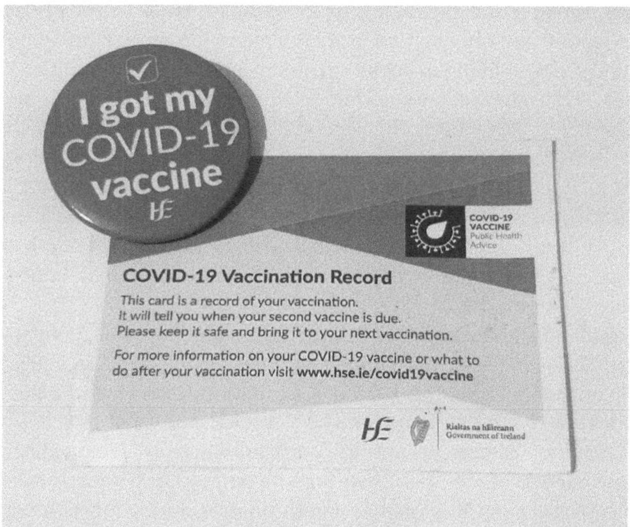

PICTURE 3.8.3 Vaccination Record. Maisie M. Jones.

The political forces that may have contributed to the produced political upheaval in the United Kingdom and elsewhere in recent years, amongst other places, rely on such susceptibility. In fact, these suspects rely heavily on the Internet and technology for the rapid transmission of semi- or selective facts before they can be verified. Societies are collectively waking up to this trend. Perhaps, so too we may be living in such a tech-driven era of hysteria that the Irish fringe has exploited. Suppose we believe that such groups' goal is to undermine organizations such as the World Health Organization and the supranational constructs, and the order and checks and balances that they represent. In that case, vigilance has to start at home, for each one of us [19].

8.6 Guest essay: Is vaccine equity possible? Reflections from the United Kingdom, as an island state, with a strong health system

Regina Keith

The COVID-19 pandemic is a stark reminder that diseases do not respect borders [20]. On 6 September 2021, there were 223,022,538 confirmed cases of COVID-19 and over 4.6 million deaths globally [21]. In July 2020, the World Health Organisation (WHO) reported that health workers accounted for 10% of global COVID-19 cases, highlighting the negative impact of the disease on health systems. Even though over 5 billion vaccine doses have been administered globally, as of September 2021, the vaccination programme is failing to reach all countries and populations in an equitable way [21].

One specific difference with the COVID-19 pandemic and other infectious diseases is that its impact has been witnessed closer to home in high-income countries with developed health systems (such as China, Italy, the United Kingdom, and the United States), rather than just in low-income countries [20]. Early research into the factors which increase the prevalence of COVID-19-related morbidity (illness) and mortality (death) identified being obese and elderly, especially if living in social care settings with co-morbidities like type II diabetes [22]. High-income countries affected had large populations with these risk factors. However, high-income countries have had the capacity to respond to the emergency, putting in place health and economic strategies to curb the spread of the virus and mitigate against the impact on their population. These strategies included investing in vaccine research, establishing effective testing and contact-tracing systems, protecting health services, increasing intensive care capacity, and implementing mandatory social-distancing measures [23,24].

New Zealand and Australia reacted quickly to the pandemic, enacting urgent travel restrictions in March 2020, to reduce the risk of the virus being brought into their countries [25]. The United Kingdom and the United States closed international travel routes more slowly. It could be argued that this allowed greater circulation of new variants to countries with less developed health systems. The United Kingdom, like other high-income nations, set up national economic measures to support families and businesses during lockdowns, such as covering salaries for staff not able to work, providing vouchers for families who get free school meals, and topping up social security payments by an extra £20 every week for those on universal credit [26]. However, in countries like India, the implementation of strict lockdown measures resulted in a mass exodus from urban cities to rural homes, promoting the spread of the virus. Prior to the lockdown in India, there were only 320 reported cases [27]. By 4 August 2021, India had reported over 31 million cases and 425,767 deaths [28]. The lack of access to essential healthcare services especially in more rural areas resulted in high levels of excessive morbidity and mortality in India. Despite being the biggest producer of vaccines globally, only 10% of the Indian population had been given their first COVID vaccine by August 2021 [29]. As more people in rural areas required health care, the health system was overwhelmed, resulting in the government calling for Indian vaccine companies to use available vaccines for the Indian population [30]. One source stated that 45% of Indians had to borrow money for food during the pandemic with 96% reporting the lack of government promised food rations as a cause [31]. Another source noted that 62% of Indian households had changed from food secure to insecure during the pandemic [32]. Lockdowns also reduced the capacity of countries to make vaccines and drugs, with India reporting a 44% reduction in manufacturing due to lack of essential raw materials [33].

Why focus on vaccine equity, when 2 billion people do not have access to essential health care?

Public health strategists are urging political leaders to increase resources and political commitment to a global response to the pandemic, at a time when many countries are feeling the negative impact of 18 months of reduced economic growth due to the pandemic. Globally, poverty has increased from 15% to 24% during the COVID-19 pandemic [34]. In the United Kingdom, the economy has reduced by 6.5%, whilst in the United States, it has contracted by 5.9% [34]. Countries are keen to restart their economies, and part of the new COVID-19 recovery plans include how to cut costs to save money, rather than focusing on increasing investment in strengthening global health systems and governance. The United Kingdom merged their overseas aid programme with the Foreign and Commonwealth Office (FCO) and cut the international development budget by over 2% of GDP [35,36], a move criticized by development organizations and partners [36].

The Director General of the WHO has stated repeatedly that the pandemic will not be over until all populations are vaccinated [37]. To achieve this global vaccination strategy, prevent further preventable deaths, and halt the emergence of new strains of the virus, urgent investments are required to support the strengthening of resilient health systems to deliver essential health care for all. To date, the global response dialogue has focused on vaccine equity, implementing strategies to share vaccines with countries with limited resources as soon as possible. The ACT Accelerator programme has developed tools to support the roll-out of the vaccines, such as delivering tests and cold chain systems for the safe delivery of the vaccines. The ACT Accelerator programme needed US$38.1 billion to support the effective delivery of vaccines to low-income countries; however, by September 2021, donors have pledged around $18.1 billion, leaving a funding gap of $16.6 billion [38].

COVID and the UK health system

In the United Kingdom, 92% of adults have had or are willing to have their COVID-19 vaccinations, which are delivered free at multiple venues around the country [39]. Even with an average of 12 health workers per 1000 population, the United Kingdom has had to employ many more staff and set up hundreds of large-scale vaccination sites, to ensure that the COVID-19 vaccines are delivered effectively, without wastage and at the correct temperature [40]. Like other countries, the UK healthcare workforce has been dramatically impacted by COVID-19, with over 900 healthcare workers losing their lives due to it by February 2021 [40]. Others have suffered from an expanded workload with increasing incidence of burnout [40]. To support the UK health system becoming more resilient, The King's Fund, a UK-based think tank, has suggested five steps for health system recovery in the United Kingdom:

(1) Put the health workforce at the centre of planning strategies,
(2) Focusing on reducing inequalities,
(3) Reform the social care system,
(4) Continue to invest in further expansion of digital systems, and
(5) Strengthen the relationship between communities and health systems [40].

These strategies resemble those agreed upon in 1978, when 136 countries agreed on the right to health for all, through the principles of primary health care, in the Alma Ata Declaration. In 2018, Keith suggested that the Alma Ata principles of global governance, equity, and community engagement, through an intersectoral lens, were essential to achieve universal health coverage by 2030 [41]. These strategies are in line with targets set by world leaders at the United Nations General Assembly (UNGA) in the Sustainable Development Goals [42]. In her research, Keith supports the call from the King's Fund on the need to reduce inequity; she demonstrates that equitable provision is achieved through public services provided free at the point of access [43]. Keith also highlights the positive impact of active engagement with health workers and communities in health planning and policy determination [43]. Although the United Kingdom is a member of the G7 nations, successive governments have failed to invest adequately in reducing inequalities and addressing the growing problem of social care due to a rapidly ageing population, with new strategies recently announced by the UK government to increase investment in social and health care through tax rises [44]. The future of public health research group carried out research in England on the impact of COVID-19 on public health nutrition (PHN) services, illustrating the negative impact of 10 years of aggressive austerity measures on service capacity at community level. The lack of investment has resulted in essential services being discontinued, resulting in reduced community public health capacity, just as the Covid-related demand increased [45].

Politicians and advocacy groups are presently calling for a halt to the UK government's plan to discontinue the £20 universal credit top-up payments (social security support), and to support them through the COVID pandemic [46]. There are growing fears that this will push millions more families into poverty this year.

Despite having a strong health system with essential services free at the point of access, the United Kingdom has struggled to continue essential routine health care whilst tackling the COVID-19 pandemic, resulting in long backlogs which will require extra resources to tackle. The WHO states that 90% of countries have reported health service disruption, including reduced coverage of routine vaccinations, due to the COVID-19 pandemic [47].

The vaccination works

On 17 July 2021, the United Kingdom relaxed their travel, social, and working restrictions. Despite a return to work, education, and social interaction, the positive impact of the effective vaccine roll-out has led to reduced numbers of the population who are seriously ill or needing hospitalization. Discussions are now underway regarding the need and safety for children aged 12–18 to be vaccinated and if a booster dose is required for those who had had their two vaccinations [48].

At the 2021 G7 summit in Cornwall, the UK Prime Minister Boris Johnson called on wealthy countries to increase their support to address the COVID-19 pandemic in countries with developing economies or weaker health systems. He pledged 100 million doses, from the United Kingdom's surplus vaccines, to contribute to this effort [48]. Although this pledge is welcome, and it adds to the commitment already given by the United Kingdom to the COVAX partnership, wealthier countries need to concurrently invest in supporting low- and middle-income countries to strengthen resilient, pro-poor health systems. However, evidence illustrates that without functioning health systems, effective vaccines will not reach the most vulnerable [49].

How can health systems be strengthened?

In 2000, the WHO set out six essential building blocks to achieve a functioning health system, including governance and leadership, human resources for health, health financing, robust health information, supply, and delivery systems [50]. In 2002, Save the Children and the London School of Hygiene and Tropical Medicine carried out research into the impact of the new Global Alliance of Vaccines and Immunisations (GAVI) on four countries in sub-Saharan Africa (SSA) [51]. The research focused on the integration of new vaccines into health systems. The lessons learned from this research could feed into the present global plans to scale up the COVID-19 vaccination coverage. Factors such as investing in strengthening health systems to ensure that supply chains are kept at the needed temperature and that waste management, staff training, and support need to be prioritized [51]. Further research has added that another important factor is to ensure these public health services are delivered free at the point of access [43].

Global Health Governance, WHO, and Universal Health Coverage

The World Health Organisation was established in 1948 to support the right to health for all and to provide global health governance; however, for the last three decades, the organization has been underfunded, reducing its capacity to carry out its role effectively [41].

The WHO estimates that Africa will reach 52.5% universal health coverage of essential health services by 2023, which will leave over half of the population unable to access essential health care [37]. This is supported by the launch of the recent State of the Universal Health Coverage in Africa report [52]. When populations must pay for these services, at point of access, the poorest 20% do not seek care, whilst another 20% are pulled into poverty to pay for health care [41].

Health system strengthening also includes increasing the robust nature of health coordination and governance, ensuring that there is an effective health-financing mechanism which protects the most vulnerable populations from impoverishment, whilst ensuring that services are delivered close to populations and that there are at least 4.5 health workers per 1000 population to deliver the services. For these factors to be effective, health workers need to have the resources they need such as vaccines, fridges, needles, gloves, and waste management systems to ensure the vaccines reach the populations most in need, in a safe and effective manner [51].

Global shortage in skilled and supported health workers

In 2006, the WHO estimated that to deliver essential measles vaccinations to at least 80% of the target population, a country required around 2.5 health workers per 1000 population [53]. In 2016, that figure was reviewed considering the increase in noncommunicable diseases, resulting in an estimate of 4.5 health workers per 1000 population [54]. Globally, to reach the Sustainable Development Goals for Universal Health Coverage, there is a shortage of 18 million health workers to vaccinate 20% of the global population with the COVID-19 vaccine, and 1.1 million new health workers are needed [55,56]. 2021 has been dedicated as the year of the health worker, to increase global attention on the importance of protecting and supporting our healthcare workers [57].

> Despite health workforce being a leading expenditure of governments, only 7% of all development assistance for health went to support health workforce between 1990-2016. Most resources went for short term project-based interventions, rather than investments in long-standing health workforce challenges
>
> *[57]*

Research in Ethiopia, Nigeria, Liberia, and Sierra Leone demonstrates that when there is less than one health worker per 1000 population, essential services such as vaccinations fail to reach the most vulnerable populations [43].

What is needed and what tools are available to support countries?

COVID-19 vaccines require an effective cold chain which means chilled transport to vaccination sites and chilled storage at the clinic sites, to ensure they are still active. Twenty years ago, when the Global Alliance for Vaccines and Immunisations was created, many health systems were not prepared effectively for the new vaccines, which required more storage and different temperatures, resulting in millions of doses being lost [51]. Lessons have been learned, but health systems are still weak; less than 50% of those living in Africa have access to essential healthcare services [52]. To help offset these challenges, the COVAX vaccine programme is being supported by the ACT Accelerator partnership, which is focusing on four pillars: diagnostics, therapeutics, vaccines, and health system strengthening. The ACT partnership works with countries to ensure vaccines are delivered into systems that have effective cold chain capacity, with functioning tests, tracing, and treatment capacity for COVID-19. To date, most resources have focused on vaccines, and there is still a $18.5 billion dollar shortfall, which needs to be covered urgently. Systems need to be strengthened to ensure that the essential gains made in the last 20 years are not lost due to hesitancy in global solidarity [37]. Whilst the world continues to fight against the pandemic, other killers are still hard at work, with malaria, diarrhoea, and pneumonia still claiming the lives of over 5 million under-fives every year. Around one-fifth of these children die in Nigeria, where coverage of essential services is under 50% for the poorest populations [52].

What actions are needed?

As mentioned, reaching populations with an effective vaccine requires functioning health systems or vaccination campaigns. Using past research, there are five priority areas:

(1) Investing in the training, recruitment, retention, protection, and vaccination of health workers.
Globally, to reach the Sustainable Development Goals for Universal Health Coverage, there is a shortage of 18 million health workers [54]. To vaccinate 20% of the global population with a COVID-19 vaccine, 1.1 million new health workers are needed [55].

(2) Surveillance (testing and tracing).
We need to know where the virus is circulating and implement public health measures to protect the most vulnerable populations, including health workers [55]. This requires testing systems that are rapid, and free for all, so that we can halt preventable deaths by preventing transmission. If workers must isolate themselves, they need to have their jobs and salaries protected to ensure compliance. These social safety nets are expensive, but the international community has achieved this in the past when responding to the Ebola and the HIV pandemics.

(3) Effective cold chain system
Vaccines require chilled transport to vaccination sites and chilled storage at these sites, to ensure they are still active. We must learn lessons from previous vaccination programmes, such as the roll-out of the quadrivalent and tetravalent vaccinations, 20 years ago. Many health systems were not prepared effectively for the new vaccines, resulting in millions of doses being wasted. Safely managing the waste from vaccinations also requires system investment to prevent harm [51].

(4) Equity
To ensure that people will not delay in seeking testing and healthcare, these services need to be considered public health priorities and be offered free to all at the point of access. Research in Sierra Leone, Liberia, Ethiopia, Tanzania, Sudan, Rwanda, and Burundi demonstrates that when populations must pay for these services, at point of access, the poorest do not seek care, whilst others are pulled into poverty to pay for healthcare [43].

(5) Active community engagement and trust

If populations do not trust the services or the vaccinations, they will not get vaccinated. This vaccine hesitancy is reduced when communities are actively engaged in planning and assessing services. Regular transparent communication is essential in an emergency to rebuild trust in communities [43].

We have seen what happens when we do not invest in effective health systems and when we do not value our health workers. Vaccine equity comprises more than giving vaccines to the poor; it is about global health solidarity, perhaps requiring a new framework convention, ensuring that the right to health for all becomes more than an empty promise made in 1978. If every nation put their pledges into action, we would be able to build resilient health systems ready to tackle COVID-19 and the next disease to hit our growing global population.

PICTURE 3.8.4 Covid Passport. Maisie M. Jones.

8.7 Guest essay: A vaccination conversation

Sebastian Kevany

With a fellow surfer, we gazed out at the waves rolling in over Killiney. As we procrastinated about getting into cold wetsuits, the conversation turned to the pandemic. The exchange began with his distress at government incompetency. I agreed, as far as dissolving the National Public Health Emergency Team was concerned. He then went further, saying the media, via hype and hysteria, were running the country, and that lockdowns were a result of public sector incompetence. Ireland, he felt, was compared unfavourably to continental Europe in that regard; Irish people were, in his opinion, both more passive and more gullible than others.

I replied that maybe that was the case, though I doubted it—but, either way, Ireland was badly affected by loosening restrictions in December 2020—and that, in that context, in winter 2021, lockdowns represented the only alternative to anything less than universal vaccinations. Ireland had, I felt, been too liberal in its tolerance for those remaining unvaccinated for nonmedical reasons.

He then took a different tack—comparing COVID vaccine requirements to MMR—the latter, of course, not being legally required; vaccines, to him, therefore seemed discriminatory. My take was that the situations were not comparable—antivaccine resistance to MMR, in the modern era, has never threatened to cripple the health system or the economy. A closer comparison, I felt, was with the use of seat belts in cars—some would once have seen this as an infringement of their personal freedom of choice, regarding them as dangerous impositions. Despite this, once society reached a critical mass of acceptance and education—as a result of mandates and compulsory use—such public health policies become unquestioned, and second nature.

The choice, to me, thus became simple: more lockdowns, or more mandatory vaccinations. He replied that neither, for him, would solve the problem. Vaccines would be beaten by variants; lockdowns had too many negative side effects. In this regard, he referenced media hype again—the way that, he felt, the pandemic was being used to sell newspapers, when in many ways it was, to him, just like the flu. I agreed, to some extent, about the media, but also replied that they had done a lot to help public health, as well. Similarly, toning down media coverage, though it might be helpful, was not going to change the fundamental trade-off between mandatory vaccinations and lockdowns. Even without journalists, I said, the choice remains the same.

He then posed a third option: to learn to live with nature. As he framed it: live longer caged, or shorter and free? My reply was that, even if we do that—whatever that means, exactly—people are still going to go to hospital when they get sick. So I replied that, ideally, a long and uncaged life would be best...

The only other option, it seemed to me, was for hospitals to turn away unvaccinated people, as part of a ruthless triage system. But why should hospitals cater to smokers, he said, but shouldn't cater to the unvaccinated? Should, in both cases, the health system be more ruthless, more Darwinist? Of course, that should, we agreed, never happen. The worst that would transpire, I reckoned, would be an increase in insurance premiums for the unvaccinated, as well as of the other limitations they have made the choice to live with.

He then asked about Mexico and South Africa—why was the hype so much less? My reply was that those countries were far more accustomed—like so much of the world—to tragedy, morbidity, and early death than Ireland, or other developed and affluent places.

Eventually, we agreed that the only third option, unfeasible in the short term, was huge investment in the health system. But, even then, would that be acceptable—would we be channelling all our resources—away from roads and infrastructure and education and other necessities—to cater to the needs of the unvaccinated—as well as to the freedom of choice, as well as expectation of health care, that they expect?

Below us, the surf rolled in: like the waves breaking on the shore, we both had the feeling that we hadn't made much progress, despite the effort. Eventually, we agreed that the entire pandemic has been a wake-up call for human health: that we must, if we can, try to stop abusing our bodies with bad diets, smoking, or excessive drinking, all of which are increasingly part of the underlying conditions that make up vaccinated hospitalizations. For the unvaccinated, also, other elements of health were pivotal—mandatory vaccinations were unlikely to succeed, so the key way, perhaps to try and keep everyone out of hospital by taking more and more responsibility for their own health.

8.8 Key points

- The antivaccine movement was underestimated during vaccine roll-out planning. This was associated, also, with rampant cyber liberty and the spread of 'fake news' about vaccines.
- Many western governments realized that, as a result, their initial vaccine policies had been too lenient, and gradually imposed greater restrictions on the unvaccinated.
- Whilst vaccines were, at the time of publication, never enforced in the same way as car seat belts, the similarities in public health benefits of both are worthy of consideration.

References

[1] E. Burke, The Works of the Right Honourable Edmund Burke, vol. 5, Wells and Lilly, 1826.
[2] K. Megget, Even covid-19 can't kill the anti-vaccination movement, BMJ 369 (2020).
[3] Growing Up in the Age of Fake News, 2021, April 29. Retrieved December 6, 2021, from https://en.unesco.org/courier/2021-2/growing-age-fake-news.
[4] WHO Ten Health Issues WHO Will Tackle This Year. (n.d.). Retrieved December 6, 2021, from https://www.who.int/news-room/spotlight/ten-threats-to-global-health-in-2019.
[5] https://www.theguardian.com/world/2021/jul/17/covid-misinformation-conspiracy-theories-ccdh-report.
[6] https://www.dailymail.co.uk/news/article-9797397/Tucker-Carlson-slams-idiot-Republicans-focusing-Cuba-tyrannical-government.html.
[7] https://edition.cnn.com/2020/05/01/politics/donald-trump-michigan-gretchen-whitmer-protests/index.html.
[8] https://www.medicalnewstoday.com/articles/covid-19-has-hit-people-in-republican-led-states-hardest-study-finds.
[9] Equality and Human Rights Article 10: Freedom of Expression. (n.d.). Retrieved December 6, 2021, from https://www.equalityhumanrights.com/en/human-rights-act/article-10-freedom-expression.
[10] G. Karabulut, K.F. Zimmermann, M.H. Bilgin, A.C. Doker, Democracy and COVID-19 outcomes, Econ. Lett. 203 (2021) 109840.
[11] I. Ahmed, Dismantling the anti-vaxx industry, Nat. Med. 27 (3) (2021) 366, https://doi.org/10.1038/s41591-021-01260-6.
[12] K.M. Douglas, R.M. Sutton, A. Cichocka, The psychology of conspiracy theories, Curr. Dir. Psychol. Sci. 26 (6) (2017) 538–542, https://doi.org/10.1177/0963721417718261.
[13] The Lancet Infectious Diseases, The COVID-19 infodemic, Lancet Infect. Dis. 20 (8) (2020) 875.
[14] J.O. Lloyd-Smith, J.S. Sebastian, P.E. Kopp, W.M. Getz, Superspreading and the effect of individual variation on disease emergence, Nature 438 (7066) (2005) 355–359.
[15] N. Williams, T. Radia, K. Harman, P. Agrawal, J. Cook, A. Gupta, COVID-19 severe acute respiratory syndrome coronavirus 2 (SARS-CoV-2) infection in children and adolescents: a systematic review of critically unwell children and the association with underlying comorbidities, Eur. J. Pediatr 180 (3) (2021) 689–697.
[16] S. Sridhar, S. Begom, A. Bermingham, K. Hoschler, W. Adamson, W. Carman, et al., Cellular immune correlates of protection against symptomatic pandemic influenza, Nat. Med. 19 (10) (2013) 1305–1312.

[17] S. Balsari, P. Vemulapalli, M. Gofine, K. Oswal, R. Merchant, S. Saunik, et al., A retrospective analysis of hypertension screening at a mass gathering in India: implications for non-communicable disease control strategies, J. Hum. Hypertens. 31 (11) (2017) 750–753.
[18] A. Pope, in: J. Murray (Ed.), The Works of Alexander Pope, vol. 2, 1871.
[19] S. van Der Linden, J. Roozenbeek, J. Compton, Inoculating against fake news about COVID-19, Front. Psychol. 11 (2020) 2928.
[20] OECD, Beyond Containment: Health Systems Responses to COVID-19 in the OECD, OECD, 2021. [cited 23 September 2021]. Available from: https://oecd.org/coronavirus/policy-responses/beyond-containment-health-systems-responses-to-covid-19-in-the-oecd-6ab740c0/.
[21] Covid19.who.int, WHO Coronavirus (COVID-19) Dashboard, Covid19.who.int, 2021. [cited 23 September 2021]. Available from: https://covid19.who.int/.
[22] GOV.UK, Drivers of the Higher COVID-19 Incidence, Morbidity and Mortality Among Minority Ethnic Groups, 23 September 2020, GOV.UK, 2021. [cited 23 September 2021]. Available from: https://www.gov.uk/government/publications/drivers-of-the-higher-covid-19-incidence-morbidity-and-mortality-among-minority-ethnic-groups-23-september-2020.
[23] Who.int, Contact Tracing in the Context of COVID-19, Who.int, 2021. [cited 23 September 2021]. Available from: https://www.who.int/publications/m/item/contact-tracing-in-the-context-of-covid-19.
[24] OECD, Enhancing Public Trust in COVID-19 Vaccination: The Role of Governments, OECD, 2021. [cited 23 September 2021]. Available from: https://www.oecd.org/coronavirus/policy-responses/enhancing-public-trust-in-covid-19-vaccination-the-role-of-governments-eae0ec5a/.
[25] Who.int, New Zealand Takes Early and Hard Action to Tackle COVID-19, Who.int, 2021. [cited 23 September 2021]. Available from: https://www.who.int/westernpacific/news/feature-stories/detail/new-zealand-takes-early-and-hard-action-to-tackle-covid-19.
[26] Emcouncils.gov.uk, Economic Support Measures—Covid, Emcouncils.gov.uk, 2021. [cited 23 September 2021]. Available from: https://www.emcouncils.gov.uk/News/economic-support-measures-covid.
[27] A. Joshi, et al., India's COVID-19 Burdens, 2020, Front. Public Health 9 (608810) (2021). [cited 23 September 2021]. Available from https://www.frontiersin.org/articles/10.3389/fpubh.2021.608810/full.
[28] Covid19.who.int, India: WHO Coronavirus Disease (COVID-19) Dashboard With Vaccination Data, Covid19.who.int, 2021. [cited 23 September 2021]. Available from: https://covid19.who.int/region/searo/country/in.
[29] S. Seervai, A. Shah, India's Failure of Leadership Collapses Health System in COVID-19 Surge, The Commonwealth Fund, 2021. Available from: https://www.commonwealthfund.org/blog/2021/indias-failure-leadership-collapses-health-system-covid-19-surge.
[30] A. Upadhyay, 'Healthcare System In Rural India Is A Cause Of Concern,' Says Public Healthcare Expert Amid COVID-19 Pandemic, NDTV-Dettol Banega Swasth Swachh India, 2021. [cited 23 September 2021]. Available from: https://swachhindia.ndtv.com/healthcare-system-in-rural-india-is-a-cause-of-concern-says-public-healthcare-expert-amid-covid-19-pandemic-49759/.
[31] I. Vispo, et al., State of the Right to Food and Nutrition Report, Global Network for the Right to Food and Nutrition, 2021. [cited 23 September 2021] Available from https://www.righttofoodandnutrition.org/files/20210719_state-rtfn-report_2021_eng_v15.pdf.
[32] P. Nguyen, et al., Impact of COVID-19 on household food insecurity and interlinkages with child feeding practices and coping strategies in Uttar Pradesh, India: a longitudinal community-based study, BMJ Open 11 (048738) (2021). Available from: https://bmjopen.bmj.com/content/bmjopen/11/4/e048738.full.pdf.

[33] A. McDonnell, et al., A Path to Resiliency: Mitigating the Impacts of COVID-19 on Essential Medicines Supply Chains, Centre for Global Development, 2021. Available from: https://www.cgdev.org/sites/default/files/Mitigating-impact-covid-essential-medicine-supply-chain_0.pdf.
[34] D. Laborde, et al., COVID-19 risks to global food security: Economic fallout and food supply chain disruptions require attention from policymakers, Science 369 (6503) (2020), https://doi.org/10.1126/science.abc4765.
[35] GOV.UK, Prime Minister Announces Merger of Department for International Development and Foreign Office, GOV.UK, 2021. [cited 23 September 2021]. Available from: https://www.gov.uk/government/news/prime-minister-announces-merger-of-department-for-international-development-and-foreign-office.
[36] Government wins vote to lock in cuts to overseas aid, BBC News (2021). [cited 23 September 2021]. Available from: https://www.bbc.com/news/uk-politics-57826111.
[37] Who.int, WHO Director-General's Opening Remarks at 148th Session of the Executive Board, Who.int, 2021. [cited 23 September 2021]. Available from: https://www.who.int/director-general/speeches/detail/who-director-general-s-opening-remarks-at-148th-session-of-the-executive-board.
[38] Who.int, Access to COVID-19 Tools Funding Commitment Tracker, Who.int, 2021. [cited 23 September 2021]. Available from: https://www.who.int/publications/m/item/access-to-covid-19-tools-tracker.
[39] ONS UK, COVID 19 Latest Insights: Vaccines Office of National Statistics August 4th, 2021, 2021, Accessed: https://www.ons.gov.uk/peoplepopulationandcommunity/healthandsocialcare/conditionsanddiseases/articles/coronaviruscovid19latestinsights/vaccines.
[40] Org.uk, NHS Workforce: Our Position, Org.uk, 2021. [cited 2021 Sep 24]. Available from: https://www.kingsfund.org.uk/projects/positions/nhs-workforce.
[41] R. Keith, From Alma Ata to Astana: can renewed commitment to Health for All by 2030 be committed to in October 2018? World Nutr 9 (2) (2018) 121–126. Available from: https://worldnutritionjournal.org/index.php/wn/article/view/586/522.
[42] United Nations General Assembly (UNGA), The UN Sustainable Development Goals, 2015, [cited 2021 Sep 24]. Available from: https://www.un.org/en/development/desa/population/migration/generalassembly/docs/globalcompact/A_RES_70_1_E.pdf.
[43] R. Keith, Universal Health Coverage by 2030, or the Right to Health? The Continued Relevance of the Alma Ata Principles of Voice and Equity, Through an Intersectoral Approach, to Reducing Global Inequalities (PhD Thesis), University of Westminster, London UK, 2021.
[44] Gov.uk, Health and Social Care Levy, Gov.uk, 2021. [cited 2021 Sep 24]. Available from: https://www.gov.uk/government/publications/health-and-social-care-levy/health-and-social-care-levy.
[45] S. Noonan-Gunning, K. Lewis, L. Kennedy, J. Swann, S. Gursimran Kaur, R. Keith, Is England's Public Health Nutrition System in crisis? A qualitative analysis of the capacity to feed all in need during the COVID-19 pandemic, World Nutr 12 (2) (2021) 83–103.
[46] N.E. Wilson, Make universal credit top-up permanent—Tory MPs, BBC (2021 Aug 26). [cited 2021 Sep 24]; Available from: https://www.bbc.com/news/uk-politics-58331179.
[47] Who.int, COVID-19 Significantly Impacts Health Services for Noncommunicable Diseases, Who.int, 2020. [cited 2021 Sep 24]. Available from: https://www.who.int/news/item/01-06-2020-covid-19-significantly-impacts-health-services-for-noncommunicable-diseases.
[48] Department of Health and Social Care, JCVI Statement, August 2021: COVID-19 Vaccination of Children and Young People Aged 12 to 17 Years, 2021, [cited 2021 Sep 24]. Available from: https://www.gov.uk/government/publications/jcvi-statement-august-2021-covid-19-vaccination-of-children-and-young-people-aged-12-to-17-years.

[49] D. Werner, D. Saunders, Questioning the Solution: The Politics of Primary Health Care and Child Survival With an in-depth Critique of Oral Rehydration Therapy, Healthwrights publication USA, 1997.

[50] Handbook Of Indicators A, Their Measurement Strategies. Monitoring the Building Blocks of Health Systems, [cited 2021 Sep 24]. Available from: https://www.who.int/healthinfo/systems/WHO_MBHSS_2010_full_web.pdf.

[51] M. Starling, R. Brugha, G. Walt, A. Heaton, R. Keith, New Products Into Old Systems: The Initial Impact of the Global Alliance of Vaccines and Immunizations from a country perspective, LSHTM and Save the Children UK, London UK, 2002.

[52] AHAIC, Executive summary: The state of universal health coverage in Africa, in: AHAIC 2021 virtual conference, 2021. [cited 2021 Sep 24]. Available from: https://ahaic.org/download/executive-summary-the-state-of-universal-health-coverage-in-africa/.

[53] WHO, World Health Report: Working Together for Health, World Health Organisation Geneva, 2006. [cited 27 September 2021]. Available from: https://www.who.int/whr/2006/whr06_en.pdf.

[54] G. Cometto, et al., Policy Briefing: Health workforce demand and shortages to 2030: an overview of forecasted trends in global health labour market Background paper No, 1 to the UN High level Commission on employment and economic growth, Global Health Workforce Alliance World Health Organisation, Geneva, 2016.

[55] WHO, Global Strategy on Human Workforce for Health, 2016, [cited 27 September 2021]. Available from: https://www.who.int/hrh/resources/global_strategy_workforce2030_14_print.pdf.

[56] WHO, Access to COVID-19 Tools Funding Commitment Tracker, 2021. [cited 27 September 2021]. Available from: Access to COVID-19 tools funding commitment tracker (who.int).

[57] WHO, Key facts on the Year of the Health Worker 2021, World Health Organisation, 2021. [cited 27 September 2021]. Available from: YHCW2021 campaign: Facts on health and care workers (who.int).

Chapter 9

Military involvement in pandemics

Sebastian Kevany, Michael Baker, Deon Canyon, and Aoife Kirk

Chapter outline

9.1 The securitization of health

The COVID-19 pandemic sent a ripple effect of shockwaves through the international scientific community, quickly followed by economists, policymakers, and politicians. As a result, worldwide trade, air travel, and tourism had significant unexpected collapses, and the pandemic also highlighted the need for improved health and disease surveillance [1].

Furthermore, the global economy was brought to its knees by the unexpected spread of infection in an interlinked, globalized world, which has resulted in millions of lives lost, along with widespread, long-term illnesses and comorbidities. Therefore, it is vital to consider the pandemic and the recent rise in nationalism and political isolationism throughout the late 2010s decade. This, along with the political use of the blame game, lockdowns, frontline workers, and travel restrictions, has caused further divisiveness in the general public.

In a sense, the pandemic is a stark reminder of a national security breach. Perhaps, we can learn from such events of previous national and international security disasters, for example, Pearl Harbor and 9/11 in the United States,

Outbreaks, Epidemics, and Health Security. https://doi.org/10.1016/B978-0-323-91705-6.00010-1

which resulted in significant overhauls of national and international intelligence to prevent future surprises [2]. These security events also led to a substantial role in global diplomacy. Similarly, COVID-19 has resulted in significant health and political diplomacy, from vaccine equity matters to border closures, to ensure national and international safety for civilians and citizens.

PICTURE 3.9.1 Military personnel prepare Covid testing efforts to support local health systems. Sebastian Kevany.

The balance of scientific advisors providing political heads of state with information to help them make informed, accurate decisions without causing public outcry or panic has often been a tricky tightrope to balance during this pandemic. Hopefully, governments will now make a more significant commitment to strengthening health systems, developing new tools to pursue enhanced disease intelligence efforts and cybersecurity, and enhancing vigilance to enable epidemic response emergency plans in the future [3].

PICTURE 3.9.2 **In Ireland and elsewhere, military and police checkpoints were in place to crucial disease spread.** Sebastian Kevany.

While health and disease intelligence is a central part of any state's national and international security, other specialities often overlap closely, including counterterrorism, counterespionage, and cybersecurity. As a result, it is essential for cross-sector collaboration in these areas, and this, unfortunately, was particularly true in Ireland. In May 2021, just as the handling of the second wave of coronavirus cases was commended and applauded by the Irish National Public Health Emergency Team (NPHET), a cyberattack struck the Irish Health Services IT system.

The risk of leakage of confidential health data, including mental health records and vulnerable patient details, was an incident of an immediate threat to the Irish Government. Shutting off the IT software across the country affected thousands of patients, with at the time of writing a projected one million patients awaiting access to consultant review on waiting lists in Ireland. This untimely event also resulted in severe implications on the public health system, including the postponement of care for patients, disruption of daily duties for healthcare workers, and impact on the provision of timely care, including radiology reporting and chemotherapy regimens [4].

The cyberattack on health resources has exemplified the interlinking of health and disease intelligence. One example is the National Cyber Security Centre, which had a vacant director position during the cyberattack and minimal staff within the department in Ireland. Again, a lesson for other countries—as data becomes more valuable—sectors easily manipulate risks for their own advantage.

Building up an emergency response sector may have a high upfront cost to the health sector, in comparison to having an industry that is organized and prepared to respond to future national security catastrophes deriving from either spontaneous or deliberate spread of hazardous or infectious agents. Moreover, health and politics are becoming increasingly and inexorably intertwined. As a result, the risk of 'cross-contamination' across previously separate professional and disciplinary areas increases exponentially [5].

9.2 Health, surveillance, and intelligence

Associations between world politics, global health initiatives, and clandestine organizations such as the Central Intelligence Agency (CIA) have never before been so prominent in the public eye. State and nonstate actors have simultaneously taken the advantage of the geopolitical fallout and institutional weaknesses caused by the current viral pandemic to conduct terrorism, spy probes, disinformation campaigns, and cyberattacks [5]. There remains a significant threat of agencies using human delivery agents to spread vectors. In a hypothetical future, where there is little or no functioning diplomacy or supranational collaborative effort among nations, the links between infectious disease control and containment strategies are inevitably strained.

Perhaps, what is indeed required is an intersectoral approach involving global health assessment units, civilian and military intelligence gathering, and national security agencies to ensure that the limitations of privacy versus personal liberties are questioned, examined, and explored. Can new software, for example, which is often already pushing the privacy of individuals, assist with tracking the surveillance of health behaviours of future epidemic attacks? Political strategies, which often play out in both domestic and international fields, often result in states deciding what they are able and willing to share regarding pandemic information [3].

There remains a market field gap where rumours may spread and critical information may be missed. But, much like investors in global markets, so too do states and nonstate actors require and seek precious information, which may mitigate risk and reduce global (health, or otherwise) threats. While there has been some commentary in various forms and mediums on how this intrudes on an individual's privacy, many nations see the benefits of information outweighing the social costs and are moving ahead in this area.

9.3 Guest essay: The crucial role of the department of defence fits into early detection and response to disease outbreaks

Michael Baker

21st Century Global Health Intelligence is an increasingly important aspect of intelligence gathering to promote national security and requires a greater share of the resources and visionary leadership. US Medical Intelligence is vitally important to the security of the nation and its people.

Infectious diseases continue to evolve and potentially disrupt nations around the globe. This process is accelerated by demographic, political, and climate change pressures on populations that push humanity into habitats that were once wilderness or were considered unfit for living. The next pandemic may evolve while we remain exposed due to rapid, unsustainable urbanization, climate change, destructive food harvesting and producing practices, globalization, and reliance on other nations for essential items.

Zoonotic transmission, lab accidents, or an intentional biowarfare event can trigger an outbreak, and disease can quickly spread globally—as we have recently seen with the COVID-19 pandemic. The COVID-19 pandemic has also shown crucial US national security vulnerabilities and shortfalls in our response capability, and our adversaries have likely taken a note.

The ability to fuse that information into a useful product along with information from other sources informs government leaders promptly and potentially improves early response. Given that diseases are transboundary in nature, it is essential that United States has early detection assets. Information about disease outbreaks such as incidence virulence, transmissibility, genome data, and virulence statistics are crucial for combatting disease outbreaks but are characteristically difficult to obtain.

The US Department of Defense (DOD) and by extension the US Military is an important part of any early detection network. When the threat is detected, it is also the most capable agency to spearhead and deliver the response in a fast and reliable way.

Background

The National Center for Medical Intelligence (NCMI) serves as the lead US Department of Defense (DOD) agency for the production of medical intelligence, responsible for coordinating and preparing 'integrated, all-source intelligence' for the DOD and other government and international organizations on 'foreign health threats and other medical issues to protect US interests worldwide'. US intelligence may provide government leaders with early warning and information unavailable from any other source, particularly about foreign states. to reflect the organization's wider audience which may include the White House, Department of State, Homeland Security, other domestic customers, and allied partners.

The US DOD has forward deployed bases, forces, labs, hospitals, intelligence assets, and surveillance resources, all of which are backed up by an agency that has demonstrated success in early warning, testing, and response measures. In 1998, five different organizations within the Department of Defense (DoD) were brought together to create the Defense Threat Reduction Agency (DTRA) to better synchronize plans and actions for nuclear deterrence, weapons of mass destruction, and biothreats. DTRA quickly provided subject matter expertise, mobilized portable lab testing facilities, field vaccines, and treatments for the Ebola Outbreak in West Africa.

DOD's network of overseas laboratories perform research on infectious diseases of both public health and military importance and are key to the Department of Defense Global Emerging Infections Surveillance and Response System.

These intelligence tools demonstrated their value in November 2019 when the US intelligence community and NMCI began to warn about the coronavirus outbreak in China which could develop into a 'cataclysmic event' and briefed policymakers, decision makers, and the National Security Council at the White House. In this current pandemic, the military medical intelligence gatherers and government intelligence agencies were well ahead of the curve in raising the alarm on this growing threat.

The US Military is capable of providing crucial medical intelligence and is also best positioned to lead a rapid response as was seen in 2014. *Operation United Assistance* was a 2014 United States military mission spearheaded by The 101st Airborne Division to help combat the Ebola virus epidemic in West Africa. US DOD's DTRA quickly provided subject matter expertise, personnel, mobilized portable lab testing facilities, field vaccines, and treatments for the Ebola Outbreak in West Africa. DoD has the assets and experience to deploy necessary relief personnel and resources to all corners of the globe at a moment's notice—there is no agency better prepared to respond in times of crisis.

The US Military is unique in its ability to respond quickly with trained personnel, equipment, relief supplies, transportation, and communications.

Humanitarian assistance and disaster response are core DoD capabilities, but they are always conducted in a supporting role to assist other US Government agencies.

Military medical intelligence gathering originated with efforts in World War II. The US Central Intelligence Agency was established after World War II in 1947 and began producing medical intelligence reports focussed on Communist Bloc capabilities and trends, while the US Army Medical Intelligence and Information Agency (USAMIIA) handled the related military medical intelligence.

Given that diseases are transboundary in nature, secrets and information such as disease genome and virulence statistics are crucial to obtain from within non-transparent, closed regimes like China, Russia, Iran, and North Korea—especially if they are the source of the outbreak.

The value of these tools was evident in November 2019 when the US intelligence community and NMCI began to warn about a global epidemic, saying that the coronavirus outbreak in China could develop into a 'cataclysmic event', and policymakers, decision makers, and the National Security Council at the White House were repeatedly briefed on the issue. In early January 2020, mention of the novel coronavirus outbreak first appeared in the President's Daily Brief (PDB) of intelligence matters that are placed on the president's desk every morning. In this current pandemic, government intelligence agencies and military medical intelligence gatherers were well ahead of the curve in raising the alarm of this growing threat.

The military requires timely and comprehensive threat detection to support overall readiness, counter the spread of disease, and promote stabilization missions.

DARPA was instrumental in the development of RNA vaccines and provided $25 million in financial support to Moderna in 2013 to pursue messenger RNA-based antibody drugs and vaccines. DARPA announced that it was committing up to $56 million in additional funding to Moderna.

The programme is being run by the Defense Advanced Research Projects Agency (DARPA) and is called the Detect It with Gene Editing Technologies. The acronym, DIGET, is pronounced 'dig it'.

They use modelling and bioinformatics to choose the antibody that seems most effective at neutralizing the pathogen and then determine the genetic sequence that codes for the creation of that particular antibody. That snippet of genetic code can then be manufactured quickly and at scale, and injected into people.

Jenkins says this approach is much faster than manufacturing the antibodies themselves. Once the genetic snippets are delivered by an injection, 'your body becomes the bioreactor' that creates the antibodies, she says. The P3 programme's goal is to have protective levels of antibodies circulating within 6 to 24h.

DARPA calls this a 'firebreak' technology because it can provide immediate immunity to medical personnel, first responders, and other vulnerable people.

DARPA continues to work closely with the Department of Defense (DoD), multiple US government agencies, as well as its academic and industry partners to provide technical and scientific solutions to address the COVID-19 pandemic.

PICTURE 3.9.3 Militaries were often involved in vaccination as well as testing efforts. Sebastian Kevany.

9.4 A defence involvement checklist

Training and education

Improvement and upskilling practitioners in epidemiological basics, including recognition of evidence-based research, is an essential lesson from this pandemic. Early in January 2020, as reports were coming from China regarding the high ICU admission rates of COVID-19 patients, the disease may have been underplayed in countries with less focus on global health security [6]. As the pandemic struck Northern Italy and horror-like scenes arose from media sharing, so too did a wealth of 'research' which began to emerge. Suddenly, it appeared that many social media users and the general public became 'armchair epidemiologists' overnight [7]. The general public knowledge can be credited to the National Public Health Emergency Team (NPHET), which explained the epidemiology through diagrams and simplified examples during press briefings.

This fundamental training for the public in epidemiology has become a requirement for many countries, where staff were recruited and involved in various pandemic response measures. Recognizing terms such as 'flattening the curve', individuals understanding risk and societal actions [8], ensuring communication of handwashing, social distancing (whether one metre or two apart—remains to be seen), and the use of sharing accurate health information became important during this pandemic. Scientific and medically astute observers, academics, frontline workers, and citizens are all necessary for a response effort. In addition, the education of national leaders, governors, politicians, and other decision makers about what these warnings mean and possible consequences might require an urgent response.

Improving electronic databases

Ensuring the use and development of the newest high-tech tools and additional resources in conjunction with the enhancement of our intelligence-gathering agencies will provide a better early warning and response system and apparatus [9].

Previous outbreaks, such as SARS, Ebola, and H5N1, may have been looked upon as forewarning, but perhaps these outbreaks have been instructive in preparing for the 2020–21 crisis. Yet, many nations did not adequately prepare for previous infectious disease outbreaks or proactively respond with their total capacity as quickly as needed to avoid public health (and economic) harm [10]. Ensuring adequate electronic databases, including new tools, such as spatial epidemiology or more old-school methods of contact tracing, are in place to update data frequently and with a sense of urgency would be extremely useful for local, national, and international datasets. These datasets, in turn, would guide public policy.

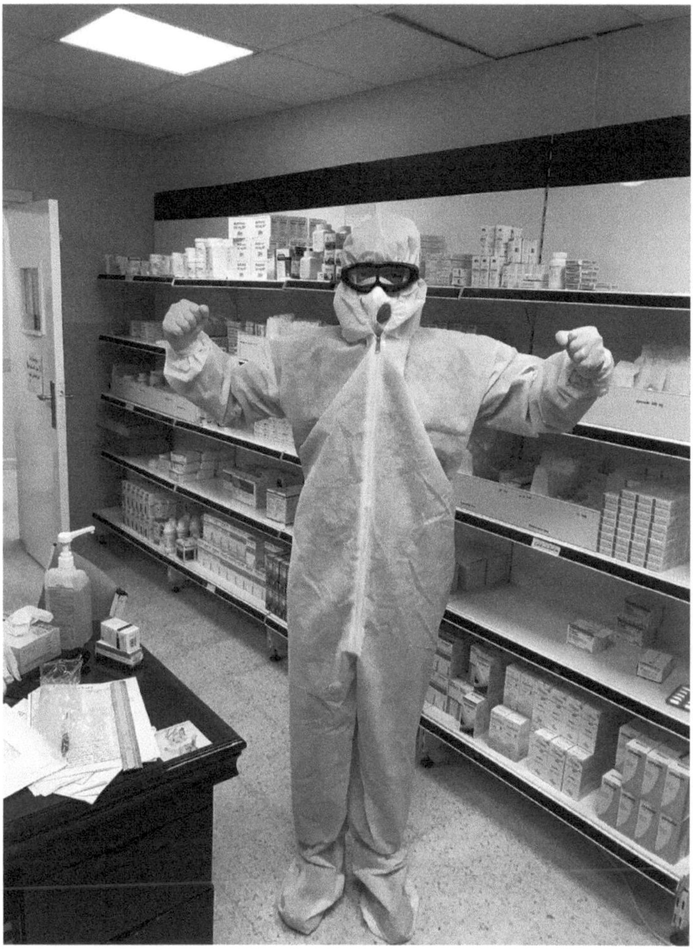

PICTURE 3.9.4 **HazMat gear often had military overtones.** Jeremy Fowler.

Standardization of data

All collaborating countries should be onboard by adopting a standard framework for transparent data sharing to facilitate early reporting. Classical data sources for outbreak intelligence include clinical reports, notifiable disease reporting, lab reports, pathology results, registries, and death records. Health intelligence originates at the local level, usually as clinical information.

Samples tested in the laboratory following the identification of suspected clinical cases often determine the hazard or pathogen. Suppose a red flag is triggered, for example, a cluster of cases, or suspicion about a particular subject arises. Then, the epidemiologic investigation is carried out to determine the source of infection and often any additional exposures, along with further laboratory investigations.

'Syndromic surveillance' is the inclusion of nonclinical data which automated nondiagnostic systems can collect [11]. Examples can include pharmacy records, ambulance call categories, personnel records, emergency department (ED) attendances, or primary diagnosis. While individual symptoms may not indicate a particular disease, detecting a cluster of symptoms or a syndrome can trigger early warning systems. Scientific and epidemic response teams may agree that the key to recognizing and managing an effective response to a pandemic is early detection and early response [12].

Early reporting

Infection rate and transmission mode were crucial factors to consider initially, as they may or may not render a pandemic uncontrollable, regardless of early warning. Unfortunately, in the current pandemic crisis, the existing early warning apparatus was often inhibited by the current state of surveillance tools, a lack of transparency and international cooperation, delays in reporting and lab confirmation, disinformation campaigns, and counterfeit equipment and drugs.

In an era with more resources and valuable tools than ever, the monitoring of COVID-19 infections can be carried out through minute-by-minute tracking systems, electronic tracking of data, rapid communication across countries, and rapid scientific molecular diagnostics. The increasing number of COVID-19 infections and their rapid global spread emphasize the importance of early detection. Moreover, it reveals humanity's complacency or outright reluctance to respond to clear signals of danger in some countries and states [13].

Intersectional outbreak response

This intersection of medicine, intelligence, and national security was evident during World War I. For example, the 1918 Influenza, often more colloquially known as 'Spanish flu', was emphasized by the suppressed war-time media to avoid blaming military training centres and enforcing xenophobia. Epidemiologically, however, the first breakouts were reported towards the end of World War I in US-based training camps, resulting in high mortality rates among younger healthy males, females, and the elderly [14].

After World War II, the Central Intelligence Agency was established to produce medical intelligence reports focussed on Communist Bloc capabilities and trends. At the same time, the US Army Medical Intelligence and Information Agency (USAMIIA) handled the related military medical intelligence programme. The USAMIIA is one of the many examples of the military acting on medical intelligence worldwide. The 'National Center for Medical Intelligence' (NCMI) grew from there to reflect the cross-sector approach, which included the White House, Department of State, Homeland Security, and other foreign partners.

9.5 How can intelligence help shape the response to future outbreaks?

As demonstrated during the 2003 SARS outbreak, the 2009 influenza pandemic, multidrug-resistant tuberculosis, and now the COVID-19 pandemic, infectious diseases are not constrained by international borders [15]. Outbreaks—whether linked to lab accidents, bioterrorism, or zoonotic causes—risk spreading rapidly across the globe. Therefore, outbreaks have a significant adverse impact on the economy, international security, and societal affairs.

In this current pandemic, government intelligence agencies, scientific advisory boards, and military medical intelligence gatherers were often well ahead of the curve in raising the alarm of this growing threat. However, the early response was lacking throughout most countries, with a lack of foresight to provide neither adequate resources nor modern tracking tools, leaving many countries to build a tracking tool from the basics with minimal resources and financial aid provided. Therefore, intelligence and medical fields must be open, transparent, and collaborative on essential developments in international disease outbreaks.

If well resourced, supported, and funded, departments such as Centres for Disease Control and Prevention, National Centers for Medical Intelligence, and Public Health departments are effective leading agencies. Disaster health agencies would benefit arguably from collaboration with technologically astute intelligence-gathering agencies to use tools that emerge in high-tech societies for contact tracing and isolation. Security and international relations at this time are evolving.

Yet, one point does seem clear: it is only through the timely and detailed collection and sharing of disease outbreaks and surveillance information between countries that pandemics will be contained or prevented. Building a wall or limiting travel will only delay the spread of disease. The earlier and more quickly surveillance notes a problem, the sooner alerts can be transmitted, the better the public health outcomes, and thus, the better the prospects for regional and international security and cooperation [16].

9.6 Key points

- Military forces are essential tools for pandemic responses, particularly in terms of military medical skills and capacity.
- However, these powerful forces need to be used judiciously and carefully, particularly when involved with international operations.
- The evolution of military forces from combat to humanitarian roles is welcomed by the public health community.

References

[1] M. French, T. Monahan, Disease surveillance: how might surveillance studies address COVID-19? Surveill. Soc. 18 (1) (2020) 1–11.

[2] R.K. Betts, Two faces of intelligence failure: September 11 and Iraq's missing WMD, Polit. Sci. Q 122 (4) (2007) 585–606. http://www.jstor.org/stable/20202928.

[3] N.K. Ibrahim, Epidemiologic surveillance for controlling Covid-19 pandemic: types, challenges and implications, J. Infect. Public Health 13 (11) (2020) 1630–1638.

[4] T. Anderson, W. Torreggiani, The impact of the cyberattack on radiology systems in Ireland, Ir. Med. J. 114 (5) (2021) 137.

[5] M. Baker, S. Kevany, D. Canyon, The Intersection of Global Health, Military Medical Intelligence, and National Security in the Management of Transboundary Hazards and Outbreaks, 2020, Apcss.org. Available at: https://apcss.org/wp-content/uploads/2020/07/N2508_Baker_Canyon_Intersection_Health_Intel.pdf. Accessed 7 December 2021.

[6] Z. Zhao, A. Chen, W. Hou, J.M. Graham, H. Li, P.S. Richman, H.C. Thode, A.J. Singer, T.Q. Duong, Prediction model and risk scores of ICU admission and mortality in COVID-19, PLoS One 15 (7) (2020) e0236618.

[7] D.E. Alexander, On evidence-based practice in disaster risk reduction, Int. J. Disaster Risk Sci. (2021) 1–9.

[8] S.G. Reddy, Coronavirus and the limits of economics, Foreign Policy (2020). March 31.

[9] Q. Zheng, F.K. Jones, S.V. Leavitt, L. Ung, A.B. Labrique, D.H. Peters, E.C. Lee, A.S. Azman, HIT-COVID, a global database tracking public health interventions to COVID-19, Sci. Data 7 (1) (2020) 1–8.

[10] R.D. Smith, Responding to global infectious disease outbreaks: lessons from SARS on the role of risk perception, communication and management, Soc. Sci. Med. 63 (12) (2006) 3113–3123.

[11] K.J. Henning, What is syndromic surveillance? Morb. Mortal. Wkly Rep. (2004) 7–11.

[12] C. Ansell, A. Boin, A. Keller, Managing transboundary crises: identifying the building blocks of an effective response system, J. Conting. Crisis Manag. 18 (4) (2010) 195–207.

[13] H.A.S. Hashmi, H.M. Asif, Early detection and assessment of covid-19, Front. Med. 7 (2020) 311.

[14] J.L. Sanchez, M.J. Cooper, C.A. Myers, J.F. Cummings, K.G. Vest, K.L. Russell, J.L. Sanchez, M.J. Hiser, C.A. Gaydos, Respiratory infections in the US military: recent experience and control, Clin. Microbiol. Rev. 28 (3) (2015) 743–800.

[15] K. Shaw, The 2003 SARS outbreak and its impact on infection control practices, Public Health 120 (1) (2006) 8–14.

[16] S. Laato, A.K.M. Najmul Islam, M.N. Islam, E. Whelan, What drives unverified information sharing and cyberchondria during the COVID-19 pandemic? Eur. J. Inf. Syst. 29 (3) (2020) 288–305.

Part IV

Public health information sharing

Chapter 10

Information sharing

Sebastian Kevany, Deon Canyon, and Aoife Kirk

Chapter outline

10.1 Who should know what?

Protected health information (PHI) is often an area of uncertainty and mistrust in a modern democratic society. However, in the wake of the devastating 2020—ongoing pandemic and previous epidemic outbreaks, public health information and health security to inform further outbreaks are necessary, to allow decision-makers to be better and better equipped to respond. In addition, there are increasingly stronger medico-legal arguments for PHI, in certain reportable public health emergency situations, to be made public or at least accessible by recognized academic and research entities [1]. Whilst this has a host of potentially complex legal and ethical ramifications—the public health and economic benefits may well outweigh these arguments for societal benefit.

Through associated legislation, individuals are assured of near-total confidentiality in reporting medical conditions to healthcare providers, independent

Outbreaks, Epidemics, and Health Security. https://doi.org/10.1016/B978-0-323-91705-6.00014-9

of their nature. However, reportable diseases are often a quasiexception. Certain conditions such as HIV/AIDS, TB, polio, and other reportable infectious diseases are registered in confidential aggregated national and international databases. With the advent of electronic medical records (EMRs) and health management information systems (HMISs), this dynamic is changing: whilst confidentiality and privacy remain assured, there are more and more ways in which PHI can be centralized, aggregated, and analysed [2].

The release of vital demographic and geographical information becomes more crucial at the early (and end) stages of an epidemic and pandemic. With greater access to demographics and risk factors associated with reportable conditions for the public, a wide range of academics and analysts can begin to make independent and informed decisions. Moreover, these decisions may assist in guiding decision-makers on the cause, effect, prospects, and other elements of epidemiology—a professionally diverse thinking exercise, rather than one that is the exclusive preserve of senior civil servants.

Thus, the analysis moves from the entire realm of bureaucrats and their consultants to a more appropriate, more diverse group of experts. Examples of these groups of independent critical analysis experts appeared during various pandemic stages in each country. Allowing the release of this information to be publicly available also, importantly, will enable resources to be concentrated in the worst-affected areas.

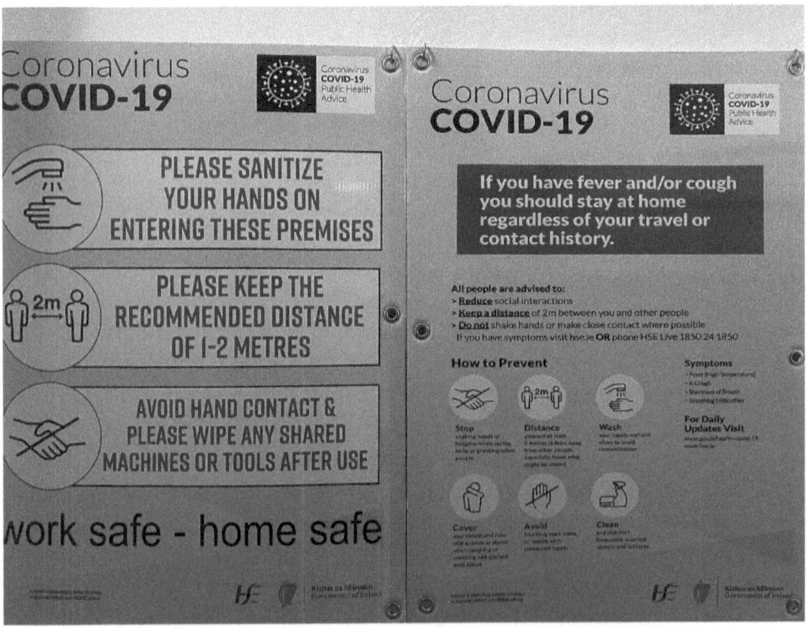

PICTURE 4.10.1 **We had the information, but did we always have the data to back it up?** Aoife Kirk.

An even more compelling rationale, though, is the concept of ensuring essential public epidemic containment information. Public health efforts invariably suffer without high granularity and resolution in outbreak data—down to the details of street addresses, ideally, but if not at least information kept on district or county levels [3]. The de-identification and anonymization of such information are often standard practices. This approach contrasts with the difficulties which may arise with vagueness in the precision of details during epidemic situations, which may leave both public health law enforcers and the general public at a disadvantage.

10.2 Guest essay: The importance of death registration (vital statistics)

Kiran Kamble

Coronavirus disease 2019 (COVID-19) is a communicable disease also known as severe acute respiratory syndrome coronavirus 2 (SARS-CoV-2) [4] and has infected approximately 202 million people worldwide claiming the lives of 4.28 million [5] as of mid-August 2021. The economic loss due to this pandemic is unimaginable with the United Nations' Department of Economic and Social Affairs projecting a global loss in output to the tune of USD 8.5 trillion over the next 2 years [6]. These figures belie the emotional and financial shocks that individuals and families had to bear across the world. The global devastation caused by this pandemic has no precedent in recorded history, not even the two World Wars.

The pandemic has exposed several policy levels, administrative and implementation systems gaps across the developed and developing countries. The severity and extent of these gaps may vary across nations, but no country has been spared. As a physician, a global health practitioner, and a health services researcher, the following is a significant gap that I would like to highlight.

This gap is the weak to nonexistent civil registration and vital statistics (CRVS) system, specifically, a system for accurately registering mortality data, especially in the developing countries. We have all witnessed the unfortunate politicization of mortality data in this COVID-19 pandemic across the world. Politicians, uniformly across the developed and developing world, were busy downplaying the mortality data and in some cases even fudging the data by undercounting and hiding in order to cover up their leadership failures [7]. Apart from these political factors, there are several other structural and technical reasons, brought to the fore by this pandemic, that over time have led to the gaps in collecting accurate mortality and morbidity data. Some of these gaps are [8–10] (1) lack of a regulatory and legal framework guiding a national vital statistics registration system; (2) absence of a vital statistics implementation strategy, with a clarity on the role of the government and the citizen, and lack of standardized tools and data collection formats; (3) weak administrative will to establish necessary infrastructure, including human resources; (4) inadequate to nonexistent standards for data coding, analysis, and sharing; (5) nonproactive systems with disjointed approach; (6) lack of quality control mechanisms; and (7) lack of political will and absent public awareness. There is a lot of work that needs to be done in strengthening vital registration systems in the developing world, and given so many other urgent population health matters to be catered to, one would wonder why it is so important now to have a strong vital registration system, especially accurate mortality data.

Mortality data is a significant indicator of the health status of a population, highlighting the disease burden and the influence of structural factors and social determinants of health on a population's health [7–13]. It is also an important aid for programme implementers and policymakers to understand the racial and ethnic inequities in affliction of a particular health disorder, as well as the disparities in provision of health services [14]. Information on mortality constitutes an important part of epidemiological and public health surveillance and is a good indicator of the efficacy and functioning of a health system [12]. Mortality data has proved to be a crucial indicator in understanding the impact of COVID-19, given the lack of standardized medical classification [15]. Excess mortality, defined as the "difference between the observed and expected number of deaths in a specified time period" [16], has been used to determine the mortality caused by the pandemic. This, however, is a comparative estimate and does not provide the true picture of the direct and indirect mortality caused by the pandemic, and such granular data is necessary for formulating precise and relevant health programmes [17]. So, undoubtedly, current and accurate registration of mortality is important, but what are the most cost-effective ways to institutionalize and implement a standardized CRVS system in a developing set-up?

As early as 1968, the United Nations' Economic and Social Council floated the idea of establishing a 'World Programme for the Improvement of Vital Statistics' [18] and, later in a resolution in 1976, recognized the civil registration system as a part of human rights [19]. Since then, the United Nations and several multilateral organizations, including the governments of France, Great Britain, and the United States, have conducted research on the gaps and actions required to establish a CRVS system in countries with varied political, economic, and cultural backgrounds. Following is a summary of a country-wise list of activities proposed through these different initiatives.

- Establish multistakeholder coordination within the country involving legal experts and representatives from public administration, civil society, medical, public health, and medico-legal fields, politicians, and religious leaders.
- Seek assistance from the United Nations and other international organizations.
- Develop legal framework and guidelines for civil registration and vital statistics through a consultative process.
- Establish the institutional infrastructure, including identification of the staffing needs, their recruitment, and training protocols.
- Institutionalize and promote methodological research systems, whilst assigning specific data collection, collation, and analysis responsibilities at all levels.

- Ensure training of all medical and paramedical personnel from the public and private sectors to the civil registration and vital statistics systems.
- Update the national civil registration and vital statistics standards based on international updates, whilst considering the local context.

The World Health Organization (WHO) and the World Bank (WB) have set an ambitious target of universal CRVS for the entire global population by 2030, and 45% of global deaths were registered by 2013 [20]. There is still a long way to go, and public awareness is necessary to ensure that the establishment of a current and dynamic civil registration and vital statistics system becomes a political and administrative priority.

10.3 A double-edged sword

There are threats, however, to public information sharing being so freely available. Stigma, suspicion and unnecessary fear, and economic decline inevitably follow as the same blanket policies are enforced across all geographical areas, industries, and demographics—regardless of disease-related differences [21]. Otherwise, the stigma associated is concerning. For example, one approach reminiscent of the Ebola outbreak, where military and police searches were often required to 'expose' the 'hidden' patients who feared internment. Usually, stigmatization of HIV/AIDS patients and people exposed to the Ebola virus has been derived from false information and beliefs about transmission and lack of control over data [22].

Therefore, conversely, the availability of more granular information results in a well-informed, more resilient public able to make more sensible, evidence-based decisions, both within and outside affected geographical areas [23]. Unfortunately, this was sadly lacking in 2020 and 2021 in most countries, many of which continued to place a greater degree of importance on personal privacy than curtailing the pandemic's devastating health and economic effects. Moreover, in the face of COVID-19, many high-income countries adopted nationalistic approaches rather than cooperation regarding information sharing and collective action despite repeated pleas and appeals from the WHO [24].

Individuals and citizens need to have up-to-date access to accurate, high-resolution information on disease outbreaks. In 2020 and 2021, this information, whilst available, was the preserve of only a select few and was released in often unhelpful, aggregated, and at times frightening forms of communication to the public. Humanity has been fortunate in recent months that the global public health emergency has not become a complex humanitarian emergency as well. Unfortunately, at one stage, even that changed in the United States and United Kingdom, where nationwide protests may have provided a new hotbed for disease transmission.

PICTURE 4.10.2 Polices such as outdoor dining were driven by statistics. Aoife Kirk.

10.4 Guest essay: Guam experience and lessons for responding to epidemic threats—Academic institution fosters collaborations/multisectoral approach for COVID outbreak and vaccine preparedness

Yvette C. Paulino[a]; Margaret Hattori-Uchima[a]; Laura E. Buback[b]; Teofila P. Cruz[a]; Annette L. Aguon[c]; Carlos R. Taitano[a]; Michael J.A. Reid[b]

Guam, an island territory of the United States located in the Western Pacific, serves as a travel hub connecting the United States to countries in Asia and to other islands throughout Micronesia. It is said that Guam is where 'America's Day Begins', located 2500 miles west of the International Date Line: 20 h ahead of Hawaii and 1 day ahead of the United States. Whilst Guam is geographically remote, it is culturally diverse—consisting of the indigenous CHamorus and other Pacific Islanders (49%), Asians (34%), predominantly Filipinos, Mixed (9%), Whites (7%), and Others

[a]University of Guam.
[b]University of California, San Francisco.
[c]Department of Public Health and Social Services, Guam.

(1%)—and globally well-connected. Guam is thus in a unique position to harness both local and regional partnerships to strengthen the COVID-19 response for its diverse population. The lessons from Guam are unique to its context; however, they may also be useful for other jurisdictions in preparing for future pandemics.

Initial response through local collaborations

The island reported its first travel-related COVID-19 case on 15 March 2020, although the outbreak on the USS Theodore Roosevelt is what attracted national and international attention. The increase in COVID-19 cases and deaths over the next few months placed an additional strain on the island's healthcare system. In all response efforts led by the Guam Department of Public Health and Social Services (DPHSS), the value of partnerships was emphasized. Guam's island sense ofcommunity and connectednesswas a challenge leading to surges and rapid spread of COVID-19, especially in underserved communities and large, often multigenerational households. However, that sense of community and solidarity also fostered a unique 'esprit de corps' that enabled partners to come together around the COVID-19 response. *Inafa'maolek* reflects the CHamoru value of cooperation and mutuality to benefit all. This *inafa'maolek* was seen in action during the pandemic in relation to testing, contact tracing, clinical support, community outreach, and vaccination. The expansive network of academic, government, private, community, and faith-based partners was a key to ensuring consistent supplies, and a well-trained, culturally sensitive workforce was in place for a strengthened and timely response.

As the sole 4-year degree-granting institution of higher learning on the island, the University of Guam (UOG) has long been a strategic partner of the local DPHSS in controlling past disease outbreaks. Faculty and students from the UOG School of Health have a long relationship of collaboration with DPHSS and were engaged even prior to the State of Emergency, during island-wide disaster management planning. The UOG School of Health actively engaged with responses to H1N1, measles outbreaks, and the 2019 emergence of dengue fever on the island, necessitating outreach and education to the villages. Then, the COVID-19 pandemic once again elicited support from the School of Health, consisting of nurses, social workers, health educators, and public health professionals. Early on, the Government of Guam launched a COVID-19 hotline known locally as 311. The hotline centre was situated at UOG and, through support from the UOG Global Learning and Engagement Center, was expanded to include multiple phone lines and adequate space for appropriate physical distancing. The 311 Hotline Center served as a Nurse Triage, run by nurses from DPHSS, UOG, and the Guam Department of Education. Meanwhile, the island was faced with a depleting supply of personal protective equipment and the needs across the island were significant. UOG hosted a telethon that was successful in raising over $43,000 for COVID-19 relief. The funds opened a mechanism for procuring personal protective equipment and necessary items promptly. The collaboration between the University and local healthcare organizations provides a support that can, in situations such as procurement and hiring of personnel, respond more quickly and efficiently to urgent pandemic-related needs.

Prior to the pandemic, the healthcare workforce in Guam faced shortages of registered nurses, nursing assistants, and other healthcare professionals. During the first months of the pandemic, the School of Health was immediately tasked with nursing workforce development activities such as assisting in acute care and community-based settings, bolstering the understaffed DPHSS nursing workforce, and responding to requests for COVID-19-related support. For example, the Guam National Guard (GUNG) requested assistance from UOG to develop an innovative hospital training programme to transition Army Medics into hospital care providers, whilst the Guam Regional Medical City requested assistance to develop and implement a 1-week training programme. GUNG medics successfully integrated into the staffing pattern of the Guam Regional Medical City, thereby encouraging the uptake of the GUNG medics at the Guam Memorial Hospital—the island's government-funded hospital. There, the GUNG medics were utilized in many patient care areas, including COVID-19 care units. The nursing students were critical throughout the pandemic, prompting the UOG School of Health to graduate the Nursing Class of 2020 early and enter the workforce immediately to support pandemic efforts. All BSN graduates successfully passed the National Council Licensing Exam (NCLEX) and are currently registered nurses working in local hospitals and healthcare agencies.

Throughout the pandemic, there was also a growing concern for the homeless population and their potential exposure to COVID-19. Factors affecting the homeless population and increasing COVID-19 risk included closure of soup kitchens, lack of basic hygiene supplies and facilities, lack of access to services due to government and private sector shutdowns, existing shelters on lockdown and not accepting new clients, poor management of chronic illness, declining access to mental health services, inability to reach COVID-19 testing centres (drive-through and clinic-based testing), and the rising numbers of homeless individuals and families due to layoffs and ongoing business closures. The Guam Homeless Coalition (GHC)—a group of government agencies, nonprofit organizations, and private sector partners—worked collaboratively to provide COVID-19 testing to the homeless population. GHC member organizations provided outreach to the homeless to inform them of the testing availability, and three successful COVID-19 testing outreach events were held for the homeless population. The outreach included social services by GHC member organizations in addition to COVID-19 testing. Positive cases were managed within 48h, and DPHSS and Coalition staff worked together to provide isolation to those who tested positive (with their consent). Another mitigation strategy was the opening of an emergency homeless shelter led by the Government of Guam. The emergency shelter opened in the first 120 days of the pandemic.

Continued response through national collaborations

The DPHSS and UOG partnership continued through the COVID-19 pandemic, especially during the second wave of COVID-19 infections in mid-August 2020 that prompted the need to increase contact tracing (CT). The University of California San Francisco (UCSF), from across the Pacific, was consulted for its expertise in CT training, and a multisectoral partnership evolved. Through support from the Governor of Guam, Lourdes Leon Guerrero, herself a registered nurse, funding was obtained to secure technical assistance from UCSF and stand up the UOG Call Center. In September 2020, UCSF and UOG conducted its first virtual CT training in Guam, covering the epidemiology of COVID-19, containment strategies, and health coaching, including skills laboratories for practising CT interviews. In October 2020, a second CT training occurred, followed by a training of 10 trainers, who would eventually deliver additional training sessions, and the UOG CT

Call Center was launched. Through this partnership, 159 CTs and 10 CT trainers were trained, including personnel from the island's education, judiciary, healthcare, faith, and hospitality sectors. The UOG CT Call Center has been operating full-time supporting contact tracing and laboratory testing and evolved to support vaccination efforts. Leveraging the technical and training expertise of local and regional academic partners, DPHSS was able to establish a robust onboarding and training programme. The unique partnership between UCSF and UOG also ensured that the workforce was optimally trained to provide an evidence-based, culturally sensitive response. The long-standing collaboration between DPHSS and UOG ensured that the new workforce was rapidly deployed at the CT Call Center.

PICTURE 4.10.3 Workers break for sunlight, fresh air, and a photo opportunity outside the UOG Call Center. Carlos Taitano.

Training in CT was sustained through support from the Government of Guam as well as the Health Resources and Services Administration grant funding the Geriatrics Workforce Enhancement Program (GWEP), which is managed by the UOG School of Health. The GWEP also received funds to promote telehealth and tele-education for students, healthcare workers, and the public. Training supported by CARES funds included COVID-19 training for healthcare staff/caregivers and health professions students. Online COVID-19 training was implemented with 355 healthcare workers/caregivers and health professions students trained.

As the United States rolled out the nation's vaccination plan, Guam was fortunate to receive its first shipment of 3900 doses of Pfizer-BioNTech COVID-19 vaccines on 12 December 2020. The vaccines were administered less than a week later to residents according to Phase 1 priority group eligibility, which included healthcare workers at high risk of contracting COVID-19 and residents of long-term care facilities.

Ongoing emphasis on vaccination

In January 2021, as hospitalized COVID-19 patient numbers decreased and more vaccine doses were received, the focus on Guam shifted to vaccination efforts. DPHSS began mass vaccination clinics in several locations, with hundreds of island residents seeking vaccinations. Efforts were expanded island-wide, and the GUNG was mobilized to stand up large-scale vaccinations. The UOG Calvo Field House was determined to be an optimal UOG-GUNG vaccination site for its central location and massive size. The UOG Call Center was leveraged for vaccine deployment and data entry of vaccines administered at the UOG-GUNG site and other community outreaches. More recently, in response to a need for the Government of Guam to provide COVID-19 vaccination verification to the public, the demand for verifications exceeded the ability of the DPHSS. The UOG Call Center responded and served as the main provider of COVID-19 vaccination verifications. Again, the Governor of Guam was instrumental in providing sustained funding to support the ongoing vaccination efforts at the UOG Field House and Call Center. Collaboration at the highest levels of government facilitated an effective and timely response to the pandemic.

By 28 July 2021, Guam became the first US territory/state to vaccinate 80% of its adult population. Guam's strategy for vaccination scale-up offers important insights in terms of coordinating an effective vaccine campaign, addressing hesitancy, and engaging all key stakeholders. Local data on COVID-19 test results was used to inform targeted vaccination, and existing partnerships once again accelerated results. One group targeted was the homeless population. Continued collaboration amongst GHC member organizations led to early vaccination for homeless service providers, residents of homeless shelters, and group homes. Additionally, the GHC, UOG, DPHSS, and other organizations held COVID-19 vaccination pods for the homeless, with outreach conducted prior to the vaccination dates, to inform the homeless about the COVID-19 vaccines. Homeless shelter clients were also included in the vaccination outreach events. These actions have led to limited COVID-19 spread amongst the homeless, with only four identified cases amongst the street homeless population, and no outbreaks in homeless shelters. The emergency shelter which opened for COVID response had several COVID-19-positive cases but were quickly identified and managed without spreading to other residents. The efforts to promote the safety of homeless individuals and families have been successful, in large part due to the collaborative work of GHC member organizations and the administration of the Government of Guam.

Another group targeted was the Micronesian populations. The GWEP staff collaborated with DPHSS and helped establish a community network for people from the Federated States of Micronesia (FSM). The FSM Community Network had the goal of conducting outreach and vaccination events for FSM residents. The Network was formed to address the problem of COVID-19 health disparities affecting Micronesians. Vaccination pods were conducted in various churches with high Micronesian congregations, as well as at the Guam Regional Medical City. The collaboration with Guam Regional Medical City and the FSM Community Network organizations enabled the DPHSS to engage with the Micronesian community in Guam. This relationship continues to be a positive force for increasing vaccination rates amongst our minority populations.

The third group targeted was UOG students to increase vaccination uptake amongst the island's young adults. A multidisciplinary team of students from the

Criminal Justice, Nursing, Social Work, Public Administration, and Public Health Programs collaborated on a COVID-19 campaign project to develop a social media toolkit. Included in the toolkit were infographics promoting vaccination, links to related resources for the most updated information, and a video on navigating the toolkit. The video was presented by the team of students at the sixth Annual HART (Health Awareness to Reach Tritons) Fair in April 2021, and the toolkit was disseminated shortly thereafter.

In addition to promoting awareness amongst young adults, many UOG students were actively involved at the UOG-GUNG vaccination site. For example, seniors from the Nursing, Social Work, and Health Science Programs were trained and given the opportunity to assist in the vaccination clinic. Students earned clinical/practicum credit for the hours spent at the vaccination site. This opportunity provided students with real-world experience with pandemic relief and specifically vaccination deployment. Prior to the vaccination implementation, students from the UOG School of Health participated in CT and UOG Call Center operations as part of their senior capstone coursework. Students were trained through the UOG-UCSF collaboration and gained valuable hands-on experience. Furthermore, the collaboration amongst partners provided the relevant experience to prepare the health professional workforce in various aspects of pandemic response.

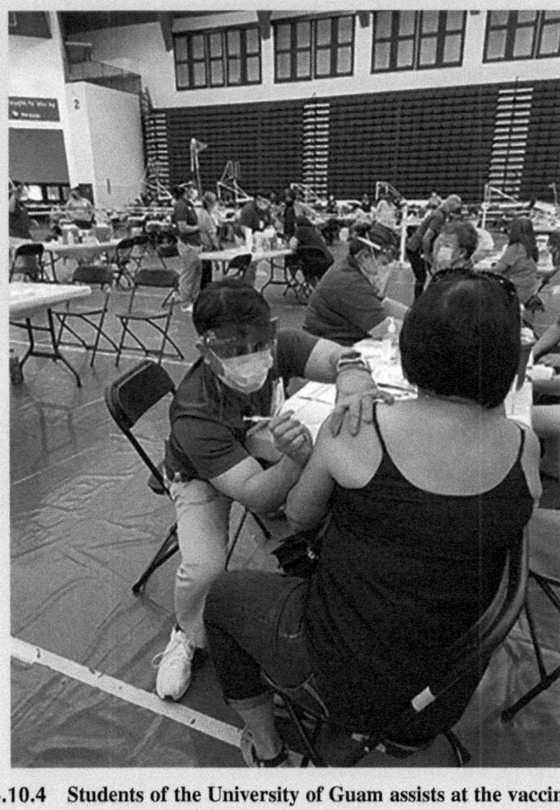

PICTURE 4.10.4 Students of the University of Guam assists at the vaccination site on campus. Margaret Hattori-Uchima.

Whilst the story of COVID-19 is still evolving in Guam, it is clear that the diverse, multisectoral partnerships are essential. Whilst we focus on academic and community partnerships, the tourist industry and military sector in Guam also played a key role in shaping the COVID-19 response on the island. Many can learn from the island where America's Day begins, which has also outpaced any other territory and exceeded most of the United States in vaccine rates. Looking to the future, Guam's island status may also well play a pivotal role as a sentinel surveillance site for future pandemics. Whilst America's Day begins in Guam, it is also potentially where new pandemic detection can begin. Guam is an important nexus for tourism, especially from East Asia, and also receives a lot of military traffic. It is therefore strategically and epidemiologically important to invest in surveillance and public health infrastructure in Guam, expanding laboratory capabilities to detect new and emerging infections, and strengthening public health expertise to respond to future emerging infectious threats.

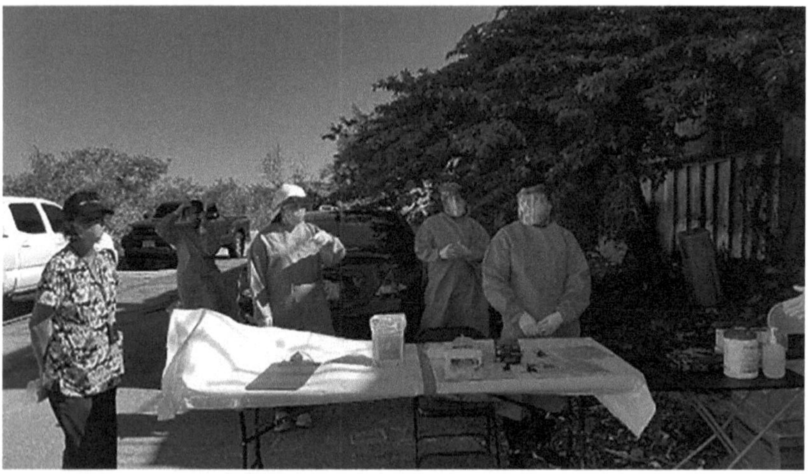

PICTURE 4.10.5 **Vaccination outreach to the homeless.** Yvette C. Paulino.

10.5 The role of tech

The need to act on PHI reform in emergency epidemic situations is, therefore, relatively urgent. Before the concatenation of circumstances that froze the world and its economy repeats itself in the short or long term with subsequent waves of infection, governments should consider enacting PHI policies that enable:

- More informed public decision-making;
- Greater academic involvement;
- Improved targeting of resources;
- More effective epidemic control;
- Relieving the public of unnecessary paranoia;
- Reducing restrictions in non/less-affected areas.

Access to more granular, detailed public health information may have improved the compliance and understanding of the general public. This detailed PHI negates inaccurate word-of-mouth messaging and reliance on social media and provides far more relevant information for policy decision-making than the (almost) irrelevant aggregated data typically provided by government statistics departments. Unfortunately, in the 2020–21 pandemic case, public understanding relied almost exclusively on word of mouth, social media sources, rumours [25], and aggregated government statistics.

Yet voluntary buy-in to even such limited efforts inevitably varied by country, by what level of privacy was culturally acceptable. On the one hand, many Westerners and sceptics perceived the actions of China and Korea to be an infringement of personal freedom, PHI, and human rights [3]. On the other hand, other countries (such as New Zealand and Taiwan) viewed such information-gathering methods as essential in regulating an evidence-based and targeted removal of lockdown efforts, enabling the reopening of society, and helping to restart the economy.

The apps released and used in Europe send notifications when users get close to other users who have tested positive for the COVID-19 virus. Advances in Bluetooth Low Energy technology enabled this to occur without the app gathering information on its users' identity or location [26]. However, these apps also had minor issues—how, for example, would this be feasible in hospitals, where frontline workers would be surrounded by positive patients? Would this app work on public transport? The apps would not replace the manual function and groundwork of contact tracing through epidemiological measures. Yet, when used correctly with access to PHI technology, they would assist with the efficiency of gathering the data. Perhaps, the best takeaway from using apps was to increase contact tracing test and trace, encourage accurate information through the app, and improve uptake of swabbing of close contacts.

10.6 Guest essay: Global solidarity: Support for equitable access to health services through strengthened health systems, free at the point of access

Regina Keith

Werner and Saunders state that *"solidarity is not an act of charity. It is an act of unity between allies fighting on different terrains towards the same objective"* [27]. As we fight against the COVID-19 pandemic, perhaps it is time that we decide that global solidarity for health is a worthwhile target. Just as diseases can spread across borders, so too can solutions. Over 40 years ago, 134 health ministers decided that health for all was a worthy and achievable goal. Inspired by an evidence-based, rational policy process, led by the World Health Organization (WHO) and UNICEF, they signed the Alma Ata Declaration [28]. This was the first-ever global health policy, a commitment that acknowledged that health was a right for all and identified primary health care (PHC), as the strategy to achieve the vision of health for all by the year 2000.

The Alma Ata defined health as "a state of complete physical, mental and social well being, and not merely the absence of disease or infirmity" and stated that health was an *"important world-wide social goal whose realization requires the action of many other social and economic sectors in addition to the health sector"* [28]. The evidence presented by the WHO and UNICEF inspired the principles of the newly agreed strategy. These included equity of access to comprehensive primary healthcare services, including the prevention, promotion, and treatment of illnesses. The services were to be locally determined, through active community engagement, and implemented concurrently with interventions addressing issues such as food security, nutrition, social protection, water, sanitation, and quality education [27,28].

The design was built on the experience of good health at low cost in countries like Sri Lanka, Cuba, Guyana and places like Kerala in India [29]. Good health was assessed through the achievement of Universal Coverage of Health services, with sustained reductions in preventable maternal and child deaths [29]. However, the countries had other common factors such as a clear political commitment to the equitable provision of health for all, supported by active engagement from communities to identify and address local problems that affected health, such as lack of access to clean water, sanitation, and food security [27,29–33].

All the participating countries had invested in strong community health systems that linked formal health services to households, with concurrent investment in education (especially for girls), nutrition, and social protection for marginalized populations [29,30,33]. Kerala, in India, invested 14% of its budget on health, and Costa Rica abolished its armed forces in 1948, spending its money on social development [29]. The participants of the Alma Ata Conference returned home with renewed hope for a future where everyone had equitable access to health care.

That hope was quickly dampened by a lack of funding to implement the ambitious policy. Funding agencies like the World Bank and the Rockefeller Foundation felt the policy was too expensive and suggested that the comprehensive PHC vision be reduced to a more manageable set of four interventions.

This resulted in a focus on four interventions to reduce childhood deaths: growth monitoring, breastfeeding promotion, oral rehydration solution, and childhood immunizations. They became known as the GOBI interventions: these GOBI interventions were supported by health education messages, using top–down approaches to the selected priorities—an approach divergent from the agreed Alma Ata principles of primary health care [27].

As countries implemented the GOBI interventions, a global recession occurred in the 1980s, which reduced the resources available to support health and other essential social spending [27]. Yet leaders in many newly independent countries had been elected on the promise of free health and education services. However, they did not have the resources to implement these promises. Many of these countries borrowed money from the International Monetary Fund (IMF), which had strict conditions, set out in structural adjustment policies. These conditions required them to reduce social sector spending and implement policy changes, such as implementing charges for accessing health care, to ensure that they could repay their loans and maintain fiscal stability [27]. Other conditionalities were also introduced, such as increasing inflation and devaluing local currencies. These measures had devastating impacts on the ability of countries to implement their new primary healthcare (PHC) policies, based on the right to health for all [27,34].

Although many countries included comprehensive PHC in their health policies, they only had resources (mostly from donors) to implement selective disease-specific interventions. These disease-specific programmes did not adhere to the PHC principles, and 'health for all by the year 2000' was not achieved [27]. The right to health focus was increasingly replaced by a new focus on child survival [35], using community compliance, disease-specific interventions, and behaviour change communication tools, rather than the development of more robust community health services, to address local health and social problems [27,36,37]. Although 12 million under-fives were dying from preventable causes each year at this time, the focus on selective, disease-specific interventions led to fragmentation of the health services and lack of investment in human resources and other sectors such as water, sanitation, and education [38].

Millennium development goals and making poverty history

Despite the failure to achieve the Alma Ata target of health for all by the year 2000, there was at that time increasing global advocacy that urgent action was needed to address growing global inequalities in health and education. At the start of the new millennium, global leaders agreed to work together to address some of these inequities. As over 12 million children under 5 were dying from preventable causes, and millions of children held little chance of formal education, the global leaders agreed on eight global goals, also known as Millennium Development Goals (MDGs), which aimed to be achieved by 2015. Three of these eight goals were focused on health: Millennium Development Goal four focused on reducing the number of under 5-year-old children dying from preventable causes, MDG five focused on reducing maternal deaths, whilst MDG six focused on halting the growing prevalence of HIV, TB, and malaria [39].

Despite setting these global commitments, there was little change to how health was being implemented, with a continued focus on disease-specific interventions [27,38,40]. In 2003, the Grow Up Free from Poverty (GUFP) Coalition carried out research which predicted that the MDGs for maternal and child health (MDG 4 and MDG 5) would not be met for another 150 years in sub-Saharan Africa, resulting in 80 million maternal and child preventable deaths by 2015 [31]. The GUFP report was launched at the UK Parliament by the then Secretary of State, Baroness Amos. The report encouraged organizations to develop more robust evidence to support the Make Poverty History Campaign (which aimed to halve poverty by doubling aid) in 2005.

Save the Children UK (SC UK) launched their Cost of Coping with Illness in East and Central Africa series in 2005, to support a global call for more action. This research was based on a peer-reviewed mixed method research programme, exploring the barriers to timely access to health care, in seven countries (Ethiopia, Sudan, Tanzania, Uganda, Uganda, Burundi, and Rwanda) in sub-Saharan Africa (SSA). The 5-year longitudinal study was commissioned by Save the Children to follow up the findings from their 2001 Bitterest Pill report, which highlighted the collapse of health systems in parts of sub-Saharan Africa [38].

The longitudinal study found that one-third of the research population could not afford to seek health care, whilst a further third of the population was pulled into poverty, due to selling capital finite assets, to pay for health care [41–53]. This research supported the increasing global recognition regarding the negative impact that health user charges were having on healthcare service utilization. Evidence had been growing since the 1990s, when the World Bank and the IMF started aggressively promoting their use as conditions linked to structural adjustment plans [54–64].

Innovative research in 2005 estimated how much it would cost to abolish these user charges in 20 of the United Kingdom's priority countries in Africa. The results were published in the British Medical Journal (BMJ), estimating that abolishing health care fees in these countries could prevent the death of 233,000 children under five, at a cost of £1.75 per person per year [62]. Tony Blair, the UK Prime Minister at the time, used the research in the Labour Party Manifesto for the May 2005 General Election. He promised, if elected, to support 12 countries to move away from health user fees by 2015 [60]. The Make Poverty History Campaign galvanized a growing coalition of civil society advocates to call for user fee abolition and to promote more equitable access to health for the poorest. In countries like Liberia in West Africa, with UK aid support for free health care, child mortality rates were halved from 235 under 5-year-old deaths per 1000 live births in 1990 to 111 under 5-year-old deaths per 1000 live births in 2009 [50].

The Make Poverty History Campaign also succeeded in encouraging world leaders to promise to double aid to half poverty. In 2005, the WHO also invited partners to a consultation on the development of their new World Health Report: Making Every Mother and Child Count. The WHO Director General raised concerns that approximately 100,000 people were forced into poverty each year through paying for health [65]. The presentation was supported by academic and NGO research [31,42–50]. When the 2005 World Health Report was released, it advised that countries should return to the principles of Alma Ata and move away from user fees at point of service [65].

The importance of health systems, proper health financing, and trained supported health workers were also included in the report: cited as fundamental for achieving health-related MDGs and echoing findings from the Fustukian et al. research [65]. At the May 2005 World Health Assembly, 189 ministers of health agreed to 2 new resolutions which focused on strengthening maternal and child health systems, working with communities more, and moving towards pro-poor health financing systems [66,67]. However, most low-income and fragile countries did not have the resources or the capacity to move towards more equitable health financing mechanisms without external support [53].

Fragile states are those countries that will not, or cannot, provide basic services for their populations [68]. These countries are either too poor to invest in local services, or they are rebuilding their country following conflict or disaster. Some such leaders of fragile states may decide not to support health and social spending for political reasons, or sometimes as a tool in conflict. Although this renewed interest in global health led to new resources, much of these resources were spent through new global private-public partnerships, which again were often disease-specific and vertical in nature. This resulted in less than 3% of the new resources strengthening national health systems [69,70].

In 2010, there were 320 million children living in fragile states, but resources were not being invested in the countries with the greatest need. In 2009, donor aid for maternal and child health accounted for only 3% of international aid [71]. The impact of these internationally led vertical programmes (often delivered as disease-specific single intervention programmes, not integrated into routine health services), along with growing national debt and recession, resulted in the collapse of rural health systems in many low-income countries, especially those struggling with high levels of poverty, debt, and conflict [72,73].

Resources at that time were often not reaching the health systems, and the impact of the HIV pandemic was placing more strain on health and economic systems in the poorest countries, especially in Africa [31,70]. There was also a global shortage of health workers, with a minimum of 4 million extra needed [74,75]. In 2008, on the 30th anniversary of the Alma Ata Declaration, WHO released the World Health Report, focusing on the need to return to the principles of primary health care [76].

As anticipated by the Grow Up Free from Poverty Coalition research, 2015 arrived, and the health-related MDGs for maternal and child health were not achieved. Although the MDGs were not met, some progress was achieved, with 721 million fewer people living in poverty, 2.6 billion people having access to improved water, and 53% fewer children dying from preventable causes [77].

Sustainable development goals and the 2018 Astana declaration

In 2015, 17 new global targets were agreed upon—the Sustainable Development Goals (SDGs). Many of these goals are linked to health and nutrition, with SDG 2 focusing on food security and nutrition whilst SDG 3 focusing on improving global health. SDG 3 also includes a commitment to achieving Universal Health Coverage (UHC) by 2030. Many right-to-health advocates were concerned that the focus on UHC would halt the progressive realization of the right to health for all [34,78,79]. On the 40th anniversary of the Alma Ata Declaration in 2018, a global conference on PHC was held in Astana, and the Astana Declaration was signed. The final Astana Declaration included many of the PHC principles, including the right to health and the importance of reducing inequities [80].

However, Universal Coverage of Health Services do not fully address the social determinants of health or the full progressive achievement of the right to health as set out in the Alma Ata Declaration [78].

With under 10 years to go until 2030, the world is not on track to meet the SDGs for health and nutrition. Malnutrition levels are increasing globally, with 2 billion people who do not have access to essential health services when they need them [81]. There are still 5 million under 5-year-olds dying from preventable causes, so there is still a long way to go to achieve health for all [82]. Unless we learn lessons from the past 40 years and reorient global health systems in line with the Alma Ata principles, we may be reflecting on another failed set of targets in 2030.

The present COVID-19 pandemic has highlighted the devastating consequences of health service inequities throughout the world. Health systems, and economies, are stretched—even in high-income nations. We are thus at a pivotal moment in time. This is a good time to recommit to implementing the evidence-based Alma Ata principles of equity and voice, through an intersectoral approach, to ensure the progressive realization of the right to health for all [78]. To achieve this vision, we need to globally ensure:

(1) Political commitment to provide essential healthcare services free at the point of access, to promote equity and reduce delays in timely access of health care [78].

(2) Concurrent investment in sectors such as food security, nutrition, social protection, water, sanitation, and quality education (especially for girls and ensuring gender equality), with investment in essential community-based health services. These services could prevent 2.5 million under 5-year-old deaths annually [83].

(3) Investment in active community engagement, prioritizing the most vulnerable populations to reduce inequalities and increase timely health service utilization.

(4) Investment in global health governance systems, strengthening the World Health Organization and national health in all policies approach to governance [83].

Global health solidarity and governance must be invested in, and pledges need to be captured in laws such as a framework convention, to achieve the health and nutrition SDGs and the progression of the realization of health for all [57,78,83].

10.7 An evidence base

Everyone now knows the benefits of having an evidence-based data pool to inform decision-making: the better the information, the better the decisions [84]. For example, an informed government is better positioned to increase or decrease cordon sanitaire provisions for lockdown and reopening. Likewise, a more informed public is more likely to take appropriate steps to self-regulate, accept government decisions and comply with them to reduce the spread of disease. The only questions that may then remain are "Do you trust your government and your population not to abuse the data?" and "How can the security sector ensure that the data is not abused?"

The availability of public health information to responders, affected individuals, and decision-makers is of crucial importance and, in future, should be high on the agenda of policymakers in future epidemics [85]. Governments should consider enacting PHI policies that enable more effective epidemic control, thus relieving the public of unnecessary paranoia and reducing restrictions in non/less-affected areas.

10.8 Key points

- In Ireland, as in many other countries, protection of personal health information (e.g. COVID infection) was placed at a higher level of importance than infectious disease control.
- However, in future pandemics, greater transparency and information sharing, combined with geospatial maps of outbreaks, may result in much better containment efforts.
- In such cases, the need to protect individuals from being stigmatized because of their infection has to be balanced against the broader needs of society and public health.

References

[1] A. De Carli, M. Franco, A. Gassmann, C. Killer, B. Rodrigues, E. Scheid, D. Schoenbaechler, B. Stiller, WeTrace—a privacy-preserving mobile COVID-19 tracing approach and application, arXiv (2020). preprint arXiv:2004.08812.

[2] D.O. Argente, C.-T. Hsieh, M. Lee, The Cost of Privacy: Welfare Effects of the Disclosure of Covid-19 Cases, No. w27220, National Bureau of Economic Research, 2020.

[3] S. Kevany, D. Canyon, Urgent Policies Required to Grant Public Access to Protected Health Information During Emergency Disease Outbreaks and Pandemics, Daniel K. Inouye Asia-Pacific Center for Security Studies, 2020.

[4] WHO, Naming the Coronavirus Disease (COVID-19) and the Virus that Causes It, 2021. https://www.who.int/emergencies/diseases/novel-coronavirus-2019/technical-guidance/naming-the-coronavirus-disease-(covid-2019)-and-the-virus-that-causes-it. (Accessed 6 August 2021).

[5] Johns Hopkins University and Medicine, Johns Hopkins Coronavirus Resource Center, 2021. https://coronavirus.jhu.edu/.

[6] United Nations, COVID-19 to Slash Global Economic Output by $8.5 Trillion Over Next Two Years | United Nations. https://www.un.org/en/desa/covid-19-slash-global-economic-output-85-trillion-over-next-two-years. (Accessed 13 August 2021).

[7] S. Annaka, Political regime, data transparency, and COVID-19 death cases, SSM Popul. Health 15 (2021), https://doi.org/10.1016/j.ssmph.2021.100832, 100832.

[8] G.A. Kumar, L. Dandona, R. Dandona, Completeness of death registration in the Civil Registration System, India (2005 to 2015), Indian J. Med. Res. 149 (6) (2019) 740, https://doi.org/10.4103/ijmr.IJMR_1620_17.

[9] D. Cobos Muñoz, D. de Savigny, R. Sorchik, et al., Better data for better outcomes: the importance of process mapping and management in CRVS systems, BMC Med. 18 (1) (2020) 67, https://doi.org/10.1186/s12916-020-01522-z.

[10] K. Hill, A.D. Lopez, K. Shibuya, P. Jha, Interim measures for meeting needs for health sector data: births, deaths, and causes of death, Lancet 370 (9600) (2007) 1726–1735, https://doi.org/10.1016/S0140-6736(07)61309-9.

[11] B.K. Sartorius, K. Sartorius, Global infant mortality trends and attributable determinants—an ecological study using data from 192 countries for the period 1990–2011, Popul. Health Metrics 12 (1) (2014) 29, https://doi.org/10.1186/s12963-014-0029-6.

[12] J.B.R. Hernandez, P.Y. Kim, Epidemiology Morbidity And Mortality, StatPearls Publishing, 2020. https://www.ncbi.nlm.nih.gov/books/NBK547668/. (Accessed 27 August 2021).

[13] R. Gibson Parrish, Measuring population health outcomes, CDC Emerg. Infect. Dis. J. 7 (4) (2010). https://www.cdc.gov/pcd/issues/2010/jul/10_0005.htm.

[14] M.R. Benjamins, A. Silva, N.S. Saiyed, F.G. De Maio, Comparison of all-cause mortality rates and inequities between black and white populations across the 30 most populous US cities, JAMA Netw. Open 4 (1) (2021), https://doi.org/10.1001/jamanetworkopen.2020.32086, e2032086.

[15] S. Elezkurtaj, S. Greuel, J. Ihlow, et al., Causes of death and comorbidities in hospitalized patients with COVID-19, Sci. Rep. 11 (1) (2021) 4263, https://doi.org/10.1038/s41598-021-82862-5.

[16] National Center for Health Statistics, Excess Deaths Associated with COVID-19. COVID-19 Death Data and Resources, 2021. https://www.cdc.gov/nchs/nvss/vsrr/covid19/excess_deaths.htm.

[17] R. Baker, E. Sullivan, J. Camosso-Stefinovic, et al., Making use of mortality data to improve quality and safety in general practice: a review of current approaches, Qual. Saf. Health Care 16 (2) (2007) 84–89, https://doi.org/10.1136/qshc.2006.019885.

[18] UN, Index to the Proceedings of the Economic and Social Council—Forty-fourth Session, United Nations, 1968. E.68, I.20 https://www.un.org/sites/un2.un.org/files/1430-20140410140246505662_0.pdf.

[19] United Nations Fund for Population Activities, Actions for the Improvement of Civil Registration and Vital Statistics, 1983. https://www.cdc.gov/nchs/data/isp/027_action_for_the_improvement_of_civil_regist_and_vital_stat.pdf.

[20] C. AbouZahr, M.W. Bratschi, D.C. Muñoz, et al., How can we accelerate progress on civil registration and vital statistics? Bull. World Health Organ. 96 (4) (2018), https://doi.org/10.2471/BLT.18.211086. 226–226A.

[21] T. Ramaci, M. Barattucci, C. Ledda, V. Rapisarda, Social stigma during COVID-19 and its impact on HCWs outcomes, Sustainability 12 (9) (2020) 3834.

[22] T.R. Wojda, P.L. Valenza, K. Cornejo, T. McGinley, S.C. Galwankar, D. Kelkar, R.P. Sharpe, T.J. Papadimos, S.P. Stawicki, The Ebola outbreak of 2014-2015: from coordinated multilateral action to effective disease containment, vaccine development, and beyond, J. Global Infect. Dis. 7 (4) (2015) 127–138, https://doi.org/10.4103/0974-777X.170495.

[23] V. Vasudevan, A. Gnanasekaran, B. Bansal, C. Lahariya, G.G. Parameswaran, J. Zou, 100+ COVID-19 Websites and Apps: Yet a Lack of Granular Data From India, 2021.

[24] World Health Organization, Sustainable preparedness for health security and resilience: adopting a whole-of-society approach and breaking the "panic-then-forget" cycle, in: Meeting Report: Side Event, 75th Session of the United Nations General Assembly, 1 October 2020, 2020.

[25] F. Tagliabue, L. Galassi, P. Mariani, The "pandemic" of disinformation in COVID-19, SN Compr. Clin. Med. 2 (9) (2020) 1287–1289.

[26] T. Sharma, M. Bashir, Use of apps in the COVID-19 response and the loss of privacy protection, Nat. Med. 26 (8) (2020) 1165–1167.

[27] D. Werner, D. Saunders, Questioning the Solution: The Politics of Primary Health Care and Child Survival With an In-Depth Critique of Oral Rehydration Therapy, Healthwrights Publication USA, 1997.

[28] World Health Organization, Declaration of Alma Ata, in: International Conference on Primary Health Care. Alma-Ata, USSR World Health Organisation Sept 6–12, 1978, 1978.
[29] B. Scott, J.A. Halsted, K. Walsh, S. Warren, Good Health at Low-Cost, Rockefeller Foundation, New York, USA, 1985, pp. 42–43.
[30] S. Mehrota, C. Jolly, Development With a Human Face: Experiences in Social Achievement and Economic Growth, Oxford University Press, Oxford, 1998.
[31] S. Fustukian, R. Keith, A. Penrose, 80 Million Lives: Meeting the Health Related MDGs for Child and Maternal Survival, Save the Children UK and CAFOD for the Grow Up Free from Poverty Coalition, 2003.
[32] R. Keith, Special report: WHO should focus on improving health systems, Lancet 361 (2003) 5.
[33] D. Balabanova, M. McKee, A. Mills, Good Health at Low Cost 25 Years on: What Makes a Successful Health System?, London School of Hygiene and Tropical Medicine, 2011.
[34] F. Baum, The New Public Health Fourth Edition, Oxford University Press, Australia and New Zealand, 2015.
[35] J. Grant, Marketing Child Survival, Assignment Child 65/66 (1984) 3.
[36] G. Kent, The Politics of Children's Survival, Prager, New York, 1991.
[37] C. Schuftan, The child survival revolution: a critique, Fam. Pract. 7 (4) (1990) 329.
[38] Save the Children, Medact, The Bitterest Pill of all: The Collapse of Africa's Health Systems, Save the Children UK, 2001.
[39] United Nations, Millennium Declaration. General assembly, in: Resolution 55th Session Document A/RES/55/2. September 8th, 2000, United Nations, New York, USA, 2000.
[40] Save the Children UK, Poor in Health, Save the Children, 1996.
[41] S. Russell, Ability to pay for health care: concepts and evidence, Health Policy Plan. 11 (3) (1996) 219.
[42] N. Cadge, R. Keith, The Cost of Coping with Illness: Burundi, Briefing Paper Save the Children, Save the Children UK, 2005.
[43] N. Cadge, R. Keith, The Cost of Coping with Illness: Ethiopia, Briefing Paper Save the Children, Save the Children UK, 2005.
[44] N. Cadge, R. Keith, The Cost of Coping with Illness: Uganda, Save the Children, Briefing Paper, 2005.
[45] N. Cadge, R. Keith, The Cost of Coping with Illness: Sudan, Briefing Paper Save the Children, Save the Children UK, 2005.
[46] N. Cadge, R. Keith, The Cost of Coping with Illness: Rwanda, Briefing Paper Save the Children, Save the Children UK, 2000.
[47] N. Cadge, R. Keith, The Cost of Coping with Illness: Tanzania, Briefing Paper Save the Children, Save the Children UK, 2005.
[48] N. Cadge, R. Keith, The Cost of Coping with Illness in DRC, Briefing Paper Save the Children UK, Save the Children UK, 2005.
[49] N. Cadge, R. Keith, East and Central Africa Synthesis Briefing Paper: The Cost of Coping with Illness in ECA, Save the Children UK, 2005.
[50] N. Cadge, R. Keith, Policy Briefing: Fragile States, Save the Children UK, 2010.
[51] S. Witter, The Unbearable Cost of Illness: Poverty, Ill-Health, and Access to Healthcare Evidence From Lindi Rural District, Tanzania, Save the Children UK, London, 2005.
[52] S. Witter, Ability to Pay for Health Care in Khartoum State: Results of a Household Survey, Save the Children UK, Khartoum, 2005.
[53] S. Witter, R. Keith, An Unnecessary evil? User fees for health care in low-income countries, Make Poverty History, Save the Children UK, 2005.

[54] L. Gilson, S. Russell, K. Buse, The political economy of user fees with targeting: developing equitable health financing policies, J. Int. Dev. 7 (3) (1995) 69–401.
[55] A. Creese, J. Kutzin, Lessons from cost recovery in Health, in: Marketing Education and Health in Developing Countries, Claredon Press, 1996. Oxford Department for International Development, Fragile States Briefing Paper, 2005.
[56] L. Gilson, The lessons of user fee experience in Africa, Health Policy Plan. 12 (1997) 273–285.
[57] L. Gilson, A. Adyepong, Strengthening health system leadership for better governance: what does it take? Health Policy Plan. 33 (2008) 111–114.
[58] S. Russel, L. Gilson, User fee policies to promote health services access: a wolf in sheep's clothing, Int. J. Health Serv. 27 (2) (1997).
[59] S. Russell, K. Abdella, Too Poor to Be Sick: Coping with the Costs of Illness in East Hararghe, Ethiopia, Save the Children UK, London, 2002.
[60] R. Keith, P. Shackleton, Paying With Their Lives: The Cost of Illness for Children in Africa, Save the Children UK, 2006.
[61] D. Woodward, User Charges for Health Services in Developing Countries: An approach to Analysing the Effects on Utilisation and Health Outcomes, Overseas Development Agency, London, 1997.
[62] C. James, S. Morris, R. Keith, A. Taylor, Impact on child mortality of removing user fees: simulation model, BMJ 331 (2005) 747–749.
[63] C.D. James, K. Hanson, B. McPake, D. Balabanova, D. Gwatkin, I. Hopwood, C. Kirunga, R. Knippenberg, B. Meessen, S.S. Morris, A. Preker, Y. Souteyrand, A. Tibouti, P. Villeneuve, K. Xu, To retain or remove user fees?: reflections on the current debate in low- and middle-income countries, Appl. Health Econ. Health Policy 5 (3) (2006) 137–153.
[64] R. Rowden, The Ghosts of user fees past: exploring accountability for victims of a 30-year economic policy mistake, Health Hum. Rights 15 (1) (2013) E175–E185.
[65] World Health Organisation, Making Every Mother and Child Count: World Health Report 2005, World Health Organisation, Switzerland, 2005.
[66] World Health Assembly, Resolution 58.31: Making Every Mother and Child Count and Resolution Notes From the 58th World Health Assembly, World Health Organisation, Geneva, 2005.
[67] World Health Assembly, Resolution 58.33 Sustainable Health Financing, Universal Coverage and Social Health Insurance World Health Assembly, WHO, Geneva, Switzerland, 2005.
[68] R. Yates, International Experiences in Removing User Fees for Health Services: Implications for Mozambique, DFID Health Resource Centre, 2006. June 2006.
[69] A. Heaton, R. Keith, A Long Way to Go: A Critique of Global Alliance of Vaccines and Immunisations (GAVI)'s Initial Impact Save the Children UK Briefing Analysis Paper Based on New Products into Old Systems Research, 2002.
[70] M. Starling, R. Brugha, G. Walt, A. Heaton, R. Keith, New Products into Old Systems: The Initial Impact of the GAVI From a Country Perspective, LSHTM and Save the Children UK, 2002.
[71] R. Keith, N. Cadge, Unlocking Progress in Fragile States, Save the Children UK, 2010.
[72] Action Aid, Where is the Money? Towards Transparency in UK Aids Expenditure London UK, 2005.
[73] Action for Global Health, Healthy Aid: Why Europe Must Deliver Aid Better Spend to Save the Health MDGs, Action for Global Health, 2008.
[74] K. Blanchet, R. Keith, P. Shackleton, One Million More: Mobilising the African Diaspora Healthcare Professionals for Capacity Building in Africa, Save the Children UK, 2006.
[75] World Health Organisation, World Health Report: Working Together for Health, WHO, Geneva, 2006.
[76] WHO/UNICEF, Countdown to 2015-Decade Report (2000-2010) Taking Stock of Maternal Newborn and Child survival, World Health Organisation, Geneva, Switzerland, 2010.

[77] United Nations, The UN Sustainable Development Goals, United Nations, 2015. United Nations (UN), Sustainable Development Goals, United Nations (UN), New York, USA, 2020. UNICEF, For Every Child a Fair Chance: The Promise of Equity, UNICEF, New York, USA, 2015.
[78] R. Keith, The importance of the Alma Ata principles of equity and voice through intersectoral investment and leadership in achieving the health and nutrition sustainable development goals by 2030, World Nutr. J. 9 (2) (2018) 121–126.
[79] Peoples Health Movement, Final Call to Action, People's Health Movement, Cape Town, 2012.
[80] World Health Organisation, Astana Declaration from the International Conference on Primary Health Care, WHO, 2018.
[81] World Health Organisation, Global Nutrition Report, World Health Organisation, Geneva, 2020. Accessed: Home - Global Nutrition Report.
[82] World Health Organisation, Universal Health Coverage: Key Facts, WHO, Geneva, Switzerland, 2020.
[83] R. Keith, Improving Global Health Governance: Summary Paper for the WHA, World Vision International, UK, 2011.
[84] J.A.M. Gray, Evidence-Based Healthcare and Public Health: How to Make Decisions about Health Services and Public Health, Elsevier Health Sciences, 2009.
[85] H. Cho, D. Ippolito, Y.W. Yu, Contact tracing mobile apps for COVID-19: Privacy considerations and related trade-offs, arXiv (2020). preprint arXiv:2003.11511.

Chapter 11

Public access to public health information

Sebastian Kevany, Aoife Kirk, and Deon Canyon

Chapter outline

11.1 The primacy of privacy

Ireland made extraordinary advances on surveillance and data management to provide the general public with crucial information about the 2020–21 epidemic. However, whether or not many information elements were communicated correctly and were genuinely universal and user-friendly remains uncertain. Moreover, relying upon sensitive health information often requires careful navigation and a working understanding of epidemiology [1].

Yet, either way, one key piece of information continued to be hard to determine during the Irish response. For many, an ideal data environment

Outbreaks, Epidemics, and Health Security. https://doi.org/10.1016/B978-0-323-91705-6.00003-4

would have been a centralized location, one in which we would be able to see new cases as they emerged each day, by region or area [2]. In this way, if numbers were portrayed as more exact and arguably more relatable for the general public, they would have had a routine understanding of where outbreak hotspots were. But, unfortunately, this information too may have been biased because of testing centre locations, time lags, overlooking asymptomatic cases, or other reasons.

Often a typical response to epidemics relies on access (early and frequently) to public information and understanding of the general public. For example, in the time of scarlet fever, a red ribbon was placed outside the door of an infected house—warning others that current quarantine was in progress and not to approach, but if at all, with caution [3].

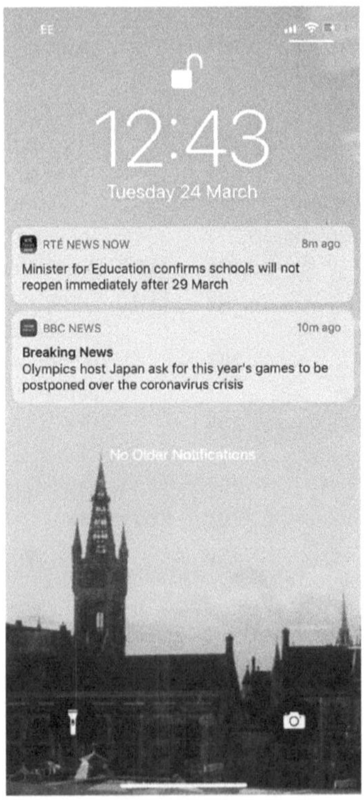

PICTURE 4.11.1 Did we know enough about what was happening in such a time of uncertainty? Eilis Conneally.

Likewise, Ebola outbreak control depended on a detailed public understanding of outbreak locations [4]. Yet, at the same time, in the modern developed world, we have become—generally speaking, rightly—more and more concerned with ensuring that health information is kept private. In this way, as with HIV/AIDS, it has become more challenging to stigmatize or discriminate against anyone because of their health condition(s). However, it also facilitates the tracking and diagnosis of reportable diseases.

11.2 Guest essay: The future: Pandemic information for the global community

Zachary Hitchcock

At the turn of the 21st century, the advent of the Internet and World Wide Web became accessible to many who could afford it. Fast forward to the present day, and we now have these tools in the palms of our hands, in the form of mini computers and in the guise of cellular phones. I can recall the early outbreaks of SARS, H1N1, Avian Influenza, and Swine Flu and having the ability to follow these outbreaks online throughout the 2000s.

Scientists declared it was just a matter of time before one of these viruses would develop the ability to infect humans through aerosol droplets and potentially create a global pandemic. I would frequent the World Health Organization (WHO) website in those early days and whilst some of the reports were easy to understand for me—because I have a background in science and emergency medicine—the common global citizen would have a difficult time making sense of these frequent epidemiological reports. Not least, because of the many languages, these reports would need to be coherently translated into to have universal appeal.

When the first reports came from Wuhan about a new strain of Coronavirus, I found the information to be muddled in ambiguities. Earlier reports, on SARS and other outbreaks, were much better presented on the WHO website. Presently, the media has taken over reporting from WHO officials—and unfortunately, this media coverage is often fraught with bias and partiality.

In this world, unbiased, impartial, and objective information is very hard to find.

Science and scientists try to bring us a foundation of their research and their theories—but they are reliant on remaining true to the unbiased, hard, scientific evidence. So how can we create a Global Information Source for information on which the Common Global Citizen can rely—a place to read, hear and see truthful reports connected to disease outbreaks and to educate ourselves about the possibility of a future pandemic?

Do we need a television network, like Bloomberg News or the Weather Channel, to give us this information through printouts; or do we need individuals that make this subject interesting to viewers who would benefit from the information? Who or what can bring this information in a parlance that can be translated to the global community?

It seems that the resources can be made available. Funding from governments, pharmaceutical companies, and many related healthcare industries, with the leadership of the WHO, could give us the ability to keep the global community up to date on any information pertaining to regional, cross-border, and global threats to our health. They can make it interesting, educational, entertaining, objective, and informative.

The average person needs to be able to grasp the potential for any threat to their community's health. We do it with the weather, and it is relatively objective: the main objective is to communicate truthfully to the world any threats and to help prepare us in the event of a hazard.

I relied on the WHO for information when it became more accessible through technology. I was a little dismayed at their responsiveness at the onset of this pandemic, but knew it was just a matter of time, so I read between the lines taking into consideration the location and political structures involved.

Now, it is time for the WHO to regain its leadership role—and partner to create a trusted information network to keep the global community up to date and informed through the technology all humans have easy access to. Make it simple and easy to understand, interpret, and take action if needed.

The world needs a vehicle to give us up-to-date, easy-to-understand information that anyone can interpret—and take necessary precautions related to the health of this planet's populations. I could imagine if something like this related to our obsession with pets, for example, it would be a huge hit. If only we regarded each other like we care for our furry, scaled, or feathered friends. How do we all become universally informed? We need another tool at our disposal.

Global health information that is both trusted and easy to understand, interpreted, and implemented if action needs to be taken. This requires effective and efficient virtual and actual communications platforms.

Accessible to everyone. We have the technology…

11.3 Geographical detail

This information sharing and guarding has been, under normal circumstances, a welcome and humanitarian development. But, understandably, the disclosure of street addresses, personal information, or other details of infected people in an outbreak situation is ruled out on privacy, confidentiality, stigma, and anonymity grounds. And yet, one wonders whether there was a threshold at which such ethical and legal ideals should have had to bow to public health needs.

There were other compromising options available. For example, Ireland has small area maps, called electoral or census divisions, used to display cumulative cases of infections since March. This data was publicly available, though officials often did not regularly update it. In a further precaution—one that further reduced resolution and usability of the information and improved data protection—cumulative cases were grouped according to a number range, for example, from 0 to 5, or 20 to 40.

11.4 Guest essay: The pandemic, black health, and systemic racism

Nichol Randall

The coronavirus: I first learned about it in 2020 whilst I was on my morning walk, listening to KQED, a local news programme in the San Francisco Bay Area. The report originated from Wuhan, China, which is a world away from the United States, so I reassured myself that I was safe and that a new virus from China would not make its way to the Bay Area. Shortly after the initial report, I heard more news about the coronavirus, but this time, it seemed more relevant and real. The first case surfaced in Washington State. It was mysterious. My family and colleagues were unsure what to make of the situation, so we paid close attention to messages on the news, statements from public health officials, and reports from epidemiologists and medical professionals with whom we work closely.

Another early event that solidified the pandemic's arrival in the United States was the stranding of the Grand Princess cruise liner in the San Francisco Bay after staff members and passengers tested positive for COVID-19. Thousands of passengers were offloaded at the Port of Oakland, with port employees wearing protective gear to prevent the spread of the virus in the Bay Area.

The world was forced to shut down in the first quarter of 2020 as COVID-19 became a threat to human life and civil society. As positivity rates and deaths began to rise, it was clear to me that there were healthcare injustices in the Black community and significant racial disparities in health outcomes amongst Black patients suffering from COVID-19 in comparison with their white counterparts.

There are notable health disparities amongst Black and Minority communities throughout the United States which have become all the more visible during the pandemic. Since Black people make up the majority of essential labour, they are more likely to contract the virus. They are also more likely to die from COVID-19 as a result of a higher prevalence amongst them of underlying illnesses like diabetes, cardiovascular disease, asthma, and other lung ailments [5]. What causes these underlying illnesses and comorbidities in Black people?

Consider the environment: Black communities are overburdened by exposure to air pollution. For example, in West Oakland, California, Black people make up the majority of those who reside near freeways that are traversed by large trucks running on diesel fuel. By California state law, trucks are routed on certain freeways that happen to be where Black people live [6].

Consider housing: Due to housing discrimination and the repercussions of redlining legislation,[a] Black individuals are more likely to live in crowded dwellings. When one lives in such close quarters with individuals from the same family, it is difficult to maintain social distance.

---

[a] Redlining is a process by which banks and other institutions put their services, such as mortgages, out of the reach of residents of certain neighbourhoods based on their race and ethnicity.

Consider essential workers: Black people work in grocery stores, transportation industries, and lower end hospital jobs that all require one to work in-person. These types of jobs put one at higher risks for contracting COVID-19.

Consider access to health insurance: Black people are more likely to be without insurance or with inadequate insurance. This is amongst the reasons that chronic conditions are not properly managed, and, when a comorbid condition exists, one is more likely to become sicker and die from COVID-19 [7].

The goal is to now think of ways to prepare for future pandemics and determine how racial disparities will be mitigated and resolved. To promote a healthier Black community, public health practitioners need to push for better health care, housing opportunities and legislation, and better job opportunities. As a result, when pandemics like COVID-19 emerge, we would not see such health disparities, with Black people becoming disproportionately sick and dying at a far greater rate than people in other communities. These socio-economic issues must be addressed, and new policies must be developed to improve the health of Black people in the United States and around the world.

What can health professionals around the world do?

Health professionals can have conversations not just about the social determinants of health, but structural and systemic racism, and other types of alienation and disenfranchisement that has had a direct impact over time on the health of our communities of colour. They can define what the issues are around COVID-19 for Black communities and think about the types of interventions that need to be implemented and sustained to protect Black lives from future pandemics.

11.5 A question of trust

Yet to see a cumulative map, the general public was often to see a de facto representation of a regional situation in March or April of 2020—which may have no connection at all to what was happening later in the year in a particular area on a day-to-day basis. Moreover, the cumulative figures may have given a wrong impression of places that may, ironically, now be the safest in the country, as the epidemic had already taken its toll and worked through them. Such information, therefore, needs to be interpreted with caution.

The government and other state bodies provided a range of reasons why it is not safe to trust the general public with any more detailed information [8]. First, as noted above, asymptomatic cases were invisible [9]; there may have been biases in areas with more testing centres. The tests were not accurate; people are given a false sense of security if they live in low-incidence places. There were also time lags in testing results, so the figures given to the public on a particular day do not necessarily reflect the day before. Finally, and perhaps above all, social unrest may have resulted from specific communities revealed as high-incidence areas [10].

11.6 Guest essay: Lessons from a Yogi's perspective

Marisa Dallabetta

The following are simple opinions that come from my field of experience as a PE and a yoga teacher.

A lesson from the pandemic, which has been upsetting people's lives for months, is a great opportunity for all of us to know better our bodies, minds, and brains, and their functioning, overall for the richest countries that have more possibility to choose and study in life. Nowadays, statistics are showing us how deadly this virus is, the overage of age and illnesses that people have before catching it. Fortunately, the vast majority of them recover from it, but at the same time, it underlines the importance of staying healthy for a better result.

At the beginning of the pandemic and during the first lockdown, in Italy, we could only exercise within 200 m of our houses, the media did not speak about how to boost the immune system with the right habits of life, healthy food, physical exercise, meditation for the mind to control fears and dejection, and avoiding alcohol and cigarettes, but only showed numbers of infected and victims. Many people and my students too were scared from all the news and uncertainty. Luckily, in the following lockdown, we could train inside the municipal district, resulting in less of a negative impact on mental health. Few people are trained to keep the mind quiet and clear.

First of all, I would like to underline that our wonderful body with its special brain and mind is our first home, where we go to sleep is the second house, and it is for that reason we should, or rather, it is our duty to take care and keep fit with healthy food and exercise almost daily. It seems that it is easier to have addictions to alcohol, drugs, cigarettes, junk food, and laziness than to tackle the situation to remain healthy with a lifestyle well-suited to our body. Yoga teaches us how to keep our body fit, a stable mind, a distance from every thought and giving a gap between an event and our personal reaction.

The data of this virus and its variants show us that healthy people can be infected but rarely die. This could be a good reason to start a different kind of lifestyle, with more awareness of our precious but underestimated body.

Secondly, we have to consider the environment. A small percentage of human beings are destroying the natural environment forgetting that we are connected with it. The air that we breathe, our first energy, is polluted by industries, and intensive farms, fruits, and vegetables are polluted too by pesticides as well as the meat and fish industry with hormones and antibiotics. All of this is damaging our lungs, cells, and life, and the body in the long run becomes weaker. In the meantime, thanks are due to scientists and researchers that have discovered many drugs that may prevent or treat disease and mental illnesses. But we have to remember that our special body is also trained to help us with antibodies and an excessive use of medicines does not allow it to work at best.

On the planet Earth, animals, bacteria, viruses, plants, and so on are necessary for a good stay, but in the last century, human beings are focused on producing many worthless material things, consuming much more than what they actually serve, most of them harmful to poor countries and the environment. This gives rise to the question: Does the worst virus come from wildlife or human beings?

Too much money for researching new kinds of weapons! Why?

This is the same question that many people ask themselves. Where is the compassion for the poor, in rich and poor countries as well?

We can see now with the percentage of vaccinated people in developed and poor countries.

Will the richest nations be able to overcome the avidity and prevarication over poor people and wildlife?

This virus has changed the habits of all countries. Maybe now, it is time to think not only for ourselves but also for the environment. If we continue to exploit the planet's resources for the mere need for personal enrichment, it means that it is not yet clear the real reason for our brief stay on this wonderful planet. In my personal opinion and experience, yoga practice can improve the awareness of ourselves and the necessity of others, to share love, compassion, and peace, going beyond personal selfishness.

It is high time to take action! We can no longer put off all these problems, every single person can start to take care of himself, people, animals, and plants and then with a wise ruling class that really want to cooperate for the well-being and the unity of the entire world to reach the same aim: the happiness of people and animals and to preserve the beauty of the planet Earth; meanwhile, we have to remember that since we were born, life has walked together with death as day and night.

11.7 Information enhances responses

Yet, with all of these limitations and qualifications, public understanding of the distribution across (and, much more importantly, within) counties of new cases each day may have been the only verified information that we had to guide our response, individually and nationally. Whilst public health information's ethical and legal tenets are lauded, there is also a need for the perfect not to be the enemy of the good: there was, perhaps, a need for decision-makers to acknowledge that we were not living in normal times, whereby standard rules applied.

The benefits of having all of the factual information—however numerous the limitations—about the epidemic's geographical distribution being made available to the general public may thus have vastly outweighed the downsides. But, unfortunately, the public's alternative relied on tabloids and rumours to give us news of outbreaks—which cannot possibly be a preferable option [11].

11.8 Guest essay: Oceans and planetary health

Rita Issa

When I was approached to contribute to this book, I had just returned to land after a 3-month mission aboard the Greenpeace Boat, the Arctic Sunrise. With a world that was battling rising COVID-19 cases and the emergence of new variants, my time on the ocean waves—and the insights I learnt from communities and ecosystems along the way—provided a particularly pertinent perspective for how to tackle not only future pandemics, but the other global health threats we face.

COVID-19 emerged—we suspect—as a product of a nonharmony between humans and 'the natural world', and served as a frank reminder of the interdependent relationship between humans and the environment that was somewhat forgotten in cities and behind smartphones. On the ship however—at times weeks from land in any direction, at others visiting low-lying islands—it was difficult to not be confronted with how fragile, small, and dependent humans are on the natural systems within which we are enmeshed.

Much of what we saw in the Indian Ocean was symptomatic of the broken relationship between humans and the natural world: plastic waste, industrial fishing practices, coastal communities under threat, and ecosystems in decline. However, the journey consolidated my unwavering belief that humans are both ingenious and resilient, particularly when under pressure. During 2020, humans achieved what was previously unthinkable—vaccine development to delivery in the space of a matter of months. So too, communities living in close relationship to oceans are adapting, mitigating, and rebuilding their relationship to their ecosystems. In such communities on the frontlines is wisdom that will be necessary for preventing and tackling emergent and future global health threats—from pandemics to climate change.

Below are excerpts from my writing whilst at sea, and the lessons and reflections gathered from the oceans' frontlines.

28th March—Embarking

A couple of weeks ago, I received an email from Greenpeace International asking if I could join the ship Arctic Sunrise as onboard medic. The 2 days that I had to accept the post had me experiencing a range of emotions that had been largely suppressed during a year of pandemic lockdown: excitement and anticipation, to deep-belly anxiety running through the worst-case scenarios that I might manage as a solo-doctor for a crew isolated on the high seas. The ship was in the Indian Ocean documenting and exposing the threats our oceans face. This area is home to incredible wildlife and ecosystems—from pygmy blue whales and dugongs to colourful coral reefs and the largest seagrass meadow in the world. Such ecosystems play an important role in supporting local communities and in our global fight against the climate crisis, but they are threatened by industrial fishing, and less than 1% of global oceans are effectively protected from destructive fisheries and other harm. We were heading to an area uniquely rich in life, to witness and highlight the failures in safeguarding our oceans.

1st April—The Saya de Malha Bank

A few days into our trip, we arrived at our first destination, on the surface indistinguishable from the miles of ocean we had covered. The Saya de Malha Bank is an underwater plateau the size of Switzerland, and the world's largest seagrass meadow. Most seagrasses fringe coastlines, but the shallow depth of Saya de Malha allows sunlight to reach the seabed, and in doing so, the meadows here provide shelter and feeding ground for the rich marine life of the region.

Seagrass meadows account for less than 0.2% of the world's oceans but sequester approximately 10% of the carbon buried in ocean sediment annually. The United Kingdom's Office for National Statistics has *calculated* that seagrasses, muds, sands, and saltmarshes already capture at least 10.5 million tonnes of carbon dioxide equivalent a year in the United Kingdom alone: greater than Britain's woodlands. In addition to storing over twice the amount of carbon per square mile as land forests, seagrasses also help reduce the acidity of ocean waters. The process of ocean acidification has been driven by increasing levels of carbon dioxide in the atmosphere that in turn is absorbed into oceans. In the past 200 years, oceans have taken on *a third of* the CO_2 produced by human activity, and the vast majority of the extra heat caused by emissions. This comes at a cost to marine and human well-being. Increased ocean acidity makes it *more difficult* for certain organisms, such as corals and shellfish, to build their skeletons and shells. Rising ocean temperature leads to increased ocean stratification, changes in ocean currents, sea-level rise, and extreme weather events and impacts ocean ecosystems. Oceans are the world's largest store of carbon—a process that has been millenia in the making. During the Devonian period, 300 to 400 million years ago, plants grew rapidly and, in doing so, drew down carbon from the atmosphere, which was then buried in death as soils and sediments. In a complex process involving rock minerals, plants, fungi, and carbon dioxide, marine organisms were able to make shells, piling up over millions of years on the ocean floor as an 'enormous burial ground for carbon'. This long, slow process reduced carbon dioxide levels in the atmosphere and led to a period of global cooling. What took millenia to achieve has been jeopardized within a century.

14th April—Industrial fishing

Industrial fishing practices—widely practiced across the world—are a threat to the seafloor and seagrass meadows like Saya de Malha, marine diversity, and coastal communities. Bottom-trawling—a widely adopted fishing practice that drags heavy nets along the seafloor—releases as much carbon dioxide as the aviation industry. We came across longliners, a type of commercial fishing vessel, where a 'long line'—sometimes up to 100 km—is laid with incremental baited hooks placed in estuaries off its length. At times, before even catching sight of the vessel, we spotted buoys in the water, keeping the fishing line afloat. Longliners are effective killing machines. Along with their target catch of tuna, they bring in a high number of already threatened sharks as 'bycatch'. Approximately, 35% of longline catch is nontarget species, and many vessels engage in illegal activities such as shark-finning, with impacts that ripple through the whole ecosystem. Sharks are a species vulnerable to overfishing because they can take between 10 and 15 years to reach maturity and may bear just one pup a year. As *top* predators, they play an important role in maintaining the balance of the ecosystem, taking out larger predatory fish, and allowing herbivore fish species to maintain a balance between macroalgae and corals.

30th April—Plastics

A few weeks into our travels, we passed through an exquisitely still stretch of ocean known to sailors as the doldrums. Around the solar equator, this region is notoriously absent of currents and wind, and historical accounts tell of sailboats stuck for weeks in the vacuum. As we transited northwards into areas of currents and wind, I noticed a significant increase in the amount of plastic floating on the water. We were still far from land, and the sight of it felt jarring against the pristine blue: a reminder that human impact is extensive and few places—even the far reaches of the Indian Ocean—remain untouched. Plastic rubbish like this enters the seas through a number of mechanisms. About 80% originates on land and runs into the sea mostly as a result of poor waste management and driven in no small part by single-use items such as water bottles, plastic bags, and food containers. Discarded nets and lines known as ghost gear continue to haunt fish in the ocean and make up some 10% of all ocean plastics. The problem is extensive: the equivalent of a garbage-truck of plastic is dumped in the sea every minute, and the most dire projections suggest that our seas may contain more weight in plastics than fish by the year 2050.

Plastic does not decompose, both a design feature that drives its extensive use and a major flaw for environmental stewardship. According to Greenpeace, every piece of plastic ever made still exists, and many make it into our waters. In our relationship to plastics, like much of our relationship to natural resources, the approach of extractive capital has been to act as if we live in an open system of limitless supply with an endless capacity to deal with waste. Clearly, that is not the case. Our planet mirrors the finely tuned balance of our bodies—with feedback loops and homeostasis attempting to maintain balance. But we are consuming at too high a rate for the planetary limits we live within, and in our small, cushioned world, waste products accumulate and knock our planet into a critical state.

4th May—Low-lying islands states

Our first sighting of land after a few weeks aboard was a welcome change from the unbroken horizon of the Indian Ocean. The Republic of Maldives was first visible as if a mirage; as we got closer, it was clear that the islands were flush to the ocean without hills or mountains—unsurprising, as the flattest country on earth—and for many of the islands, the tallest points were the tips of palm trees, holiday cabins, or a handful of high rises on the capital Malé.

Over 80% of the Maldives' 1000+ islands are less than 1 m above sea level. The Intergovernmental Panel on Climate Change has predicted that the seas will rise by up to 1.1 m by the year 2100 meaning that the very existence of these islands is threatened. Rising sea level also interacts with other climate factors such as more intense storms, leading to eroding coastlines, reduced freshwater supply, and loss of habitats. The inhabitants of the Maldives are of course acutely aware of this risk, with ex-President Mohammed Waheed Hassan stating "we are one of the most vulnerable countries on Earth." There are two approaches to addressing the impacts of the climate crisis that go hand in hand: mitigation and adaptation. Mitigation is acting now to limit global heating and its impacts. For countries like the Maldives who do very little to contribute to the climate crisis, there is limited scope for mitigation at the country level. Instead, in the interconnected systems of our world which do not respect country borders or the nation state, there is a global responsibility to act. Actions which reduce the heating (and therefore the expansion of water) of the oceans, and melting ice sheets and glaciers, could limit sea-level rise to just 30 cm under best-case scenarios. This requires concerted, coordinated, and committed action by nations who are driving these processes through excessive use of fossil fuels: action which is currently lacking.

As climate breakdown is already impacting communities and ecosystems across the planet, adjusting to current and expected risk is necessary: this is adaptation. Since the late 1990s, the Maldives have been pumping sand from the seafloor to create the artificial island of Hulhumalé, in response to the threat posed by the climate crisis. Currently, around 50,000 people live on the island, with ambitions to house a quarter of a million by the mid-2020s. Climate-resilient design has been incorporated from the start, including wider streets to increase wind flow and reduce air-conditioning use, electric buses, and bike lanes and urban planning which reduces car use by placing schools and parks in close proximity to homes, a significant proportion of which are social housing. The city also aims to utilize extensive solar power and to harvest rainwater.

Faced with social, ecological, political, or economic pressure, migration is also a reasonable adaptive response to the climate crisis. Projections suggest between 25 million and 1 billion environmental migrants will move within and across borders by 2050, the majority of whom will be within Latin America, the Caribbean, South Asia, and Sub-Saharan Africa. Those who cannot move away from an area of climate risk—known as 'trapped populations'—remain highly vulnerable to the direct and indirect impacts of the climate crisis, such as extreme weather events, health risks, and food insecurity.

Thinking global, acting local

You might wonder why a text on global health is discussing oceans and ecology. The climate crisis has been identified as the '*greatest threat* to global health of the 21st century' of which our oceans play a great part: much of this is driven by complex, intersecting systems which have been exposed by the pandemic to be fragile and precarious. When facing such enormous challenges, it can be difficult to know where to start or how to adapt. The oceans reminded me of the adage to 'think global, act local'. My time aboard Arctic Sunrise reminded me that solutions for global issues cannot just come from the top. Innovation and resilience is drawn from the grassroots and frontlines, in the way that we saw when communities and hospitals rallied together to deliver mutual aid, PPE, and urgent medical care in the early days of the pandemic. To tackle future crises, it is crucial that top–down and bottom–up be resourced and respected in tandem. To mitigate effectively—be it against pandemics or climate change—we need coordinated, decisive action to be taken by all nations. But to learn to adapt, we will all do well to look to island nations whose longevity is so closely tied to ocean health, many of whom are already being smart and taking innovative approaches in order to not just survive, but flourish.

PICTURES 4.11.2 AND 4.11.3 **Beach clean before and after.** Rita Issa.

11.9 A question of disclosure

Thus, if the government can adapt its data-sharing rules to share surveillance information back to the public, we may have a much stronger chance of controlling future outbreaks. A further consideration is how this information would have significantly strengthened the national contact tracing app. For example, recent cases in one's immediate locale, anywhere in the country, could have been displayed at the tap of a screen.

In the long term, this would have also fed into a regional lockdown approach that involved avoiding the entire country grinding to a halt for what may often have been highly localized issues [12]. What if all new cases occurred in single electoral divisions on a specific day instead of equally spreading throughout the country? Therefore, disclosing this information would also have facilitated public understanding of why services in one part of the country could continue to operate as usual whilst others were closed down. Meanwhile, 'self-enforced contact tracing', in which those in outbreak areas become more vigilant and mindful of their movements, would also have been greatly facilitated.

The alternative was not pleasant and may not have been worth the risk: a second wave situation under a surveillance system that did not share all meaningful information with the general public. It is essential that funders—whether private or public—should create economic incentives for the early reporting of outbreaks, by committing to disburse emergency funds rapidly to assist countries, and perhaps even compensating for economic losses that might result [13]. Yet, as we saw earlier in 2020, China's reticence could have perhaps avoided the severe public health repercussions. Ireland must continue to learn, and act on, these vital lessons [14].

11.10 Guest essay: Water distributions and withdrawals: The link to pandemics

Enku Kebede-Francis

There were reasons why the cradles of civilization emerged around water sources. Now, a great majority of the world's population enjoy access to safe water and sanitation, which in turn has led to an increase in global life expectancy. According to the WHO, the average life expectancy grew by an unprecedented 5 years since the year 2000. Looking at the pandemic another way, the coronavirus which first appeared in 2019 (COVID-19) would have caused similar harm to the 1918 pandemic if 71% of the world's population did not have access to safe water and sanitation. Since the turn of this century, according to World Bank data, mortality rates due to unsafe water, sanitation, and lack of hygiene have decreased, leaving only a handful of countries susceptible.

At the turn of the last century, the average life expectancy hovered around 35 and the world's population was approximately 2 billion. Amongst the diseases that afflicted the world were plague, cholera, yellow fever, smallpox, typhus, poliomyelitis, syphilis, encephalitis, epidemic la grippe, typhoid, and paratyphoid fevers. Those were years of pestilence, but they also gave us the Pan American Sanitary Code of 14 November 1924, signed in Havana, Cuba. Amongst the protocols established, Article 5 is still valid and is being used by WHO and other agencies:

The Pan American Sanitary Code—Article 5

This notification is to be accompanied, or very promptly followed, by the following additional information:

(1) The area where the disease has appeared.
(2) The date of its appearance, its origin, and its form.
(3) The probable source or country from which it was introduced and the manner of introduction.
(4) The number of confirmed cases and number of deaths.
(5) The number of suspected cases and deaths.
(6) In addition, for plague, the existence amongst rodents of plague, or of an unusual mortality amongst rodents; for yellow fever, the Aedes aegypti index of the locality.
(7) The measures which have been applied for the prevention of the spread of the disease and its eradication.

It is this protocol that established health as a global common and established the need for surveillance and standardized guidelines in the management and prevention of diseases including pandemics. The process of notification for eradications of diseases and the appearance of new diseases later became the responsibility of the WHO, which has regional offices that also include the Pan American Health Organization (PAHO), which existed before 1945.

The achievements of PAHO are due to the fact that the organization had established rolling funds for vaccines, a willingness of member nations to work together, and the understanding that water is the lifeline and the centre of all development. The PAHO region is also a microcosm of the global community as members include small island nations surrounded by salt water and deserts, and low-, middle, and high income countries. Nevertheless, average life expectancy in the regions is above the world average. This can only be attributed to the Pan American Sanitary Code and the commitment of member nations, which has been replicated globally.

When the Pan American Sanitary Code was established, experts knew that water was a closed cycle and a unique natural resource. They knew that the amount of global water cannot be altered, which means we always have the same amount of water either in liquid, vapour and/or solid forms. In other words, the amount of water is the same regardless of how and where it was used.

We have been referring to water usage, but in fact we are only withdrawing water. According to the World Bank data, in 2017, the global population collectively withdrew 3,881,000,000 m^3 of water and the same amount was returned. This is similar to leasing but not purchasing natural resources, resources that belong to all and must be shared. It is also important to note that regardless of the amount of water we use annually we do not waste water. Further analysis shows that countries where per capita water withdrawals are higher and mortality rates are lower. It is also true that where less water is withdrawn, mortality rates due to unsafe water were higher. In other words, in high-income countries the mortality rate due to unsafe water in 2017 was 1% or less—but where less water was withdrawn, in the same year, in poorer countries, mortality rate due to unsafe water and sanitation was as high as 80%. It is also important to note that all governments in this group have already implemented programmes to expand water distribution in their respective countries. If this trend continues, we can expect access to clean water to become the barometer by which we judge government policies.

It is not possible for the private sector to regulate natural resources, such responsibility needs to rest with governments. Although the private sector has been allowed to manage water distribution, it is only through government regulation that safety and access can be guaranteed. It is, therefore, safe to conclude that where good governance in water management is practiced, mortality due to lack of access to safe water is low. Conversely, where there is an absence of good governance, more people suffer from preventable morbidity and mortality.

Human history has been punctuated by major changes and transformations. From time to time, the global community has been tested but has pivoted to survive. Humans moved successfully from hunter-gatherer, to agrarian, to present-day services and industry-based economies. COVID-19 once again alerted the world community to the fact that pandemics are global problems; regardless of where they originate, they have the capacity to impact the entire world.

The shrinking category of extremely impoverished countries, however, should give the world a sigh of relief. It means some of the hard work has come to fruition and the credits go to:

- The 1978 Alma Ata Declaration that advocated primary health care for all and the eradication of smallpox;
- The 1945 UNESCO's Declaration of primary education for all;
- The UNICEF's Water, Sanitation and Hygiene programme that (1) promoted access to safe and affordable water sources at reasonable distance from where people live and also (2) worked with governments to highlight the dangers of human-made pollution that contaminates drinking water sources; as well as (3) the commitment to childhood vaccines;

- The efforts of national governments to provide access to clean water to their respective populations that included the establishment of 1924 Pan American Sanitation Code; and
- OECD member countries, for not only completing their development successfully but also for their contributions to build infrastructure elsewhere and for sharing their skills and knowledge.

Water used to be attached to privilege, but a long time ago governments and citizens realized universal access to safe drinking water was a right, and not a privilege. These citizens knew that once these infrastructures were built, they are permanent, which is why so many countries now provide 90%–100% access to clean water to their citizens. However, even the most advanced countries must continue maintaining and updating this infrastructure, which makes water management programmes perpetual. Most importantly, safe distribution must be closely monitored and regulated by governments.

Acknowledging the success made to date and learning from those countries that have invested in permanent infrastructure for water and sanitation, a handful of countries must make the necessary adjustments. It is not possible for citizens to provide clean water for themselves, and it is not possible for outside agencies to regulate water safety. Providing universal access to clean water is the responsibility of all governments. This includes establishing proper regulatory processes extending access to water and sanitation for all. Indeed, many governments have given access to clean water priorities. The six countries listed provide universal access to clean water whilst life expectancy remained low.

Why do we need uninterrupted clean water and improved sanitation in the pandemic prevention context? According to WHO, waterborne diseases are highly infectious, instant, and severe. For example, cholera, malaria, and trachoma still impact several countries. Trachoma causes blindness, malaria causes chronic health conditions, and cholera kills most of its victims.

Governments also understand the importance of having clean water, but the success we have achieved could be undone if portions of the population are deprived of safe water and sanitation. There is currently no shortage of water, but closing the equitable distributions and withdrawal gaps that are causing morbidity and mortality for approximately 12% of the global population will help curtail fast-spreading infectious diseases in future.

11.11 Key points

- In pandemic situations, there may be a need to overrule conventional personal data protection laws in order to control outbreaks.
- In turn, there is a need to anticipate such situations with associated emergency legislation that can be feasibly and rapidly enacted.
- The counterpoints to greater sharing of personal information in cyberspace are associated with concessions about sharing personal health status.

References

[1] A. Zwitter, O.J. Gstrein, Big data, privacy and COVID-19–learning from humanitarian expertise in data protection, Int. J. Humanitarian Action (2020) 1–7.

[2] K.E. Jones, N.G. Patel, M.A. Levy, A. Storeygard, D. Balk, J.L. Gittleman, P. Daszak, Global trends in emerging infectious diseases, Nature 451 (7181) (2008) 990–993.

[3] G.F. Dick, G.H. Dick, The etiology of scarlet fever, J. Am. Med. Assoc. 82 (4) (1924) 301–302.

[4] P. Vinck, P.N. Pham, K.K. Bindu, J. Bedford, E.J. Nilles, Institutional trust and misinformation in the response to the 2018–19 Ebola outbreak in North Kivu, DR Congo: a population-based survey, Lancet Infect. Dis. 19 (5) (2019) 529–536.

[5] M. Vasquez Reyes, The disproportional impact of COVID-19 on African Americans, Health Hum. Rights 22 (2) (2020) 299–307.

[6] P.A. Gonzalez, M. Minkler, A.P. Garcia, M. Gordon, C. Garzón, M. Palaniappan, et al., Community-based participatory research and policy advocacy to reduce diesel exposure in West Oakland, California, Am. J. Public Health 101 (S1) (2011) S166–S175.

[7] G.A. Millett, A.T. Jones, D. Benkeser, S. Baral, L. Mercer, C. Beyrer, et al., Assessing differential impacts of COVID-19 on black communities, Ann. Epidemiol. 47 (2020) 37–44.

[8] P. Hyland, M. Shevlin, O. McBride, J. Murphy, T. Karatzias, R.P. Bentall, A. Martinez, F. Vallières, Anxiety and depression in the Republic of Ireland during the COVID-19 pandemic, Acta Psychiatr. Scand. 142 (3) (2020) 249–256.

[9] L.A. Nikolai, C.G. Meyer, P.G. Kremsner, T.P. Velavan, Asymptomatic SARS coronavirus 2 infection: invisible yet invincible, Int. J. Infect. Dis. 100 (2020) 112–116.

[10] C. Carroll, R. Conway, D. O'Donnell, C. Norton, E. Hogan, M. Browne, C.M. Buckley, P. Kavanagh, J. Martin, S. Doyle, Routine testing of close contacts of confirmed COVID-19 cases–National COVID-19 Contact Management Programme, Ireland, May to August 2020, Public Health 190 (2021) 147–151.

[11] M. Salathé, S. Khandelwal, Assessing vaccination sentiments with online social media: implications for infectious disease dynamics and control, PLoS Comput. Biol. 7 (10) (2011), e1002199.

[12] D.O. Argente, C.-T. Hsieh, M. Lee, The Cost of Privacy: Welfare Effects of the Disclosure of Covid-19 Cases, No. w27220, National Bureau of Economic Research, 2020.

[13] T.R. Wojda, P.L. Valenza, K. Cornejo, T. McGinley, S.C. Galwankar, D. Kelkar, R.P. Sharpe, T.J. Papadimos, S.P. Stawicki, The Ebola outbreak of 2014-2015: from coordinated multilateral action to effective disease containment, vaccine development, and beyond, J. Global Infect. Dis. 7 (4) (2015) 127–138, https://doi.org/10.4103/0974-777X.170495.

[14] B. Salzberger, T. Glück, B. Ehrenstein, Successful containment of COVID-19: the WHO-Report on the COVID-19 outbreak in China, Infection 48 (2020) 151–153.

Chapter 12

Health security and strengthening health systems

Sebastian Kevany and Aoife Kirk

Chapter outline

12.1 The need for strong public health systems

Ireland's successes in epidemic control should not be understated: the shape of the country's mortality curve in 2020–21 without interventions, compared to what has been achieved so far, would make for exciting viewing, were it possible to hypothesize such a counterfactual. Nonetheless, perhaps there are a series of lessons learned from Ireland and other countries' mistakes—if they could be called that—in response to the epidemic threat. Like almost every other initially affected country (and far less than some), Ireland was unprepared in critical ways from the onset of the pandemic.

Yet, in future, such situations will inevitably occur under different circumstances; the world has changed dramatically since February 2020, in an 'unprecedentedly' short space of time. Thus, what countries 'should' have done then is no longer relevant. Instead, any nation's future health security may be determined by applying and implementing a separate set of significantly more streamlined, controlled, and focused interventions. These aim to ensure all the hard work (on both national and individual levels) so far is not undone.

The future health security of Ireland, and other island states, is our focus in this chapter. As we begin to look through a 'post-COVID-19' lens towards rebuilding an economy—perhaps more focused on health and well-being, or

Outbreaks, Epidemics, and Health Security. https://doi.org/10.1016/B978-0-323-91705-6.00009-5

maybe not—it is essential to realize that the threat of further strains, pathogens, and infectious disease burden looms in the near future. So how can we build up infrastructure to maintain health—including primary care, secondary care, and tertiary care? How can we ensure supply chains, procurement, and distribution—systems and supply chains which have become globalized over the past 50 years—are in place to provide necessary protection for frontline workers without a last-minute scramble?

Unfortunately, or fortunately, it is not just about stockpiling: Ireland, as with much of the Global North, was caught out in terms of medical supply preparedness, and there was no way that this particular manifestation of an epidemic could have been realistically predicted in the public health procurement context. Future epidemics may also not necessarily be pulmonary or respiratory, so stockpiling is vital for these events (face masks, oxygen supplies, and ICU preparedness) and the rapid conversion of other facilities to medical uses. Unfortunately, such strategies may not in future be enough in themselves, nor are they necessarily the most cost-effective options.

Health security is also linked intrinsically to technology and IT infrastructure. Ireland's health system significantly lacks up-to-date protection measures and an overreliance on old communication mechanisms, including faxing, printing, and posting letters [1]. In May 2021, Ireland's Department of Health and the Health Service Executive (HSE) were struck by double lightning. As the second wave began to look more 'under control', and with measures lifting in the United Kingdom and Northern Ireland, a Russian cyberattack Darknet group attacked the healthcare organization's intranet and email, which caused significant disruptions to communication channels with significant cancellations of outpatient services [1].

The Russian-based cyberattack criminals were seeking a ransom, to which the Irish government rightly refused to engage in paying the bail. Unfortunately, this event resulted in massive disruption nationwide, with limited access to diagnostics and medical records and slower response times to healthcare appointments across the country [1]. The HSE secured an emergency High Court order preventing any organization from sharing or disseminating the information across websites or social media; however, this poses a serious question: Are countries investing enough in IT infrastructure and health and technology security? Where do these cyber-attacks leave more vulnerable countries?

Health promotion and public health have, in truth, been underfunded throughout the past 20 years in Ireland. But, interestingly, in April 2021, a little over one year since the first lockdown in March 2020, the current Health Minister Stephen Donnelly announced the launch of public health reform in Ireland [2]. Given that the previous year highlighted the critical 'national importance of a public health workforce with capability to provide a robust and resilient health protection response', [2] the announcement of the reform was crucial timing.

The announcement was welcomed by the Irish Medical Organisation, the trade union of medical doctors, after many years of negotiation on behalf of equity of pay for public health consultants and with recognition of their consultant status. The Crowe Howarth Report in 2018 highlighted that status, pay, and understaffing were significant issues for the public health workforce [3]. This, in turn, leads to 'brain drain', both within public health and on the broader hospital and community health teams—a term synonymous with Irish doctors leaving the country.

12.2 Borders and travel

Closing air and sea ports: In terms of fast, practical, and cost-effective responses, the immediate limitation on the use of air and seaports, particularly from affected areas, was not enough of a priority (within EU constructs). In retrospect, this step, combined with enforced and monitored quarantine for repatriated citizens—which often was introduced much later in Ireland—would unquestionably have caused a significant reduction in infection rates. Of note, such efforts might also have been combined with specific travel policies related to sporting events. The disastrous consequences of the Cheltenham Festival and ski holidays, for example [4], on Irish public health only became clear in retrospect, even though the medical community had warned of the risks.

Border, air, and seaport testing: Port-of-entry testing should have been prioritized for those ingress and egress pathways that remain open during lockdown periods and preceding them [5]. This might also have been a consideration for prioritizing the roll-out of limited testing resources, with such locations taking precedence.

Priority countries: A further international consideration might have been identifying priority countries based on the comprehensiveness of their response—though this may be complicated by stopover issues. Countries such as Belarus, for example, might be placed on minimal ingress and egress policies at an earlier stage in the pandemic. In contrast, those journeying from countries or other nonaffected countries might enjoy greater travel flexibility. This method of 'red zones' was eventually brought into place, albeit at a stage when the virus was internationally in wide circulation in the community [6].

Closing borders: Determining an effective way to manage borders, for example, the Northern Ireland border, the Gaza Strip, and other politically tense and fragile border areas in the health security context, has been clearly shown to be a priority in health security. It is unlikely that Irish legislative influence will extend to Northern Ireland in the immediate future, or even the medium term regarding the Northern Irish border. Yet, a cross-border working group on epidemic preparedness and policy cohesion may go some way towards addressing this to ensure cohesive policy in future [7].

Pass system consideration: Whilst it was vital for health services that public transport continued to function to enable healthcare and frontline workers to get to work, in many cases, this leniency was abused by those making nonessential journeys. A pass system for essential staff to use on buses and trains was proven effective to help to address this issue.

Adding health attaches to embassy staff: Many other countries, such as the United States and Norway, have made health attaches integral to international embassy staff. Their work on transmission of information on epidemic risks and assistance of the repatriation of epidemic-affected citizens, amongst many other day-to-day responsibilities, including those related to aid programmes and HIV/AIDS efforts—helped the diplomatic corps to meaningfully contribute to public health efforts. Ireland, and other smaller island nations, may wish to consider initiating such a strategy forthwith [8].

12.3 Macro-level public health security policies

Short-term, intensive responses: We learned that too few restrictions leave society with a sense of fatalism and lack of control; for example, Swedish approaches in this regard have been questioned in just the same way as American and British policies. Conversely, too many regulations, particularly over long periods, are unsustainable. The solution on future occasions might be to balance sustainability and reassurance: the imposition of more Draconian, immediate, and wide-ranging restrictions—but for proportionately shorter periods. This might include giving, for example, the Irish Defense Forces a significant initial role, rather than their recent provision of late-in-the-day support to overstretched law enforcement resources [9].

Immediate geographical distance restrictions—despite their shortcomings: Mobility restrictions formed a key and highly effective element of the national response; however, for these to work comprehensively, they also may need to be accompanied by hard borders beyond which no movement, except for urgent or emergency purposes, can take place. Otherwise, challenging to implement limits such as 2-km boundaries, as implemented in Ireland, may form a spatial chain linking the entire country and the risk of further isolation of rural populations.

A focus on healthcare worker and vulnerable population safety: Future epidemics—should they resemble this one—will hopefully have the advantage of responses informed by previously identified risk factors. These risk factors, such as age, comorbidities, and other considerations such as smoking or respiratory health issues, had such data been successfully gathered [10]. In all of the populations most affected in 2020–21, perhaps significantly added protective measures should be allowed for—whether these are specifically for healthcare workers or tailored towards the case of old-age retirement homes [11].

Contact tracing: Contact-tracing strategies may be described at times as 'chasing shadows' and are indeed highly unlikely to have any natural effect during the peak of a widespread and out-of-control epidemic in the age of media saturation. However, we have learned that through the use of tech to guide contact-tracing strategies, even with the myriad problems they face (and the frequent failure of such methods in 2020). The latest technology will undoubtedly help form a different basis for early-stage epidemic containment in future public health efforts.

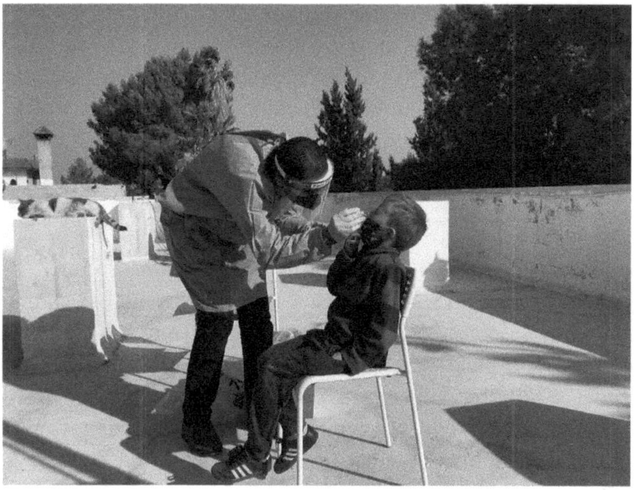

PICTURE 4.12.1 Covid testing on the Iraq border. Jeremy Fowler.

Law enforcement authority: The uncertainty and vagueness with which law enforcement was equipped with public health and epidemic control policing powers caused extensive confusion. Only with a specific legal mandate to prosecute behaviours that threaten public health can the police and others be expected to perform their duties with efficiency and accountability, should such events repeat themselves at any point in future [12]. Activities such as mask-wearing, space between passengers on public transport, and vaccine passports are tricky scenarios unless there are legislative measures to avoid unnecessary prosecution and successfully aid protection of the general population's health.

Varying coping capacity of different areas and groups: For many luckier population groups or other social demographics, the weeks and months of unreality in 2020 amounted to inconvenience, fear, and frustration—but little more. Rather than blanket policies, far greater attention and support, in future, should perhaps be focused on those living in apartments, lower socio-economic populations, along with areas of high-density housing, and those unable to access green spaces or beaches.

Whilst certain cities in the lockdown and 'recovery' phases implemented initiatives to temporarily provide individuals with improved accommodation, more permanent measures were required. The urban housing crisis across Europe in major cities suggests that further reforms are needed sustainably and equitably to avoid overcrowding of lower socio-economic groups and encompass water, sanitation, and hygiene measures for the future [13].

Virtual schooling and working: The speed with which Ireland successfully transitioned to homeschooling and working could scarcely be improved upon; nonetheless, there exists an opportunity for more formalized contingency plans in this regard, covering supervision, curricula, and formal advice for parents and employers. With luck, many elements of the new status quo in this regard will persist, marking a much-needed paradigm shift in commuting (and therefore pollution and congestion) patterns.

Not looking to the United Kingdom: Ireland has been, for so long, used to looking to the United Kingdom for leadership in global matters. Perhaps the COVID-19 epidemic has, along with Brexit, brought down the curtain once and for all in this era. UK policies have proved to be highly fallible; the groupthink of the European Union, whilst also often flawed, might prove to be a far more effective paradigm in future.

Distinguishing decision-making from advisory boards: Whilst centralized and authoritative national response control is often essential in epidemic situations [14], the distinction between advisory and decision-making bodies should be considered. In the panic of the early stages of epidemic onset, unelected bodies without a formal decision-making remit may inadvertently become highly influential. Likewise, governments may mistakenly ignore their advice, as Ireland learned to its high cost with the upward steeping curve in December 2020 when deciding to overrule the National Public Health Emergency Team's advice. Accordingly, future roles should be analyzed, not least to ensure that they do not eclipse preexisting national emergency response entities.

12.4 Guest essay: Global health security is ultimately about the individual

Jeremy Fowler

In order to achieve global health security, each individual in a society must have health security through the ability of access to a high-quality, compassionate health system. During times of increased illness like the COVID-19 pandemic, this is even more important.

When the coronavirus pandemic began in early 2020, I was working as a clinician and medical director in northern Jordan at a respiratory diseases centre that specializes in the treatment of tuberculosis (TB) and is a referral centre for complicated inpatient TB and multidrug-resistant TB. I had worked there for almost 10 years and had weathered, along with my family and colleagues, the Arab Spring and Syrian Civil War that flooded our town and the nearby Zaatari camp with refugees.

When the first reports of the novel coronavirus pandemic reached Jordan, our executive director called a meeting with the medical team to discuss how to respond. At that time we heard reports of the virus spreading, and I had seen an image of an emergency room in Cremona, Italy, filled with patients on oxygen. I watched a video of Italians warning their 'past selves' about the outbreak that was about to reach them, and admonishing the world to heed the warnings of public health officials.

As I saw the photo of the ER full of hypoxemic patients, I recalled the 2009 H1N1 flu outbreak whilst in residency training in Denver, Colorado. I was on an emergency medicine rotation when a man in his mid-40s came to the ER with the flu. He was diagnosed, treated appropriately, and sent home. The following week he returned, except this time gasping for air, face purple, and O_2 saturation in the low 80% range. The attending physician intubated and transferred him to the nearby ICU. A few days later, I began my next rotation in the intensive care unit and discovered that he was on my ward. I followed along with the ICU team caring for him throughout the entire month. He was not able to come off the ventilator during that time. I don't know what happened to him later, but I was struck by the severity of illness in such a relatively young man.

Years later, whilst working in Jordan, I admitted a 20-year-old Sri Lankan man who worked in a local garment manufacturing company. He was referred to us as a case of 'miliary TB[a]'. As I treated him, I watched with dread whilst the small pinpoint interstitial infiltrates in his lungs coalesced into large patches of thick, white pneumonia and he swiftly deteriorated, requiring higher levels of oxygen to breathe. Finally, I was forced to transfer him late that night to the local government hospital's ICU during a freak desert snowstorm that sent the ambulance sliding into the mud. Although miliary TB can lead to a similar deterioration, I suspected that he had viral pneumonia rather than TB. The MERS coronavirus followed soon after in the region, and I warned our medical providers about the clinical presentation and chest X-ray findings of viral pneumonias. By now my few experiences had instilled in me a healthy fear and respect for these viral respiratory pathogens that could so quickly end a patient's life. Although I regularly cared for patients with multidrug-resistant TB, I knew that severe viral pneumonia was out of our league and cautioned our medical providers to transfer these patients to a higher level of care before it was too late for us to help. Now, as I imagined not just one of these patients, but a whole ward of them, I realized that we may soon face something much worse than H1N1 or MERS.

Early on in the pandemic, public health guidance emphasized the importance of social distancing and business closures to prevent the unchecked spread of coronavirus. The Jordanian government enacted these recommendations in full force with lockdowns and curfews that were announced by air raid sirens and military patrols on the emptied dusty streets.

Since the sanatorium's priority patient population was those suffering from TB, we discussed closing the outpatient clinic to everyone except known TB patients in order to both protect the inpatients and allow outpatients with known TB to access care without risking exposure to COVID-19. Still, I had seen the strain that the Syrian refugee crisis had put on Jordan's health system, and had no idea how the public health system would hold up, and knew that Jordan would need as many partners as possible to address this new challenge. In addition, the majority of our outpatients had non-TB chronic respiratory diseases such as asthma, and relied on us. 'We are a respiratory diseases clinic'. I argued, 'and this is a respiratory disease. We need to continue to care for our TB patients, but should stay open and prepared to help out if needed'. We chose to keep the clinic running whilst modifying the workflow to screen for COVID.

We assessed our available PPE, ordered more, added 'bunny suits' and goggles to our stock, and prepared COVID isolation rooms in case a potentially infectious patient required emergency care before their transfer to a government-designated treatment centre. In order to preserve the focus of our centre on TB, we chose to

---

[a] Miliary TB is a form of TB in which the tubercle bacilli travel through the blood and lymphatic vessels to disseminate all over the body, causing tiny lesions that appear similar to millet seeds on a radiograph.

avoid administering COVID tests or becoming a treatment centre, but prepared to handle an ill COVID patient in such a way as to prevent spread to staff and other patients. Since there is a significant overlap with COVID and the presentation of tuberculosis and other clinical syndromes we treat, we chose to remain open for our community.

The Jordanian approach was effective during the first wave of infections that swept across the globe in March and April of 2020. During total lockdowns, no cars were allowed on the road except those with special permission. To maintain patients' access to care during the lockdowns, Jordanian medical professionals were given documents that allowed them to drive to work. Lockdowns, curfews, quarantine, and aggressive contact tracing all meant that Jordan 'flattened the curve' and kept daily new cases to less than 10 in the spring and summer of 2020. Jordan was hailed as a model country in international news syndicates. Despite the measures to keep healthcare providers working, many clinics shut down, whilst limitations on movement accompanied by fear of exposure to COVID decreased patients' access to treatment. On the one hand, I was grateful that the Jordanian Ministry of Health enacted this strong and unified response to the pandemic. It meant that the potential tidal wave of severely ill and contagious patients never reached us, and our centre did not face the crisis that immersed other communities and hospitals. But unfortunately the misapplication of this approach also led to many unintended consequences.

Soon after one lockdown, a friend called frantically for advice as his relative had been admitted to a peripheral hospital with renal failure; transfer to a higher level of care in the government system was unavailable due to COVID restrictions. His relative was unable to receive the care he needed and unfortunately passed away. Later on in the pandemic, after movement had been frequently restricted for several months, a 19-year-old woman from an outside province was admitted cachectic and bed bound with advanced TB that had destroyed much of her lungs. She had been sick for 6 months, and because of decreased access to care, had never been evaluated appropriately. She decompensated after starting treatment and did not survive. If she had been made aware of our clinic sooner, she could have been diagnosed early enough for it to make a difference.

When Jordan faced another wave of COVID infections accompanied by more lockdowns and curfews to curtail the spread, I interacted with other patients who had disastrous outcomes due to the misapplication of public health policy. One of the most upsetting cases involved the infant granddaughter of one of our hospital cleaning staff. She was born in the largest government hospital in the capital city of Amman and suffered from an anoxic brain injury during the delivery. Subsequently, she was admitted to the NICU, where she stayed for 40 days before being discharged to her home. Her mother reported to me that she was not allowed to visit her newborn daughter the entire time she was in the NICU because of the hospital's COVID visitation policies. When she received her daughter, she was still suffering from poor feeding and inadequately controlled seizures, and the paediatric neurology clinic that was supposed to see her after discharge was closed as part of a plan to decrease the spread of COVID. When I was alerted to her situation, I tried to arrange alternative follow-up in the private healthcare sector, but it was too late to help this precious little child.

Distressingly, detrimental hospital visitation policies during COVID have not been limited to the developing world or countries like Jordan that has absorbed, by some estimates, almost 20% of its population in Syrian refugees over the course of the Syrian civil war to its north. In such situations, it is very difficult, due to the crushing burden on the government health system and the potential for rapid spread of infection, to come up with a nuanced response. But even in developed countries like the United States, visitation policies have been enacted in a harmful way.

I spoke with a young medical doctor in training who wrote about his grandfather's difficult hospital stay that was exacerbated by COVID hospital visitation policies in California. And a South Korean friend was only allowed, after much petitioning, a 10-min fully masked and gowned visit to her 'dying' father who is hospitalized with Alzheimer's and end-stage congestive heart failure (CHF) in a large South Korean regional hospital, despite her having been fully vaccinated and testing negative twice for coronavirus. Often, these hospital visitation policies, whilst initially put in place to protect patients from suffering, when applied bluntly and without thought to their original intent, lead to increased patient suffering and likely poor outcomes.

What is the solution? Obviously, COVID-19 has devastated many patients and their families, and attempts to prevent the spread of the virus through physical distancing, mask-wearing, lockdowns, and limited hospital visitation policies are all appropriate measures when applied thoughtfully. However, I wonder if policy makers and writers on this topic, whilst emphasizing the importance of preventing the spread of illness by enacting general public health policies, neglected to emphasize the equal importance of continuing to give compassionate life-saving care to all patients. If we are to have a life-giving response in current and future pandemics, both of these elements must be in harmony/balance in order to hold on to one without letting go of the other. We must have clear public health policies in place, and at the same time make sure those policies exist to serve patients and medical providers, rather than medical providers and patients serving those policies. That means that, in addition to creating well-thought out public health policies, we need to maintain, encourage, and create cultures of compassion for patients and their families.

Since our clinic was used to treat a stigmatized and often feared respiratory infectious disease pathogen, we took a more nuanced approach than that which played out in the rest of the country. One case demonstrated to me the importance of our approach. A young man presented to the clinic and was screened at the gate for coronavirus symptoms. He was found to have a cough and fever with illness of 2 weeks' duration. My colleague who had designed the screening process came to ask for my advice. 'Should I send him to the government COVID testing facility without further evaluation since he is high risk for COVID?' I encouraged him to move this young man swiftly through the clinic for a chest X-ray whilst avoiding contact with other patients or ancillary medical staff. When his X-ray returned, it showed classic TB findings of apical cavitary infiltrates; a subsequent sputum AFB smear was also positive. If we had turned him away at the gate because his symptoms placed in our COVID risk category, we would have missed the diagnosis of TB, the very disease that we are prepared to treat. Whilst this seems obvious, unfortunately, many patients during the pandemic reported to me how they were sent elsewhere without evaluation because their symptoms overlapped with COVID symptoms and the treating clinic or hospital was not a designated centre. As governments and healthcare services enforced a 'one-size-fits-all' total lockdown, the patients that these decisions were meant to protect were denied care.

Prior to the pandemic, part of my work as a medical director was to help create systems that keep the medical work running smoothly and safely in order to effectively serve our patients. For many years, patients crowded at the gate of our medical facility every morning to be seen, and the number of patients we saw in the clinic corresponded to the number that showed up that day. Wait times could be up to 6 hours, and on busy days, shouting and fighting could take place as people defended their place in line. Every system we tried was circumvented by our patients; they are used to rules that block their access to services and know how to find ways to get around them. However, this situation was unsustainable both for our medical providers and for our patients. We needed a system that could bring order, with the right people to enact it. We implemented processes starting with computerized patient registration that eventually led to an electronic medical record. Staff were trained to help, and a clinic manager from the local community who understands the local culture and is a gifted organizer and communicator was employed and trained.

At the same time, I encouraged an institutional culture that advocates for individual patients and gives attention to patient safety. This meant helping ancillary staff like our pharmacy and laboratory technicians to understand the diseases we treat, encouraging them to ask questions about medications, and allowing them to remind medical providers on testing or treatment protocols if they diverted from the norm. It also meant helping staff like our clinic manager to understand our ethic of advocating for our patients, whilst still empowering him to create systems to organize care. When the COVID-19 pandemic began, we knew we had to implement a fully functioning appointment system to keep staff and patients safe and used the opportunity that the pandemic created to require patients to call ahead for an appointment, get screened for COVID symptoms, and be given appointment times rather than crowding at the gate to be seen. Since fewer patients showed up due to travel restrictions, we could adapt our system without being overrun. But as the country opened up more, we saw more patients, and I wondered if our new system was turning people away who needed help.

One morning, the triage nurse brought a young woman to me with a cough, difficulty breathing, and lower extremity swelling. She had a low-grade fever, O_2 saturation around 91%, and heart rate of about 120 beats per minute. She had recently given birth. The evaluation took only seconds as I identified her at high risk for pulmonary embolism and made the diagnosis through a quick Doppler ultrasound to the back of her right knee which showed an uncompressible popliteal vein filled with a thrombus. As I alerted my clinic manager to call an ambulance for transfer, the nurse administered low molecular weight heparin and obtained a chest X-ray that showed a large wedge-shaped opacity extending from the pleura towards the hilus of the right lung. The ambulance arrived in less than 5 minutes and whisked the patient to a hospital that could continue care. Afterward, I debriefed the clinic manager and congratulated him on his help in transferring her so quickly; he was sitting in his chair shaking his head as he recounted to me that this patient had come to the gate just a few minutes prior and did not have an appointment. As the front gate staff asked whether or not to admit her for evaluation, he felt a sense that he needed to let her in even though she was not signed up.

'Why did you let her in?' I asked him. 'I don't know', he replied, 'I just had a sense that God was telling me we needed to take care of her. If I had followed the system, she could have died at the gate'. After that experience, I felt even more convinced that our clinic manager was an excellent person to have in place. He not only knew how to create and communicate a strongly functioning system, but also knew when to bypass his system when it stood in the way of a patient getting care. Whilst it is fairly straightforward to create a system and policies that function on paper, what is equally important is creating a culture that encourages staff to provide excellent patient care, and be allowed to make decisions or petition the system to that end.

Ultimately, a public health response that serves both the public as a whole and the individual is a just response that serves to create global health security. A cookie-cutter response that discourages discernment and compassionate decision-making leads to injustice and health instability. Public health policy that loses sight of the individual it seeks to serve will become unjust and harm those it was created to heal. And injustice destabilizes systems and ultimately whole societies and countries. Consequently, global health security is really achieved by creating both thoughtful public health policy and compassionate healthcare cultures that serve each person. If we lose sight of this, we will implement systems that place barriers to care, serve systems rather than creating systems that serve us, and compromise individual patients' health security, thus compromising society's health security. Policy makers should keep a close eye on how their decisions affect the health of each individual in society, and make sure to emphasize the importance of this even when making broad public health policy decisions.

It may help to pause and ask simple questions like, 'If this was my aging mother...or my teenage son...or my aunt that doesn't speak English well...how would this policy affect them?' Along with public health strategies on masking, physical distancing, and lockdowns, policy writers and media contributors should use their influence to encourage compassionate care and advocacy for individuals in the midst of the pandemic.

PICTURE 4.12.2 Social distancing guidelines in Arabic. Jeremy Fowler.

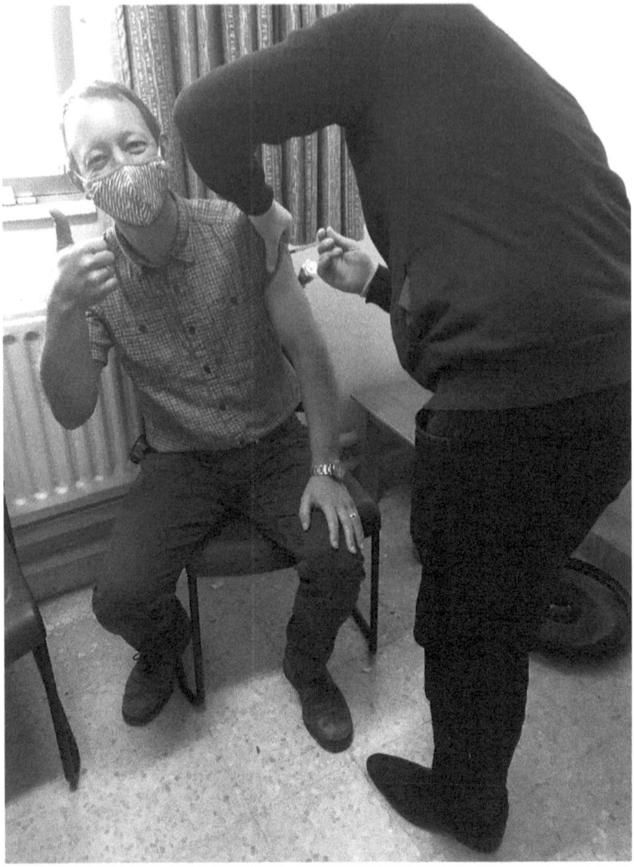

PICTURE 4.12.3 Many physicians felt a huge sense of relief when they got vaccinated—and also set the example for broader society. Jeremy Fowler.

12.5 Media and public health information

Immediate release of geographical locations of infections: Like so many other countries, Ireland discovered that, in public health emergencies, it risks becoming a victim of its own bureaucracy. This consideration, linked to the loosening of restrictions on protected health information (PHI) in emergency situations, is of critical importance.

Another lesson for the future includes releasing locational information. This information may consist of down to high levels of resolution and granularity, including the street level, the region, and the district. These lessons will help the national effort on several groups through:

- Enabling those within these areas to take added care.
- Enabling those outside those areas to avoid them.
- Allowing the government to focus both resources and restrictions on specific areas.

Perhaps, stigma risks are more than compensated for by reductions in fear, suspicion, and 'false stigma' driven by rumours of outbreaks.

Releasing PHI to academics: One of the most problematic elements of the Irish response has been the shortage of analytical information (as opposed to the seemingly endless stream of commentary, expert, or otherwise). For many epidemiologists, their efforts might have been far more rewarding, productive, and valuable if focused on immediate initiation of research into risk factors, demographic data, and spatial epidemiology—to name but a few avenues of exploration. In future, it is perhaps noteworthy that the government consider allowing much faster data mining by Ireland's academic community, perhaps via a voluntary rapid response team, with associated assistance in cutting the red tape and bureaucracy involved.

Surveillance and knowledge transfer: In many Ebola-affected countries, surveillance and knowledge transfer from field to centralized areas were crucial [15]. Establishing a civilian and a medical surveillance system for epidemic outbreak reporting may further help ensure that future responses are conducted in a targeted and focused manner.

Preventing panic hoarding: The necessary suddenness of governments' lockdown announcements was often accompanied by much unnecessary fear, not least the undignified and demeaning scenes of panic buying and conflict at supermarkets during March 2020. Fortunately, societies soon came to their senses and most avoided the allure of hoarding (and therefore shortages). Nonetheless, should similar circumstances arise in future, government public health messaging should perhaps include a specific 'nonhoarding' component as well [16].

Broadcasters, journalists, and bloggers are not epidemiologists: The question of how to control the plethora of false theories, fake news, fearmongering, and paranoia generated by mainstream media and the Internet is almost impossible to address. Better, perhaps to in future advise society to direct its attention to a limited number of trustworthy news sources; in 2020, uninformed and internet or social media fueled theories and speculation abounded, which led to much unnecessary distress and fear [17]. Too often, also, epidemic reporting was conflated with political agendas. Whilst many public figures have a role in terms of awareness building and messaging regarding the importance of adhering to government guidelines, they should also perhaps be asked to recognize that they are rarely qualified to make informed opinions on specialist medical subjects.

Avoiding excessive concern with economic consequences: There appear to be two economic options in an epidemic situation—neither of which is attractive. One is to allow businesses to continue operating, even though their trade will inevitably suffer, and face higher death and infection rates (for example, the route that the United Kingdom took). The second is to suspend businesses, invest more in public health, and face similar sets of economic consequences [18]. The key lesson from 2020–21 is that, in both scenarios, economic losses are inevitable: those demanding more laissez-faire approaches, or documenting the economic opportunity costs of restrictions, would do well to temper their criticism of government actions in this context. Indeed, there is a case to be made for suspending, distracting, and distressing economic impact analyses at such times [18].

A health benefit focus: The closure of pubs may have led to many health co-benefits; similarly, two of the main risk factors associated with this respiratory epidemic are cigarettes and alcohol: the former because of behaviour change and loosening inhibitions, as well as close confinements in small spaces, and the latter because of effects on respiratory health. Perhaps the events of 2020 may also serve to loosen the hold of the Licensed Vintners Association on Irish Society, helping the country to move away from the absurdities and disadvantages of its heavy-drinking stereotype. Of added relevance is the evidence of the role of off-license sales during pub closures—inevitable contributors to the social disorder associated with public drinking.

12.6 Feasible and enforceable individual-level efforts

Blanket and immediate face mask policies: Where possible, this might be combined with publicity efforts to de-stigmatize face mask use in future. In recent months, the lack of such policies should be attributed far more to a lack of national preparedness and fear of supply shortages than to any doubts about their effectiveness. Even a 50% transmission reduction is well worth the investment, but only if face mask use can be enforced on a routine basis.

Investing in tech: Contact-tracing apps, which are likely to be essential in controlling future community-level outbreaks, should be made accessible to all in future [19]. But, if adherence to public health guidelines is a government mandate, should the necessary technology be made available to all. This need not involve the provision of smartphones to every member of the population but could—in theory—include some kind of low-cost personal and portable alert or monitoring system that does not impinge on privacy. In the same way, efforts to allay fears about personal surveillance implications should, of course, accompany any such step.

Marshalling volunteers, businesses, and communities: Volunteer organization involvement was, for many (including for one of the co-authors of this book), inhibited or delayed at a critical time in Ireland due to vetting procedures and red tape. Whilst such safeguarding efforts are essential, their use during national public health emergencies should perhaps be reconsidered. In the same way, a mandatory emergency preparedness system for businesses, communities, and individuals—even extending to the predesignation of community wardens—might help alleviate much of the frustration and powerlessness that many felt in 2020. For example, many civil servants were delegated into beginner courses for contact-tracing services—yet during the second lockdown, there was a shortage of contact tracers to keep up with the outbreaks. Emergency preparedness systems must therefore be explored through all workplaces to ensure adequate preparation for future events.

The above inevitably excludes many other considerations, such as managing economic benefits—should automatic payments avoid the blanket reimbursements and assurances that so troubled the economy via banking guarantees in the past? There is thus a vast amount more than Ireland, and many other countries have learned in recent months; the prospective issue is to ensure that the events of 2020–21 not be dismissed as one-off anomalies. In an increasingly globalized world, only longer-term efforts to impose health security, ingress and egress policies can ensure that the mistakes of the past are not repeated.

A critical distinction that may be easy to overlook is epidemic elimination and eradication: whilst the former may have been achieved, the latter is a significantly different prospect—both in the current case and for future epidemic preparedness from any infectious disease [20]. Events such as annual lockdown preparedness mobilization and mock tests of national responses should therefore not be dismissed: it is only through rapid preparedness on the individual and the national level that future epidemic containment will depend.

The vaccine's arrival in early 2021 was a sigh of relief thanks to its protection against severe illnesses and death. However, COVID-19 and perhaps many future epidemics may be respiratory viruses, which can often circulate, mutate, and progress, similar to influenza. Nevertheless, the outstanding coordination of scientists and researchers in producing mRNA vaccines has resulted in significantly impressive results via the prevention and future elimination of COVID-19. Yet we should not speak too soon on elimination; it depends on the uptake of vaccines, along with the timing and severity of breakthrough reinfections.

Perhaps the virus will get stronger with time? Or maybe, our immune systems will transform to allow for 'catching COVID-19' without induced fear and images of ICUs? Sadly, even the common cold or flu is not benign, as it results in thousands of deaths in the elderly each year. But, of note, the effects of these everyday viruses have been dampened significantly worldwide due to mask-wearing, school closures, fewer socializations, and the lockdowns of 2020 and 2021.

Finally, timing may also be critical: as noted above, the immediate application of intense restrictions for shorter periods, involving entities such as not only the defence forces but also the coast guard and civil defence, will inevitably curtail the pandemic in a much more effective way—rather than the gradual phasing up of efforts, which allowed the infection to spread in a manner that was, for many months, beyond the control of governments.

12.7 Key points

- Strengthening health systems, which is a long-term goal, can also be achieved with surprising rapidity in the short term (e.g. investments in ventilators and PPE).

- Health security is inherently connected with health system strengthening. Countries with stronger public health systems are more resilient to health security threats.
- In the long term, public health is a key contributor to national and international security and stability. Investments in public health pay security dividends.

References

[1] Hse.ie, Conti Cyber Attack on the HSE Full Report, 2021, Available at: https://www.hse.ie/eng/services/publications/conti-cyber-attack-on-the-hse-full-report.pdf. (Accessed 14 December 2021).

[2] Gov.ie, Minister for Health Announces Implementation of Fundamentally Reformed Consultant-Led Public Health Model in Ireland, 2021, Available at: https://www.gov.ie/en/press-release/38f3d-minister-for-health-announces-implementation-of-fundamentally-reformed-consultant-led-public-health-model-in-ireland/. (Accessed 14 December 2021).

[3] Gov.ie, Crowe Horwath Report on the Role, Training and Career Structures of Public Health Physicians in Ireland, 2021. https://www.gov.ie/en/publication/ae088f-crowe-horwath-report-on-the-role-training-and-career-structures-of-p/. (Accessed 14 December 2021).

[4] J. Aron, J. Muellbauer, Measuring excess mortality: the case of England during the Covid-19 Pandemic, in: Institute for New Economic Thinking, Oxford Working Paper, (11), 2020.

[5] World Health Organization, Online Global Consultation on Contact Tracing for COVID-19, 2021. 9-11 June 2020.

[6] A. Patriti, E. Eugeni, F. Guerra, What happened to surgical emergencies in the era of COVID-19 outbreak? Considerations of surgeons working in an Italian COVID-19 red zone, Updat. Surg. 72 (2) (2020) 309–310.

[7] E. Medeiros, M. Guillermo Ramírez, G. Ocskay, J. Peyrony, Covidfencing effects on cross-border deterritorialism: the case of Europe, Eur. Plan. Stud. 29 (5) (2021) 962–982.

[8] M.D. Brown, T.K. Mackey, C.N. Shapiro, J. Kolker, T.E. Novotny, Bridging public health and foreign affairs: the tradecraft of global health diplomacy and the role of health attachés, Sci. Diplomacy 3 (3) (2014) 1–12.

[9] E.L. Kinsella, S. Hughes, S. Lemon, N. Stonebridge, R.C. Sumner, "We shouldn't waste a good crisis": the lived experience of working on the frontline through the first surge (and beyond) of COVID-19 in the UK and Ireland, Psychol. Health (2021) 1–27.

[10] T.V. Giannouchos, R.A. Sussman, J.M.M. Odriozola, K. Poulas, K. Farsalinos, Characteristics and risk factors for COVID-19 diagnosis and adverse outcomes in Mexico: an analysis of 89,756 laboratory-confirmed COVID-19 cases, Medrxiv (2020).

[11] A. Tabah, M. Ramanan, K.B. Laupland, N. Buetti, A. Cortegiani, J. Mellinghoff, A.C. Morris, et al., Personal protective equipment and intensive care unit healthcare worker safety in the COVID-19 era (PPE-SAFE): an international survey, J. Crit. Care 59 (2020) 70–75.

[12] J.E. Marco-Franco, P. Pita-Barros, D. Vivas-Orts, S. González-de-Julián, D. Vivas-Consuelo, COVID-19, fake news, and vaccines: should regulation be implemented? Int. J. Environ. Res. Public Health 18 (2) (2021) 744.

[13] Y. Rydin, A. Bleahu, M. Davies, J.D. Dávila, S. Friel, G. De Grandis, N. Groce, P.C. Hallal, I. Hamilton, P. Howden-Chapman, K.M. Lai, Shaping cities for health: complexity and the planning of urban environments in the 21st century, Lancet 379 (9831) (2012) 2079–2108.

[14] R. Loewenson, K. Accoe, N. Bajpai, K. Buse, T. Abi Deivanayagam, L. London, C.A. Méndez, T. Mirzoev, E. Nelson, A.A. Parray, A. Probandari, Reclaiming comprehensive public health, BMJ Glob. Health 5 (9) (2020) e003886.

[15] C. Fähnrich, K. Denecke, O.O. Adeoye, J. Benzler, H. Claus, G. Kirchner, S. Mall, R. Richter, M.P. Schapranow, N.G. Schwarz, D. Tom-Aba, Surveillance and outbreak response management system (SORMAS) to support the control of the Ebola virus disease outbreak in West Africa, Eurosurveillance 20 (12) (2015) 21071.

[16] D.C.K. Tse, V.W. Lau, Y.-.Y. Hong, M.C. Bligh, M. Kakarika, Prosociality and hoarding amid the COVID-19 pandemic: a tale of four countries, J. Community Appl. Soc. Psychol. (2021).

[17] C. Chew, G. Eysenbach, Pandemics in the age of Twitter: content analysis of Tweets during the 2009 H1N1 outbreak, PLoS One 5 (11) (2010) e14118.

[18] J. Bibby, G. Everest, I. Abbs, Will COVID-19 be a Watershed Moment for Health Inequalities, The Health Foundation, 2020.

[19] R.C. Khanna, M.V. Cicinelli, S.S. Gilbert, S.G. Honavar, G.V. Murthy, COVID-19 pandemic: lessons learned and future directions, Indian J. Ophthalmol. 68 (5) (2020) 703.

[20] A.E. Heywood, C. Raina Macintyre, Elimination of COVID-19: what would it look like and is it possible? Lancet Infect. Dis. 20 (9) (2020) 1005.

Chapter 13

Biodefense shields

Sebastian Kevany, Deon Canyon, Michael Baker, and Aoife Kirk

Chapter outline

13.1 A questionable track record?

There is a clear and simple correlation between population and infectious disease [1]. This places many countries in a natural but dangerous and precarious position. Not only do such countries have vast international networks, but they are also connected by trade and travel to virtually every location on the planet.

A clear link also exists between a growing population and the exploitation of the environment. Along with the exposure to wildlife, there remains a high risk of emergence and spillover of pathogens into human populations. Many developing countries have experienced vast, recent, and rapid economic growth over the last few years. As a result, there remain high numbers of people who live and work in close proximity to animals, which, unfortunately, is an ideal scenario for zoonotic diseases to jump species.

Nations such as China have a history of disease containment failure [2]. They have been linked previously to the origin of numerous global pandemics, from severe acute respiratory syndrome (SARS) to H1N1 influenza and now COVID-19, which was first reported in Wuhan, China. Thus, it is natural for such nations to be perceived as the most prominent global disease incubators and preeminent disease disseminators. Various governments have attempted to

Outbreaks, Epidemics, and Health Security. https://doi.org/10.1016/B978-0-323-91705-6.00011-3

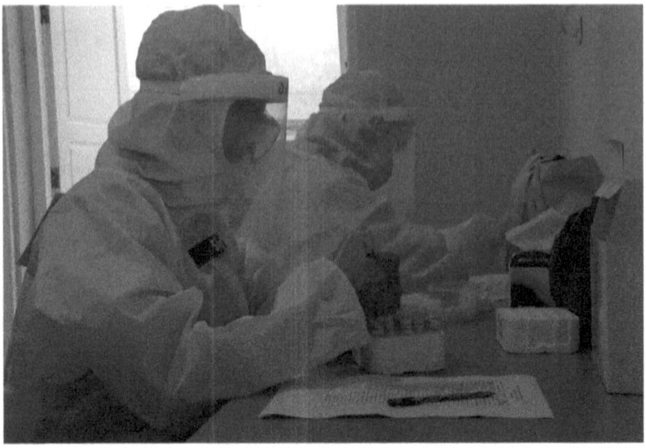

PICTURE 4.13.1 Is this what it takes to form a biodefense shield? Yodi Mahendradhata.

demonize and politicize this perspective, yet one cannot easily spin and reframe an obviously natural phenomenon.

To compound matters, national responses to related predicaments have historically been fraught with insincerity, nontransparency, and a lack of accountability. For example, in a RAND report on the COVID-19 pandemic, Chinese officials were described as having routinely delayed acknowledging the existence of a public disease threat and downplaying the severity of that threat. Moreover, the government's disease notification system, from medical observation to government reporting and public notification, is 'designed to make timely acknowledgement of an epidemic to the public ... almost impossible'.

The same behaviour was observed in January 2003, when some Asian governments downplayed the severity of the SARS coronavirus outbreak for two months. Then, in November 2002, the first human cases of SARS-CoV were found in Guangdong Province, in Southern China, with the pandemic eventually claiming 774 lives after over 8000 people were diagnosed as infected in 26 countries.

Avian influenza (H5N1) is also famous for its Guangdong Province origins, where it was first detected in geese in 1996. While the first human case was found a year later in Hong Kong, this Bird Flu persisted, with sporadic outbreaks occurring in China and the international community until 2015. In this regard, Southern China is the source of almost all poultry in the country, and the conditions under which this farming takes place were believed to have been responsible for the emergence of new bird flu virus agents. Even the 1957 Asian Flu (H2N2) has suspected origins in the southwestern province of Guizhou, while the World Health Organization (WHO) suspects that the 1968 Hong Kong Flu may have originated in China. Combined, these two pandemics killed over 3 million people worldwide [3].

13.2 Guest essay: The United Kingdom, global health diplomacy, and geopolitics

Lyndsay S. Baines

On Monday 26th July 2021, the Government of the United Kingdom lifted the majority of coronavirus disease 2019 (COVID-19) restrictions and opened the economy at a time when there was still uncertainty over the Delta variant of COVID-19. This was a bold move, as the rest of the world looked on from its position of locked economies and societal restrictions. It remains to be fully seen how the UK's gamble will play out. If the United Kingdom succeeds in reaching a plateau without hospitals becoming overwhelmed in the face of an upward surge in reported cases, then the virus may be repositioned as a manageable seasonal health issue, such as the flu virus.

The UK government hopes that its approach of focusing on its own population first will move the country towards a position of herd immunity, thereby positioning the country as a net supplier of vaccines to other countries. The United Kingdom has based its intrepid approach on the success of the national vaccination programme: 68.9% of adults in the United Kingdom are fully vaccinated at present, on par with 68.7% in the United States [4] (late 2021). Furthermore, the UK government has commenced vaccinations in 16- to17-year-olds, which will enable return to normalcy in high schools. In addition, the UK government is placing its trust in public caution and personal calculated risk taking versus lockdowns, masking, and social distancing. This approach has not necessarily been supported by the scientific community, which wrote in a highly cited publication that the strategy of herd immunity by mass infection was 'unscientific and unethical' [5]. The increase in alerts from the National Health Service *'test and trace'* system referred to as the 'pingdemic' does not seem to support the strategy either [6].

In our opinion, herd immunity may never be fully achieved due to new COVID-19 variants. On the other hand, the government risks economic depression and a breakdown of civil society if it continues to pursue a relentless policy of lockdowns, social restrictions, and a freeze on elective medical interventions. There is clearly a need for middle ground—which may, however, remain elusive as both sides of the debate are deeply entrenched in their positions.

While the challenge to discover an effective vaccine and establish a national vaccination programme dominated the early months of the COVID-19 pandemic, attention has now turned to equity and the role of vaccine diplomacy [7]. Countries in the Global South are eagerly looking to schemes, such as COVAX, as a solution to vaccine inequity [8]. Adequate supplies of vaccines may not be enough, however, as it has recently been reported that several countries do not have funds to buy fuel to transport doses to clinics, train people to administer shots, or persuade people to get them. In addition, some multinational manufacturers of vaccines have avoided direct negotiations with COVAX but reached agreements with the US administration—which hurts COVAX's credibility [9]. This is health vaccine diplomacy in its most basic form as the Global North improves their image and gains leverage over countries to which they are supplying vaccines. China and Russia have taken the lead in such schemes, and on the surface appear to be promoting health equity.

However, in our opinion, vaccine diplomacy has a role beyond the mere supply of vaccines, which should extend to quality and safety. Sinopharm and Sinovac vaccines are unregulated in the Global North, whereas AstraZeneca, Pfizer, and Moderna are approved in the Global South [10]. China, for example, has not prioritized its own population first (as the United Kingdom did), before exporting vaccines. Health diplomacy works best from a position of strength: a case in point is India, which exported significant amounts of the vaccine, but quickly reversed course after the upsurge in COVID-19 variants. Even when India resumes exporting its vaccine, it is likely to favour its neighbours and allies to buttress its policy of counteracting China [11].

Furthermore, in our opinion, the pandemic and the advent of effective vaccines has fundamentally changed the role of global health diplomacy. Even before the vaccines were formally approved, health diplomats in the Global North had negotiated bilateral prepurchase agreements with pharmaceutical companies, leading to an increase in the cost of vaccines. The Global South had neither the diplomatic skills nor the networks to afford vaccines in such a competitive market. We urge stakeholders in the Global South to prepare themselves in the art of global health diplomacy—for the next pandemic, or the so-called third wave of COVID-19.

While the Global South may appear to have lost out in obtaining large amounts of vaccines, this may have resulted in further strengthening of their primary and public health resources. Government leaders and opinion makers in the Global South have realized, and in some cases convinced their people, that public health should not be neglected in favour of glamorous tertiary health facilities [12]. The future will see countries in the Global North, Russia, India, and China engage in aggressive global health diplomacy to pursue their national objectives. North–South cooperation should be the way forward as it is now clear that the pandemic will not be over anywhere, until it is over everywhere [9,13].

PICTURE 4.13.2 **Biodefense takes on many forms, including a form of health border protection.** Yodi Mahendradhata.

13.3 Deficits in capacity and communications

Despite having some of the best researchers and laboratories focused on these types of diseases, many governments still refuse to change how they respond to these outbreaks. For example, during the original COVID-19 outbreak in Wuhan, the scientific quality assurances by the Chinese government were suspected. Nevertheless, their expert team investigated the outbreak and concluded that there was 'no person-to-person transmission and the outbreak was well controlled after closing the seafood market'.

Although the Chinese equivalent of the Center for Disease Control (CDC) had access to the most current bat virus research, it likely had 'gaps in coronavirus' expertise. Unfortunately, this gap in research perhaps resulted in no cases being reported in Wuhan for two weeks. Subsequently, many experts have cast suspicion on the reporting during the outbreak; despite the situation rapidly escalating into a crisis, the government remained reticent regarding the latest outbreak information. It was released only once the information had passed through all the bureaucratic channels and vetted at the highest levels. Again, this sets unfortunate precedents for many countries in the region and worldwide in terms of information-sharing [14]. Nevertheless, one lesson from this chain of events is that there is, potentially, a danger to human life if primary outbreak management institutions become political tools around the world.

13.4 Guest essay: New lessons or old hurdles?

Viveka Guzman,[a] *Ori Solomon,*[b] *Kim Nordmann,*[c] *Mehr Muhammad Adeel Riaz*[d]

> For too long, the world has operated on a cycle of panic and neglect. We throw money at an outbreak, and when it's over, we forget about it and do nothing to prevent the next one.
>
> *Dr. Tedros Adhanom, World Health Organization (WHO) Director General*

Long before the COVID-19 pandemic took hold, many calls for action were issued for international epidemic preparedness. One could argue that such warnings came from a mixed bag of sources of authority, ranging from observational studies in ecology to Hollywood movies. However, prominent public health experts,

[a]Viveka Guzman, MBBS, MSc, is a SPHeRE PhD scholar at the Royal College of Surgeons in Ireland. She is also the Projects team lead at the Global Health Next Generation Network (GHNGN).
[b]Ori Solomon, MGH, is a PhD candidate in Microbiology at McGill University. He is currently serving as the Secretary-General of the GHNGN.
[c]Kim Nordmann, MGH, BSc, is a medical student at the RWTH Aachen University in Germany. She is part of the Projects team and the internal affairs team of the GHNGN.
[d]Mehr Muhammad Adeel Riaz, MBBS, is a junior doctor working in Pakistan. He is currently serving as part of the international team of Projects team at the GHNGN.

intellectuals, and policy makers are amongst those who predicted an upcoming infectious disease outbreak that would pose a massive threat for global health security and every aspect of human society. Emergency preparedness is defined by the WHO as 'the knowledge, capacities and organizational systems developed by governments, response and recovery organizations, communities and individuals to effectively anticipate, respond to, and recover from the impacts of likely, imminent or current emergencies' [15].

Evidence of the modern approach to infectious disease outbreaks can be traced back to the emergence and spread of HIV/AIDS [16]. Since then, other infectious disease outbreaks like the severe acute respiratory syndrome (SARS), Ebola, the Middle East respiratory syndrome (MERS), influenza A (H1N1), and Zika have been stark reminders of the remaining gaps in emergency preparedness. After each one of the massive infectious outbreaks of the last two decades, a post hoc assessment has followed. Such assessments have typically involved a process of collective diagnosis, through which official commissions draw lessons from the outbreak that point to the need to anticipate and prevent similar future crises. Critically, no warnings or 'lessons to be learnt' analysis generated the changes needed to prevent the devastation of COVID-19.

At the time of writing, the world is still amid the COVID-19 pandemic. As of August 2021, more than 4.5 million people have died of the disease across the globe [17]. Tens of millions of people are at risk of falling into extreme poverty, with a deep recession looming on the horizon [18,19]. Under these conditions, vulnerabilities amongst older people, slum dwellers, migrants, and displaced people have significantly multiplied [20]. The diversion of critical resources from other public health challenges may result in exacerbated maternal mortality and morbidity, increased rates of adolescent pregnancies, HIV, and sexually transmitted diseases [20]. Emerging data also shows that domestic violence against women and girls has intensified since the beginning of the pandemic [21]. However, unequal access to vaccination remains rampant across low-income countries.

The global response to COVID-19 will inevitably influence our reaction to upcoming threats. The WHO has commissioned the first assessment of the initial response to the pandemic, aiming to identify the experiences gained and make recommendations to improve capacities for the future [22]. This and previous outbreak response evaluations offer actionable insights and valuable solutions but also risk perpetuating long-known issues by ignoring fundamental challenges [23]. More than ever, it is vital to reflect on the deep barriers to implementing the lessons garnered from previous outbreaks, and reflect on how we can build sustainable and resilient global systems that will ensure health, economic, and social security in the light of future outbreaks.

Evidence of a 'déjà-vu'

Back in 2005, 196 countries agreed to abide by the International Health Regulations (IHR), which require national governments to develop core capacities to detect, assess, report, and respond to 'public health emergencies of international concern' (PHEIC) [24]. Since the PHEIC system's inception, the WHO has declared this type of emergency on six occasions: in 2009 for H1N1, in 2014 for the re-emergence of Polio and for Ebola, in 2016 for Zika, in 2019 for another Ebola outbreak, and in 2020 for COVID-19. Experts have retrospectively evaluated each of these outbreaks and produced calls to action for 'the next pandemic that is around the corner'. For instance, after the 2009 H1N1 pandemic, a review of the response carried out by the WHO concluded that 'the world is ill-prepared for a severe pandemic or for any similarly global, sustained and threatening public health emergency' and called for 'research, reliance on a multisectoral approach, strengthened healthcare delivery systems, economic development in low- and middle-income countries and improved health status' [25]. Five years later, in the aftermath of the Ebola epidemic in West Africa, another commission established by the WHO established that 'the world cannot afford another period of inaction until the next crisis' [26]. Again in 2017, members from the World Bank, several universities, and other international agencies worked together and called for 'capacity building in health and financial services as well as national and international emergency simulation exercises' [25]. A closer review of post hoc assessments from some of the latest infectious disease outbreaks reveals further similarities between the lessons to be learnt and calls to action (Table 1).

Nonetheless, assuming that we have not learnt any lessons from previous infectious disease outbreaks is misguided. Many international organizations, national governments, and other stakeholders have made significant progress in preparedness to face health emergencies based on the lessons of history [32]. However, responses to PHEICs have shown to be insufficient in preventing outbreaks such as COVID-19 and highlight significant vulnerabilities that expose us to increasing devastation and harm from the next pandemic—unless we tackle underlying issues and take up the necessary actions now. In the next section, we reflect on four mechanisms that have consistently hindered our path to effectively prevent and respond to massive infectious outbreaks and attempt to unravel why it has been so despite the incentives to do otherwise.

Funding

Increased funding towards global health security is often identified as a key component to be improved in overall pandemic preparedness. Through sustainable and reliable funding, measures can be taken to ensure swift reaction mechanisms in case of a PHEIC, for instance, the acquisition of medications and vaccines, training of healthcare professionals, development of surveillance systems, amongst many other resources needed in pandemic preparedness. The needs for funding have been identified at multiple levels ranging from regional and national governments to international and multilateral agencies [33]. However, it remains a challenge to secure and maintain appropriate resources in the long term. After pandemics, allocation of funds for pandemic preparedness tends to spike due to pressure from experts, media, and the public, but galvanized resources quickly fade. As executive director of the Big Cities Health Coalition, Chrissie Juliano, pointed out: 'Something happens, we throw a ton of money at it, and then in a year or two we go back to our shrunken budgets and we can't do the minimum things we have to do day in and day out, let alone be prepared for the next emergency' [34]. This is known as the cycle of 'panic, neglect, panic, neglect' [35].

TABLE 1 Overview of calls to action from previous infectious outbreaks and COVID-19.[a]

	SARS [27]	H1N1 [28]	Ebola [29]	Zika [30]	COVID-19 [22,31]
Infectious disease outbreak					
Year	2002–03	2009–10	2014–16	2015–16	2019–Ongoing
Countries affected	Global (29 countries in Africa, America, Asia, Australia, and Europe)	Global (nearly all countries affected)	Mainly West-Africa with few cases in the United States and Europe	Mainly America and Oceania	Global
Illustrative calls to action and/or lessons to be learnt[b]					
Detection and surveillance	Strengthen interconnected systems for international surveillance and outbreak alert; implement screening protocols at entry points (i.e. at airports)	Reinforce evidence-based screening protocols on international travel and trade based on WHO guidelines	Define an international surveillance system for monitoring and outbreak alert	Develop, implement, and appropriately strengthen interconnected systems for international surveillance and outbreak alert	Establish a global surveillance system that generates data that is complete and comparable
Care and support	Strengthen health systems to protect at-risk populations from emerging pathogens; improve system resilience and surge capacity	Establish a more extensive global, public-health reserve workforce to increase surge capacity	Improve patient management; training healthcare workers and volunteers for patient management, improve system resilience, and surge capacity	Strengthen health systems and other relevant stakeholders at the national and community levels to provide support for affected and at-risk populations	Strengthen health systems to improve system resilience and surge capacity

Research and development	Facilitate open access and sharing of information between states and the WHO; strengthen research in new technologies, diagnostic tests, treatments vaccines, and digital systems for surveillance and case tracking	Enable and accelerate a research and evaluation programme; reach agreement on the sharing of viruses and access to vaccines and other benefits	Enable and accelerate research and ensure access to the benefits of research through a framework of norms and rules operating during and between outbreaks	Generate data and evidence needed to strengthen public health and community guidance and interventions to prevent, detect, and control outbreaks and manage disease complications	Ensure technology transfer and commitment to voluntary licensing are included in all agreements where public funding is invested in research and development [22]·
Coordination and governance	Establish multinational collaboration along with the WHO in implementing surveillance strategies, and guidelines	Plan a framework for equitable distribution and access of essential pharmaceuticals and medical devices	Establish an interagency mechanism for collaborative responses to health crises; ensure clarity on roles and responsibilities, including robust measures for accountability and transparency	Establish a coordination mechanism for collaborative responses to outbreaks	Strengthen national coordination mechanisms for collaborative responses to outbreaks Elevate leadership for global health and strengthen the WHO [22]
Finance	Invest in international surveillance cooperation	Create a contingency fund for public-health emergencies	Establish multilateral partnerships to mobilize financial support for emergency preparedness and response	Establish a revolving fund for emergency preparedness	Raise new international financing for the global public goods of pandemic preparedness and response [22]·

[a] *Several official commissions have carried out additional assessments of pitfalls and lessons in each infectious disease outbreak. The ones selected are used for illustrative purposes.*
[b] *List is not comprehensive; however, action points were chosen to illustrate some of the main calls to action.*

Time and time again, it has been demonstrated that combatting a full-blown public health emergency and its economic fallouts is considerably more costly than investing in preventative measures and emergency preparedness. However, as the value of preparedness lies in preventing a possible outbreak, its evidence of success is less tangible and remains difficult to measure [35]. Thus, funding preparedness is largely unattractive for national governments that are struggling to balance their limited resources with a plethora of conflicting agendas [35]. Instead, during the past years, most health spending of OECD countries was incurred in individual care with only a small fraction dedicated to emergency preparedness [36]. Likewise, development assistance to low- and middle-income countries widely focused on vertical programmes for HIV, tuberculosis, and malaria, with less than 1 percent dedicated towards pandemic preparedness in 2019 [37].

Part of the systemic challenge in funding is the lack of economic incentives for private companies to invest in research and development of vaccines and other medical supplies that contribute to prevent or mitigate the effects of a pandemic. As revenues for antimicrobial goods are typically low and development costs high, the private sector is little incentivized to invest into research and development. But it is more than that. If a pandemic is successfully averted through a product, the market shrinks and companies—though having been successful with their product—might face a negative return due to high investment costs [38]. South Korea recognized this effect after the MERS and Zika virus outbreaks, and started investing into infectious disease diagnostic technology in 2017 by giving domestic companies access to advanced technologies and establishing public–private partnerships [39,40]. As a result, South Korea was well placed to develop COVID-19 vaccines with two commercial companies starting research and testing before an official request by the Korean government [40]. South Korea further encouraged private companies by guaranteeing minimum purchase quantities and speedy approval from government regulators [40,41]. Similarly, in the United States, the PASTEUR act attempts to introduce a subscription model into the pharmaceutical landscape, to incentivize private companies to invest in research and development. Under this scheme, companies will be rewarded for new antimicrobial developments, depending on relevance and impact to public health as well as novelty, with yearly revenues preemptively agreed, rather than based on economic and market potential [42]. A strong and reliable political commitment to invest in pandemic preparedness and incentives to private companies are not the only contributing factors for the lack of funding but, if tackled, might help to implement the calls for action in a financially sustainable way.

Lack of a systems approach

Several approaches have been developed to assess countries' level of pandemic preparedness [24,43]. The COVID-19 pandemic has highlighted stark inconsistencies between their findings and the countries' success in responding to or containing the spread of the virus. For example, while the Global Health Security Index (GHSI) results published in 2019 rate the United States as best prepared for a pandemic, at the time of writing it has recorded the highest death toll from COVID-19 [43]. The other 10 countries with the highest COVID-19 mortality rates per million are also amongst the top 20 GHSI-scoring countries. On the other hand, countries in Southeast Asia and sub-Saharan Africa that had much lower GHIS scores were more successful in containing and responding to the first wave of COVID-19 [43]. A more recent exposure to infectious disease outbreaks can partially explain better preparedness measures. However, the most significant factor in these incongruences seems to be a precarious conceptualization of preparedness, which only considers a narrow set of measures without adequately weighing the importance of political leadership, trust in science, the role of the social determinants of health, and interlinkages between these. By analyzing infectious disease outbreaks through this reduced spectrum, the dynamic relationships between multiple stakeholders are not given the necessary consideration, and the actions for adequate responses remain fragmented. Advancing beyond this conceptual failure requires a systems approach that considers the inherent complexity of health challenges.

Systems thinking is a 'perspective that considers connections amongst different components (of a system), plans for the implications of their interactions, and requires transdisciplinary thinking' [44]. This approach is by no means a novelty in global health security. However, in most cases, it remains a discursive element that is rarely used to develop or evaluate emergency preparedness. World leaders still have to apply pandemic preparedness approaches that engage multiple sectors and deal with complex networks of relevant stakeholders with diverse priorities. In doing so, it is necessary to challenge the existing paradigm of fast economic growth that benefits only a few in the short term, towards a long-term multisectoral approach to prevent the emergence of zoonotic diseases. This requires in-depth consideration of the human relationships with the environment, particularly human dominance above other species and our consumption patterns including what we eat and how/where we live.

Additionally, a core dimension necessary to apply systems thinking in global health security is to empower local communities through education, health literacy, and leadership opportunities that support their involvement in policy and emergency response development. The inclusion of gender, ethnic, and climate-sensitive opinions in decision-making and implementation processes will help in creating synergies and sustain a long-lasting impact on local systems' resilience to tackle future pandemics. An additional consideration into putting system thinking in action is that developing new strategies from scratch after each emergency is rarely the best course of action, and existing frameworks should be utilized to catalyze support. For instance, the Sustainable Development Goals (SDGs) already provide a holistic approach that connects health to other spheres of human development. Making clear, shared objectives between the SDGs and emergency preparedness will equip people with a vision, tools, and the vocabulary to consider global health security in all its complexity, to prepare for unanticipated consequences, and to strategize for ways to galvanize a wide range of resources to address emerging challenges.

Governance and leadership

Putting into practice all the infrastructure that has been built in preparation for a pandemic, reacting quickly and decisively, and introducing reforms to better prepare for the future given a limited number of resources, comes down to two important factors: cohesive leadership and transparent governance. Emergency preparedness is a multistakeholder endeavour that involves collaboration between multilateral agencies, nongovernmental organizations, local governments, community leaders, and other stakeholders. National governments are expected to guide the direction of emergency preparedness and showcase unity and cohesion by streamlining resources and displaying strong conviction. Meanwhile, transparency earns governments trust and exposes policy and implementation gaps, so these may be addressed effectively. An adequate level of trust from relevant stakeholders in the communities affected and amongst the general population is vital to achieve compliance with public health restrictions and positive health behaviours.

In outbreaks and pandemic preparedness, a faster response is imperative for limiting new entries of the pathogens into the community, thus mitigating community transmission, and minimizing outbreak impact. Failure to scale up surveillance and laboratory diagnostic capacities together with poor funding for these services have proven to be critical gaps in outbreak preparedness. However, previous PHEICs such as H1N1, Ebola, and Zika have shown that slow political response and mobilizations are the main mechanisms leading to delays in the development of guidelines and implementation of effective measures. In fact, slow political mobilization is responsible for almost double the amount of time delay compared to surveillance delays [45]. Studies of early pandemic response have shown that in the European context, governments that deployed a more centralized response were able to mobilize federal agencies quicker and develop strategies for potential local outbreaks. One example is the United Kingdom, which was able to issue measures to mitigate the spread and impact of SARS-CoV2 much faster than other less centralized states such as Switzerland [45]. Nonetheless, decentralized systems do generally fare better in the long run due to better capability to respond and mobilize resources according to context-specific local needs, and thus more effectively decrease the R number. However, the COVID-19 pandemic demonstrated the potential faults in decentralized governments that emerge when leaders at different levels of government and agency leaders engage in power struggles and distrust each other, as was visible in the United States [46].

Examples of slow disjointed responses to outbreaks leading to uncontrollable spread in at-risk populations are plentiful. Exposing at-risk populations to novel and emerging pathogens has the dual effect of exacerbating poor conditions amongst vulnerable groups and creating opportunities for the virus to spread widely into the community. The Taiwanese response to SARS can serve as an archetype of this situation. Taiwan was one of the hardest hit countries during the SARS outbreak in 2003. At the initial entry of the SARS virus into the country, healthcare workers in hospitals were the group most affected, with a majority of cases attributed to nosocomial infections [47]. Successful infection management was ultimately due to proper isolation of infected patients and potential contacts to break chains of infection, and implementing strict infection control measures in hospitals to reduce hospital-based transmission [47]. In response to SARS and to prepare for future pandemics, the Taiwanese government invested resources and created a framework for the Central Epidemic Command Centre (CECC). When activated in response to epidemic threats, the CECC acts as a command centre for interministry action, cooperation and streamlining of resources and information. Taiwan maintained an active CECC after SARS through diverse initiatives and collaboration with other agencies. This led to a rapid, transparent, and robust government response that was the cornerstone of Taiwan's early success in response to the COVID-19 pandemic [48].

On the other side of the spectrum, Canada's experience is a cautionary tale of short-lived political will. Along with Taiwan, Canada was one of the countries hardest hit by the SARS outbreak. Canada also underwent great introspection and assessments, and implemented sweeping reform, such as the establishing of the Public Health Agency of Canada (PHAC) and building capacity for the Global Public Health Intelligence Network (GPHIN). However, contrary to Taiwan, Canada turned to complacency allowing their postoutbreak infrastructure to erode with time [49]. Notably, Canada's weakened PHAC shifted funds from GPHIN to other domestic projects and allowed stockpiles of personal protective equipment to expire without adequate monitoring. As shown in the stark contrast of the implementation of pandemic preparedness and response mechanisms between Taiwan and Canada, sustained financial investment and political will is critical. Furthermore, the cycle of collecting lessons, addressing gaps, and increasing knowledge and education does not appear to be sufficient to avoid the repetition of errors in policy making and implementation processes during public health crises. This cycle points to a greater problem than 'bad apples' or 'bad politicians', but rather a system that creates a space for different politicians to repeat such mistakes [50]. It is about striking a balance to create a system that acts as a command center—a hub where all the branches and sectors of government converge to deliver a unified approach to outbreak preparedness and response; and creating a system 'to share lessons of good and bad practices that are not about blaming but about learning' [50].

An often-overlooked impact of infectious disease outbreaks, beyond the health, economic, and emotional costs, is the erosion of trust and stability in the relationship between governments and their constituents. As pointed out by UK researchers, following 'choice-based cause of [policy] failures' (due to limited evidence, avoidance of evidence-based tools and policies, use of wrong political tools, or just simply due to the power of hindsight), there is the inevitable politics of blame, formal inquiries, and media investigations that follow [50]. The importance of these investigations is not trivial, as investigators and politicians must work together to find pitfalls decision-makers have made and learn from these mistakes for the future. However, this media coverage and political blaming creates a hostile atmosphere that instils doubt and scepticism in society, planting the seeds for political distrust and widespread cynicism in the population (not to confuse with political mistrust, which is an essential quality for the public to remain politically engaged and hold politicians accountable).[e] Cooperative and responsible reporting is needed to maintain an

[e] 'Political trust thus functions as the glue that keeps the system together and as the oil that lubricates the policy machine. Mistrust, or rather political scepticism, plays an equally important role in representative democracy. Critical citizens are more likely to engage in political activities and to keep office-holders accountable. When mistrust turns into widespread distrust and cynicism, the quality of democratic representation itself may change' [51].

informed public that will hold the government accountable to create the foundation for cooperative learning, integration of perspectives, and cooperation between stakeholders at multiple levels [52].

Context and political landscape

The nature and focus of post hoc assessments contribute to the generation of generic criticisms that often overlook crucial socio-geo-political contexts. One good example of this phenomenon is the reaction of the international community to the 2014–16 Ebola outbreak response. Following controversial criticisms related to the leadership of the outbreak response, the international health community, academic institutions, and several agencies invested resources into understanding the shortcomings of the WHO and the structural changes needed to improve future responses. One of the main aspects raised in these investigations was the need to re-establish the role of the WHO as a coordinating agency responsible for setting standards and guidelines for the international community rather than an operational agency delivering interventions in the field. It was highlighted that misunderstandings of WHO's mandate had translated into misdirected criticisms, and perpetuated a misrepresentation of the political landscape of outbreak responses that kept some of the actors most involved in the responses on the ground away from scrutiny [18]. Likewise, the consideration of politics between and within stakeholders received little attention. In drafting calls to action and lessons for future pandemics, it is crucial for official commissions to consider the politics of the response and power relationships between diverse actors in order to provide concrete answers of what we need to do differently.

Conclusion

Post hoc assessments and other comprehensive reviews of responses to infectious disease outbreaks in the past decades have identified a plethora of lessons to be learned and urgent calls for action to develop preparedness and response measures. Looking at these calls, one might experience a déjà-vu of overlapping lessons sometimes nearly identical in content. While this should not mislead us to assume that we have ignored all previous lessons, it clearly indicates that we have been neglecting systemic and underlying issues. In our analysis, we identified four core issues: (1) the lack of funding and financial incentives that would generate sustainable long-term investment and commitment to pandemic preparedness in the public and private sectors, (2) the lack of a systems approach to develop holistic pandemic preparedness and response measures that take into account the most vulnerable and is aligned with the SDGs, (3) shortcomings in existing systems of governance that weaken effective mechanisms and allow for poor decision-making, and (4) the lack of consideration to context and politics as barriers to implementing effective evidence-based solutions. Understanding and addressing these systemic issues is key to developing new guidelines, legislation and action for pandemic preparedness that spares the world from the devastation future pandemics may bring.

PICTURE 4.13.3 One positive side of the pandemic was a fitter, healthier population in other ways. Aoife Kirk.

13.5 The need for a defensive frontline in health security

In 1995, the World Health Assembly revised the International Health Regulations (IHR) to include the 'threat posed by the international spread of new and re-emerging diseases'. Revisions in 2005 redefined the scope of the IHR as 'prevent, protect against, control and provide a public health response to the international spread of disease in ways that are commensurate with, and restricted to, public health risks'. Public health experts agree that if all countries put adequate resources into IHR compliance, the occurrence of epidemics and pandemics would be very much reduced.

Unfortunately, most countries neglect this primary duty to humanity, but some are particularly negligent. Since the world cannot live up to the IHR, defensive lines, as a temporary measure, must be drawn up. Given limitations on global resources, these lines must focus on nations with a track record of disease incubation and dissemination.

Despite assurances, the track record of many countries proves that the world cannot merely take their word regarding disease outbreaks. As a first step in the complex process of preventing global pandemics [53], it is time for affluent nations to step up and create a defensive health security line to shield the rest of the world from infectious disease. Fortunately, several Asia-Pacific countries have a significant stake in disease surveillance and early warning. Current events have demonstrated that such alliances are proactive and successful at mitigating pandemic problems.

In Asia, for example, this task is perhaps best suited to Japan and Taiwan, both of which have many years of experience being on a defensive footing against high-risk countries. Yet, when one examines the security posture of Japan and Taiwan, it is clear that their role has long been limited to 'Defence of Nation'. This may be required to change and expand as these countries advance to assume greater importance on the world stage.

Japan and Taiwan are perfectly poised with great resources, scientific capacity, and ingenuity to become global pandemic shields to protect the world from disease pathogens originating in the region. Like Japan and Taiwan, South Korea and Vietnam have common health security concerns. However, amid the COVID-19 pandemic, these four countries may now be more open to participating in an early detection and disease control programme to secure the region and the world from human pathogens.

PICTURE 4.13.4 In some parts of the world, biodefense shields are more relevant than in others. Aoife Kirk.

13.6 Key points

- Participation in biodefense shields and other global health security efforts is the responsibility of all countries, even small ones such as Ireland.
- Associated international agreements are essential tools, much like the Geneva Convention, for at least partially ensuring that future pandemics are contained.
- The structure of biodefense shields and other associated efforts are closely linked to international relations and diplomacy (to be explored in the next chapter).

References

[1] B.R. Levin, M. Lipsitch, S. Bonhoeffer, Population biology, evolution, and infectious disease: convergence and synthesis, Science 283 (5403) (1999) 806–809.

[2] S.-Q. Deng, H.-J. Peng, Characteristics of and public health responses to the coronavirus disease 2019 outbreak in China, J. Clin. Med. 9 (2) (2020) 575.

[3] R. Peckham, Viral surveillance and the 1968 Hong Kong flu pandemic, J. Glob. Hist. 15 (3) (2020) 444–458.

[4] H. Ritchie, E. Mathieu, A.C. Rodes-Guirao, et al., C. Giattino, E. Oritz-Ospina, J. Hasell, B. Macdonald, D. Beltikian, M. Roser, Coronavirus Pandemic (COVID-19), OurWorldindata.org, 2021. Retrieved from: https://ourworldindata,org/coronavirus.

[5] D. Gurdasani, J. Drury, T. Greenhalgh, S. Griffin, Z. Haque, Z. Hyde, A. Katzourakis, M. McKee, S. Michie, C. Pagel, S. Reicher, A. Roberts, R. West, C. Yates, H. Ziauddeen, Mass infection is not an option: we must do more to protect our young, Lancet 398 (10297) (2021) 297–298, https://doi.org/10.1016/S0140-6736(21)01589-0.

[6] A. Polianskaya, What is the 'pingdemic'? https://inews.co.uk/news/uk/pingdemic-what-meaning-nhs-covid-app-self-isolate-how-many-explained-1109939. (Accessed 2 August 2021).

[7] B.L.H. Wong, M.S. Green, J. Reid, J.M. Martin-Moreno, N. Davidovitch, L. Chambaud, L. Leighton, M. Sheek-Hussein, R. Dhonkal, R. Otok, J.D. Middleton, Toward 'vaccine internationalism': the need for an equitable and coordinated global vaccination approach to effectively combat COVID-19, Int. J. Public Health 66 (2021) 1604077, https://doi.org/10.3389/ijph.2021.1604077.

[8] S. Berkley, COVAX: more than a beautiful idea, Lancet 398 (10298) (2021) 388, https://doi.org/10.1016/S0140-6736(21)01544-0.

[9] B. Mueller, R. Robbins, Where a vast global vaccination program went wrong, New York Times (2021). https://www.nytimes.com/2021/08/02/world/europe/covax-covid-vaccine-problems-africa.html?action=click&module=Well&pgtype=Homepage&section=World%20News. (Accessed 2 August 2021).

[10] Lancet Commission on COVID-19 Vaccines and Therapeutics Task Force Members, Operation Warp Speed: implications for global vaccine security, Lancet Glob. Health 9 (7) (2021) e1017–e1021, https://doi.org/10.1016/S2214-109X(21)00140-6.

[11] G. Paravicini, India's Halt to Vaccine Exports 'Very Problematic' for Africa. https://www.reuters.com/business/healthcare-pharmaceuticals/indias-halt-vaccine-exports-very-problematic-africa-2021-05-18/. (Accessed 2 August 2021).

[12] India's Covid Crisis Exposes Deep-Rooted Problems in Public Health After Years of Neglect. https://www.cnbc.com/2021/05/18/india-covid-crisis-shows-public-health-neglect-problems-underinvestment.html. (Accessed 2 August 2021).

[13] WHO Director-General's Keynote Speech at the 138th International Olympic Committee Session. https://www.who.int/director-general/speeches/detail/who-director-general-s-keynote-speech-at-the-138th-international-olympic-committee-session. (Accessed 2 August 2021).

[14] M. Thelwall, S. Thelwall, Retweeting for COVID-19: consensus building, information sharing, dissent, and lockdown life, arXiv preprint arXiv:2004.02793 10, 2020.

[15] World Health Organization, A Strategic Framework for Emergency Preparedness, 2017.

[16] R.A. Weiss, HIV and AIDS in relation to other pandemics, EMBO Rep. 4 (2003) (Special issue).

[17] Center for Systems Science and Engineering, COVID-19 Map—Johns Hopkins Coronavirus Resource Center, Johns Hopkins Coronavirus Resource Center, 2021. [cited 2021 Aug 25]. Available from: https://coronavirus.jhu.edu/map.html.

[18] International Labour Organization, Food and Agriculture Organization of the United Nations, International Fund for Agriculture Development, World Health Organization, Impact of

COVID-19 on People's Livelihoods, Their Health and Our Food Systems, World Health Organization, 2020, p. 3. [cited 2021 Aug 29]. Available from: https://www.who.int/news/item/13-10-2020-impact-of-covid-19-on-people's-livelihoods-their-health-and-our-food-systems.

[19] The World Bank, The Global Economic Outlook During the COVID-19 Pandemic: A Changed World, vol. 12, The World Bank, 2020, pp. 5–9. [cited 2021 Aug 29]. Available from: https://www.worldbank.org/en/news/feature/2020/06/08/the-global-economic-outlook-during-the-covid-19-pandemic-a-changed-world.

[20] United Nations Development Programme, Brief #2: Putting the UN Framework for Socio-Economic Response to COVID-19 Into Action: Insights, 2020, Available from: https://www.undp.org/content/undp/en/home/coronavirus/socio-economic-impact-of-covid-19.html.

[21] United Nations Entity for Gender Equality and the Empowerment of Women (UN Women), United Nations Secretariat, The Impact of COVID-19 on Women, vol. 2, United Nations, 2020. [cited 2021 Aug 29]. Available from: file:///C:/Users/camiv/Downloads/Policy-brief-the-impact-of-covid-19-on-women-en (1).pdf.

[22] The Independent Panel for Pandemic Preparedness and Response, COVID-19: Make It the Last Pandemic, 2021.

[23] International Rescue Committee, The Ebola Lessons Reader: What's Being Said, What's Missing and Why It Matters, 2016, Available from: https://www.rescue.org/report/ebola-lessons-reader-whats-being-said-whats-missing-and-why-it-matters.

[24] World Health Organization, Health Regulations, third ed., World Health Organization, 2005–2016. 1–84 p. [cited 2021 Aug 29]. Available from: https://apps.who.int/iris/bitstream/handle/10665/246107/9789241580496-eng.pdf?sequence=1.

[25] P. Sands, D. Casserley, R. Chaves, T.G. Evans, S. Gupta, J. Hazlewood, et al., From Panic and Neglect to Investing in Health Security: Financing Pandemic Preparedness at a National Level, International Working Group on Financing Preparedness, 2017.

[26] World Health Organization, Report of the Ebola Interim Assessment Panel—July 2015, World Health Organization, 2015. [cited 2021 Aug 29]. Available from: https://www.who.int/publications/m/item/report-of-the-ebola-interim-assessment-panel- - -july-2015.

[27] Threats I of M (US) F on M, S. Knobler, A. Mahmoud, S. Lemon, A. Mack, L. Sivitz, et al., Lessons From SARS for Future Outbreaks, National Academies Press, US, 2004. [cited 2021 Aug 29]. Available from: https://www.ncbi.nlm.nih.gov/books/NBK92465/.

[28] Review Committee, Report of the Review Committee on the Functioning of the International Health Regulations (2005) in Relation to Pandemic (H1N1) 2009, World Health Organization, 2011. Available from: https://apps.who.int/gb/ebwha/pdf_files/WHA64/A64_10-en.pdf?ua=1.

[29] S. Moon, D. Sridhar, M.A. Pate, A.K. Jha, C. Clinton, S. Delaunay, et al., Will Ebola change the game? Ten essential reforms before the next pandemic. The report of the Harvard-LSHTM Independent Panel on the Global Response to Ebola, Lancet 386 (2015) 2204–2221. Available from: https://doi.org/10.1016/S0140-6736(15)00946-0.

[30] Pan-American Health Organization, H. Ruiz, Zika Strategic Response Plan, 2016, [cited 2021 Aug 29]. Available from: http://www.who.int.

[31] Commission, Communication from the Commission to the European Parliament, the European Council, the Council, the European Economic and Social Committee and the Committee of the Regions: Drawing the Early Lessons From the COVID-19 Pandemic, European Commission, 2021. Available from: https://ec.europa.eu/info/sites/default/files/communication150621.pdf.

[32] Global Preparedness Monitoring Board (GPMB), A World At Risk: Annual Report on Global Preparedness for Health Emergencies, 2019.

[33] E. Radin, C. Eleftheriades, Financing Pandemic Preparedness and Response, 2021, Available from: https://theindependentpanel.org/wp-content/uploads/2021/05/Background-Paper-14-Financing-Pandemic-Preparedness-and-Response.pdf.

[34] M.R. Smith, L. Weber, H. Recht, Billions spent on coronavirus fight, but what happens next? abc news (2021). [cited 2021 Aug 24]. Available from: https://abcnews.go.com/Health/wireStory/public-health-seeks-steady-funding-feast-famine-77164492.

[35] M. Craven, A. Sabow, L. Van der Veken, M. Wilson, Not the last pandemic: investing now to reimagine public-health systems, McKinsey (2021). [cited 2021 Aug 24]. Available from: https://www.mckinsey.com/industries/public-and-social-sector/our-insights/not-the-last-pandemic-investing-now-to-reimagine-public-health-systems.

[36] A. Glassman, E. Smitham, Financing for Global Health Security and Pandemic Preparedness: Taking Stock and What's Next, Center for Global Development, 2021. [cited 2021 Aug 24]. Available from: https://www.cgdev.org/blog/financing-global-health-security-and-pandemic-preparedness-taking-stock-whats-next.

[37] H. Stutzman, A.E. Micah, J.L. Dieleman, Funding Pandemic Preparedness: A Global Public Good, Think Global Health, 2020. [cited 2021 Aug 24]. Available from: https://www.thinkglobalhealth.org/article/funding-pandemic-preparedness-global-public-good.

[38] J. Reder, C. Doyle Cooper, The Pandemic Preparedness Paradox, PMLiVE, 2021. [cited 2021 Aug 24]. Available from: http://www.pmlive.com/pharma_intelligence/The_pandemic_preparedness_paradox_1361186.

[39] Ministry of Science and ICT Republic of Korea, How We Fought COVID-19: A Perspective From Science & ICT, 2020, Available from: https://www.mofa.go.kr/eng/brd/m_22742/view.do?seq=28&srchFr=&srchTo=&srchWord=&srchTp=&multi_itm_seq=0&itm_seq_1=0&itm_seq_2=0&company_cd=&company_nm=&page=1&titleNm=.

[40] J. Shuren, T. Stenzel, South Korea's implementation of a COVID-19 national testing strategy, Health Aff. (2021). [cited 2021 Aug 24]. Available from: https://www.healthaffairs.org/do/10.1377/hblog20210521.255232/full/.

[41] P. Dyer, Policy and Institutional Responses to COVID-19: South Korea, Brookings Doha Center, 2021. Available from: https://www.brookings.edu/research/policy-and-institutional-responses-to-covid-19-south-korea/.

[42] S. Vorperian, S. Quake, The PASTEUR Act can help win the war against superbugs, STAT (2021). [cited 2021 Aug 24]. Available from: https://www.statnews.com/2021/06/25/pasteur-act-help-fight-superbugs-antimicrobial-resistance/.

[43] John Hopskin Center for Health Security, GHS Index (Global Health Security Index). Building Collective Action and Accountability, 2019, Available from: www.ghsindex.org.

[44] S.J. Leischow, B. Milstein, Systems thinking and modeling for public health practice, Am. J. Public Health 96 (3) (2006) 403. March [cited 2021 Aug 29]. Available from: /pmc/articles/PMC1470500/.

[45] S.J. Hoffman, S.L. Silverberg, Delays in global disease outbreak responses: lessons from H1N1, Ebola, and Zika, Am. J. Public Health 108 (3) (2018) 329–333. Available from: https://pubmed.ncbi.nlm.nih.gov/29345996.

[46] Y. Abutaleb, J. Dawsey, E. Nakashima, G. Miller, Denial and dysfunction plagued U.S. government as coronavirus raged, The Washington Post (2020). [cited 2021 Aug 29]. Available from: https://www.washingtonpost.com/national-security/2020/04/04/coronavirus-government-dysfunction/?arc404=true.

[47] P.-R. Hsueh, P.-C. Yang, SARS outbreak in Taiwan, Emerg. Infect. Dis. 10 (8) (2004) 1514–1516. Available from: https://pubmed.ncbi.nlm.nih.gov/15503404.

[48] S.-C. Chen, Taiwan's experience in fighting COVID-19, Nat. Immunol. 22 (4) (2021) 393–394. Available from: https://doi.org/10.1038/s41590-021-00908-2.
[49] G. Robertson, 'We are not prepared': the flaws inside Public Health that hurt Canada's readiness for COVID-19, The Globe and Mail (2020). Available from: https://www.theglobeandmail.com/canada/article-we-are-not-prepared-the-flaws-inside-public-health-that-hurt-canadas/.
[50] J. Gaskell, G. Stoker, W. Jennings, D. Devine, Covid-19 and the blunders of our governments: long-run system failings aggravated by political choices, Polit. Q. 91 (3) (2020) 523–533.
[51] T.W.G. Van der Meer, S. Zmerli, The deeply rooted concern with political trust, in: Handbook on Political Trust, Edward Elgar Publishing, 2017.
[52] J. Gaskell, G. Stoker, Centralized or decentralized which governance systems are having a "good" pandemic? Democr. Theory 7 (2) (2020) 33–40. Available from: https://www.berghahnjournals.com/view/journals/democratic-theory/7/2/dt070205.xml.
[53] M.S. Cohen, N. Hellmann, J.A. Levy, K. DeCock, J. Lange, The spread, treatment, and prevention of HIV-1: evolution of a global pandemic, J. Clin. Invest. 118 (4) (2008) 1244–1254.

Chapter 14

International relations implications of the pandemic

Sebastian Kevany, Deon Canyon, and Aoife Kirk

Chapter outline

14.1 Pandemics and international relations

There are a few empirical (as opposed to theoretical) studies of how epidemic transmission—an international security issue—can be linked to international relations and diplomacy. In the context of a pandemic, some have observed a rise in transparency and collaboration between nations. These unlikely collaborations have led to speculation on perhaps some significantly improved relations between nation-states. Moreover, they have led to better communication, collaboration, supranational governance, and the timely sharing of critical public health information. Others, however, are less optimistic about these changes [1].

Outbreaks, Epidemics, and Health Security. https://doi.org/10.1016/B978-0-323-91705-6.00004-6

Pan-national structures, such as the European Union, the Association of South-East Asian Nations, the United States, or the African Union, have enabled greater cooperation in the epidemic context. These structures have also led to increased border-related risk levels between member states. However, in a hypothetical situation where there is little or no diplomacy or a supranational collaborative effort between different countries, the links between infectious disease control and containment strategies are inevitably strained. The political and international implications are ongoing at the time of writing, with new strains emerging in late 2021. However, the authors emphasize for future lessons the importance of quick collaboration, sharing of science, public health information, and clear communication to maintain international relations [2].

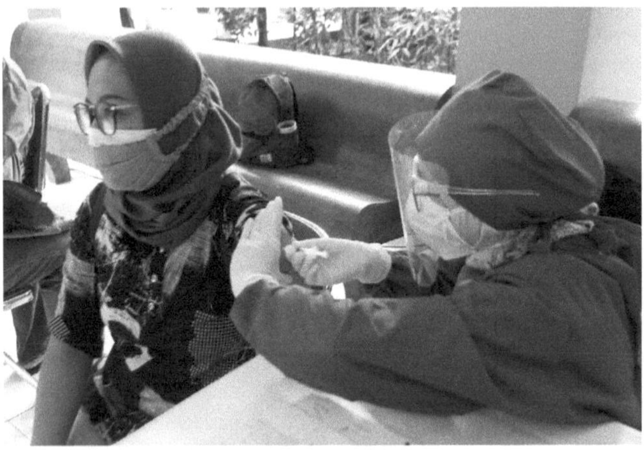

PICTURE 4.14.1 **Covid vaccines were both a barrier and an enabler to international relations.** Yodi Mahendradhata.

14.2 Limited cooperation scenarios

In an extreme case, in which no communication exists between nations, lack of information sharing will most likely lead to further spread of disease. The only way to mitigate this certainty is if limited collaborations are accompanied by corresponding temporary travel, mobility, and globalization limitations. If less harmonization is connected to the containment of vector movement, then perhaps isolationism or international antagonism may be seen as potentially helping prevent and contain the spread of infectious diseases. Moreover, in an age where rising nationalism may have promoted less cooperation between nations and supranational organizations—and often greater associated tension in some bilateral relations—information flows may weaken, with a decline in trust among states in recent years.

Certain criticism put China on the diplomatic offensive. The Chinese government requested other states and organizations to temper criticism, yet later attempted to present their efforts in the epidemic as 'ground-breaking'. Unfortunately, related problems, including timely and objective information sharing, may have made epidemic surveillance and preparedness difficult for other states, particularly for countries with more exposure to those international forces that bind them to higher levels of economic and human mobility. Moreover, straightforward cooperative elements are not necessarily easy between nation-states.

14.3 Guest essay: Diplomatic and international relations implications between countries in the epidemic era

Coral Andrews

Multilateralism and diplomacy are garnering renewed attention as essential foreign policy strategies in response to the ongoing impact of the COVID-19 pandemic. At the height of the outbreak, the World Health Organization (WHO) came under fire for its handling of the public health reporting of the international disease outbreak. WHO serves as a collaborating body between nations and is governed by the World Health Assembly (WHA) which establishes policy and performs other supervisory functions [3]. The global health agenda has grown increasingly complex, and the impact of health security and climate health have further blurred the lines associated with what constitutes the health agenda.

One suggestion is the need to harness the expertise of the former diplomats and executive leaders who led us through the historical international negotiations and establishment of a successful multilateral organization at the height of an epidemic. The organization in mind as an example is the Global Fund to Fight AIDS, Tuberculosis, and Malaria (the Global Fund). Why wouldn't we draw upon the insights from the diplomatic and global response to the HIV/AIDS epidemic as a road map for navigating the global challenges of the current pandemic? The Global Fund was established as an independent funding organization in Geneva, Switzerland, in January 2002 following a multiyear process of debates and negotiation amongst the United Nations, state and nonstate actors [4]. My reasons for supporting this position include observations gleaned from my doctoral dissertation work on global health diplomacy and an examination of the case of the Global Fund. Competing agendas between health, security, and economics need to be reconciled to promote incentives for reporting global public health threats. The pandemic has accelerated the conversations around human security and the link between climate health and human health. Cooperation between nations is essential for managing a threat that looms as largely as we have seen with this pandemic.

On 3 March 2021, US Secretary of State, Antony Blinken, delivered a speech to the American people which identified its top eight foreign policy priorities. Global cooperation was emphasized as an essential principle in navigating today's complex challenges. Diplomacy was highlighted as the mechanism by which the Secretary of State would carry out President Biden's interim National Security Strategy (NSS) and associated foreign policy [5]. In the American Rescue Plan Act of 2021, the United States included a $3.5 billion emergency funding investment in the Global Fund to support its efforts to respond to the COVID-19 pandemic impact on low- and middle-income countries [6]. On 24 April 2021, the United Nations observed the 'International Day of Multilateralism and Diplomacy for Peace'. Shaped by the ravages of the COVID-19 pandemic and its global impact, UN Secretary General, António Guterres, offered a reflective tone and highlighted our societal connectedness [7]. Advancing multilateral strategies in response to these global challenges should incorporate a framework that seeks to advance the rights of human beings within a peaceful and sustainable world [8]. On 29 June 2021, at a meeting of the G-20 in Bari and Matera, Italy, the United States' commitment to multilateralism was emphasized. This included an affirmation that multilateralism serves as an essential tool for addressing a range of global threats. To achieve this vision, it requires a reliance upon the collective international commitment of the rule-of-law, democracy, and the rights of all individuals [9].

A renewed dedication to multilateralism and diplomacy necessitates that nation-states accept some neutralization of power and a willingness to diffuse it across multilateral structures. Strengthening this requires a sorting out of global governance issues and revisiting of the International Health Regulations (IHRs), which projected itself during the pandemic as a teetering force for compelling a universal adherence to reporting global public health threats. The role of global institutions, such as the World Health Organization (WHO), will be subject to further debate in 2021. A recent survey of global health experts sought opinions about the need for a new organization focused on global health around the world. The global health experts who were polled were split (some agreeing and some disagreeing) on this question [10].

With the pandemic still upon us in many parts of the world, I believe that we need to support the existing multilateral organizations that are in place and build out diplomatic capabilities to strengthen global governance and the IHRs. Diplomacy, to be effective, relies upon the role of champions and leaders. Let's call upon those who championed the HIV/AIDS epidemic to guide us in strengthening existing multilateral organizations, such as the Global Fund, and harness the power of diplomats to rebuild nation-to-nation collaboration.

14.4 International mobility implications

From an international relations perspective, states have an incentive to cooperate if their goal is to protect and benefit their populations and the planet more broadly. Yet, the 2020–21 pandemic has created a related need to re-examine the risks of global mobility, at least in terms of the mass movement of populations for economic and political migration. Moreover, rising refugee numbers internationally, along with the overarching threat of climate change, can place high demands on housing, shelter, and welfare on host nations, and may be linked with a host of public health problems [11].

In this regard, it remains to be seen if the pandemic has created a 'new normal' in which future policies implement restrictions. These restrictions may not be imposed on international business, professional development, or diplomatic or trade-related missions, but perhaps necessarily on tourism, leisure or personal travel alongside the mass labour population movements, economic growth, and future investments worldwide. If reactions to past transnational outbreaks provide any indication, sustaining such elements of lockdown seems both improbable and unsustainable: as long as the free market is in operation in the realm of low cost and increasingly deregulated international transportation, only the imposition of stricter visa or screening requirements, for example, can be considered a valuable tool from the public health perspective.

Many nation-states also faced significant epidemic-related challenges related to their economic and population security. Declines in demand and production can displace workers and decrease or reverse economic growth, as in West Africa during the Ebola epidemic. If such threats make their way to more rural and agricultural areas, food supplies are also compromised as workers risk falling ill. In the same way, the 2020–21 pandemic exposed the danger of dependence on international supply chains for critical products; as nation-states competed for more medical equipment and supplies, global demand and prices increased for both the products and the materials needed to make them [12].

Beyond even the political and the economic, human mobility in refugees and internally displaced people (IDP) has also posed specific difficulties in the pandemic context. Sadly, the nature of refugee and IDP camps are inherent flashpoints for viral transmission. Risks of limited access to health care, crowded living conditions, and the destruction of health infrastructures in war-torn countries often leave refugees and IDPs highly vulnerable to many kinds of communicable diseases and infections.

Should epidemiology take primacy, states may, in future, have less incentive to cooperate with other countries and global institutions in absorbing displaced populations—a development that is to be feared, both on strategic and humanitarian grounds. Counter to this, perhaps it provides an opportunity to improve the living conditions of refugee camps to avoid the risk of spread to populations.

However, curtailing optional forms of international human mobility may yet be the most benign and effective action from a public health emergency perspective. Thus, except for refugees escaping persecution and war, there is a need to re-examine the risks associated with political and economic migration. The former has been a facet of many war-torn areas such as Syria, Afghanistan, Myanmar, and South Sudan. Much of the latter has been a keystone facet of both malignant and benign global economic expansion, with international labour movement occurring more rapidly—and across greater distances—than during any other historical period.

14.5 Epidemics and security

Security considerations, as well, play into this debate. Does neo-isolationism[a] increase or threaten global security; in the three-way nexus between epidemics, national, and international security, and international relations, what are the implications for each in adjusting to the other? Good international relations improve security through greater identification with our fellow humanity, exchanges of ideas, and economic, trade, political and social collaboration—in economics terms, the laws of absolute and comparative advantage. But, what happens when public health issues strain both international relations and security at the same time [13]?

For example, once nations have ensured the stability of internal health and nonhealth, will they begin to look outwards to assist their allies? Conversely, would allies with uncontrolled outbreaks receive the same treatment as enemies; to what extent do geostrategic elements come into play? And, given these considerations, what are the implications for nonpermissive international intervention epidemic control efforts?

Whatever the scenario, there appears to be a slight possibility that the 2020–21 pandemic has had a benign effect on domestic or regional security apparatuses. Internally, social stigma towards those who tested positive for the virus or those from highly affected areas has been documented; in the United States, Asians and Asian Americans were subjected to prejudice and racism [14].

These unnecessary and even paranoid views feed virulent xenophobia between and within nations that may continue to grow. In addition, large- or small-scale social groups (from communities or countries) view visitors or strangers with an increasing degree of fear and suspicion. Thus, while transnational outbreaks logically constitute a threat to international relations and regional security, they also provide an opportunistic excuse to take unjustified and unfair action.

14.6 Guest essay: Pandemic security in Papua New Guinea

Deon Canyon[b] and Joelson Anere[c]

The views expressed in this article are the author's alone, and do not necessarily reflect the official position of the DKI-APCSS, the United States Government, or the Government of Papua New Guinea.

COVID-19 in PNG

In March 2021, reported cases of COVID-19 (Covid) began to climb in Papua New Guinea (PNG) with both the Australian government and the World Health

[a]A revival of isolationism arising from a reluctance to involve the nation in further political and military commitments abroad.

[b]Daniel K. Inouye Asia-Pacific Center for Security Studies (DKI-APCSS), Honolulu Hawaii.
[c]Department of National Planning and Monitoring, Government of Papua New Guinea.

Organization suspecting widespread community infection. Official numbers, they said, did not reflect the reality, and case numbers were much higher than reported. Later in April, when there were 7000 Covid cases reported in PNG, experts estimated that the actual number was more like 100,000. Complacency, a slow response and poor testing in a nation where social distancing is minimal, were thought to be responsible.

In early July, PNG authorities confirmed the detection of a Delta variant Covid case, and the National Pandemic Response deputy controller Dr Daoni Esorom called on the populace to be proactive. At that time, less than one per cent of the population had been vaccinated, and 62% of surveyed people believed they would not be infected with Covid.

While the mechanisms and sources of Covid spread in PNG have not been researched, the population is very young (median age 22.4 years old), and extensive asymptomatic spread was suspected amongst the population. Data are especially sparse for rural and remote regions, in which 87% of the population resides.

World Bank response

In 2020, the World Bank carried out a Socio-economic Impact Assessment Survey on Covid in PNG that was jointly undertaken by the Department of National Planning and Monitoring and the United Nations Development Programme. They recommended the PNG National Executive Council focus on the following:

- Ensuring the protection of human rights and inclusion.
- Ensuring the provision of assistance to vulnerable people.
- Improving coordination amongst communities, development partners, the private sector, and international nongovernmental organizations.

The World Bank approved US$20 million in April 2021 for the COVID-19 Emergency Response Project in Papua New Guinea (PNG) to prevent, detect, and respond to the threat posed by Covid, which was also used to aid and to strengthen national systems for public health preparedness in four ways:

1. Ensuring methods to slow down and limit the spread of Covid in PNG through community engagement, strengthening provincial capacities, and developing human resources.
2. Strengthening the health system's ability to rapidly diagnose, contain the spread of Covid, and improve clinical management by improving testing, clinical management, and preparedness.
3. Management, implementation, monitoring, and evaluation.
4. Improvement of the government response capacity to emergencies.

When a sharp rise in cases occurred in June 2020, the World Bank added with another US$30 million of aid, which was aimed to build the health worker capacity, strengthen health systems, and scale up the deployment of Covid vaccines to prioritized communities, clinical management of mild cases, and nation-wide public education and community engagement delivered through UNICEF across television, radio, social media, and SMS.

Government response

With no cases in sight and the rest of the world going into lockdown, the Marape government hastily convened Parliament in 2020 and used their numerical strength to pass legislation that established the Office of Pandemic Controller

(OPC) operating out of Morata Haus, and the National Pandemic Act. The Office of the Commissioner of Police was subsequently appointed as sole 'National Health Emergency Controller' as per NEC Decision Number#: 02/2020.

This was challenged legally, as the National Health Act was not amended to cater for the Office of the Pandemic Controller. The normal administrative process to ensure compliance with the PNG Constitution requires an act to pass three readings of Parliament. This legislation occurred mid-year, after much public criticism and threats from the Opposition to seek a Supreme Court Challenge.

Initial lockdown actions included shutting all schools in March and April 2020 and then relaxing this to a recommendation until the end of the year. Some levels of schools were again shut when the cases soared in April 2021, but have relaxed again in July just as the Delta variant is entering the country. Similarly, the awareness and education of the general public was very poorly coordinated and managed from the start with some members paying large fees to get treatment. Mask wearing was mandated in late 2020 on the streets of the capital Port Moresby and has become a 'new normal' sight. The general public, market owners, and street vendors are much concerned about how to maintain the required level of hygiene and sanitation throughout their supply chains.

Economics

The Covid-induced lockdowns, accompanied by closure of workplaces and borders, hit all nations hard, but unfortunately PNG was affected even more than many due to it already being in a difficult financial economic position. Throughout 2020, PNG experienced almost no Covid cases, but the government required the closing of some workplace sectors. This was relaxed at the end of 2020, but when the cases began to rise in March 2021, workplace restrictions returned. As cases decreased in May, the government relaxed again to 'Recommend Closing' and this restriction did not change, despite cases peaking again shortly thereafter.

More attention was paid to border security with consistent quarantine and banning of arrivals from some or all regions instituted. However, extractive industries, essential to PNG's economic recovery, have continued to operate, which may have allowed Covid to enter the country undetected.

Although the government re-instated a 4-week lockdown in April 2021, it was far less restrictive than the ones imposed in 2020. The economist thus forecasted a 3% rebound in GDP in 2021 rather than increased losses.

Stay-at-home restrictions were far more lax with only a short period early in 2020 and from December 2020 to April 2021 during which residents were 'Required' not to leave home. In a country where many people have limited access to technology and many rely on walking to access services, the economic impact was too great to bear resulting in a shift to 'Recommend' not leaving house instituted in the middle of the 2021 peak of cases. The 2020 World Bank Covid surveys found that 85% of households reported using at least one of the following economic coping strategies:

- Reducing the number of children attending school (52%).
- Selling crops/harvest ahead of the planned time for sale (36%).
- Reducing their food consumption (28%).

With the national economy suffering an election coming, Prime Minister Marape expects to open international travel and stated that, 'those who choose not to be vaccinated' will have to 'face the consequences'.

Vaccines

Only 1% of PNG's population has been vaccinated with around 100,000 doses administered at around 2000 per day. At this slow rate, it will take another 3 years to vaccinate everyone in the nation. Supply is an issue, but the biggest problem has been a failure to distribute and deliver vaccines to the public. For instance, 130,000 AstraZeneca vaccines delivered through COVAX were allowed to expire, and most had to be destroyed. While this has been framed as being due to vaccine hesitancy and logistical issues, it clearly reveals a systematic failure of the health system.

Temporary vaccine clinics have improved vaccine uptake in the capital city, but the largely rural population remains largely without this service. In August 2021, 300,000 Johnson & Johnson vaccine doses arrived via COVAX, but this will have little impact in a population of nine million. Australian Doctors International reported that 'it's almost impossible to tell what's going on because there has been virtually no testing throughout last year and even now' in the provinces.

Testing results have become less reliable as the health department reallocates limited workers away from testing to vaccinating the more vulnerable segments of the population.

While many nations allocated their first doses to frontline workers and the most vulnerable, government members were amongst the first to receive vaccines in PNG. While a portion of this decision can be attributed to demonstrating the safety of the vaccine, it has the appearance of a vaccine justice issue.

Managing consequences and finding a path forwards

Many nations have reported significant public discontent with workplace restrictions, limited movement restrictions, and lockdowns, and the same is true in PNG. Efficient and effective coordination amongst all organizations and even the government were hampered by these measures. As the pandemic has raged wave after wave, these efforts have been described as poorly advised, Band-Aid solutions that have largely failed in their intent to limit the Covid pandemic.

The dire situation in Port Moresby General Hospital where patients were treated on the floor has highlighted facility limitations, human resource shortages, bottlenecks, and many other deficiencies in the health system. There has thus been a strong call for long-term international assistance to focus on improving public health infrastructure and capacity.

All best-practice evidence extols the value of clear and consistent messaging, but Médecins Sans Frontières has not observed this in PNG and locals are saying that messaging is broken. There is a stark lack of education, and enforcement of mask-wearing and social-distancing is not happening. 'It's very challenging when, on the streets, people who know there is community transmission are not wearing masks and not socially distancing'.

The pandemic has revealed the impact of social media on people's decisions, especially concerning self-protective measures and vaccination. The CEO of Australian Doctors International, Mimi Zilliacus, said, 'Lots of provincial governments have been reluctant to act, even when the response from the central government has ramped up. This is because so much misinformation has been spread, even amongst senior government officials'.

Dr. Moses Laman from the PNG Institute of Medical Research observed that misinformation on pandemic measures, treatments, and vaccines has had a significant impact on the response effort. This has been exacerbated by provincial governments not following the national government's lead in messaging and communicating the risks. When the last lockdown was in place, Health Minister Jelta Wong said government fear of a negative public response resulted in a lack of clear communication of what would happen in the coming weeks or whether further restrictions were planned. This was a legitimate concern because police and military were used to enforce lockdowns and they mistreated some people in the informal sector and markets.

Policy recommendations

1. Modernize the tax regime and banking system to support local businesses and make more efficient use of public monies. The Department of Treasury has only been able to update tax figures up to 2018, so the creation of a wider tax net is required to limit the number of citizens and businesses currently evading tax.
2. An interagency, Innovation Hub Policy Framework, would facilitate interactions between development partners and the government as they collaborate in restructuring foreign aid to support the Covid response.
3. A Health Innovation Hub that unites the University of Papua New Guinea's Health Science School and Medical School would go some way to remedying a shortage of clinical specialists. Currently, PNG has 50 medical specialists treating more than 8000 citizens each, which is grossly inadequate. The Medical School could transition to greater autonomy from UPNG, or become a stand-alone university, so that all resources are focused on graduating general practitioners and medical specialists to build national medical capacity and improve preparedness against pandemics.
4. PNG's enterprises continue to struggle to obtain start-up capital, so it is important for the government to relax barriers and streamline processes for obtaining soft loans for business start-ups. To this end, the PNG Credit Corporation, currently being developed by the Bank of Papua New Guinea to allow small-to-medium enterprises easier access to soft-loan credit financing, needs to be approved by the government. This structural improvement would reduce the high levels of unemployment that are exacerbated by pandemic-induced restrictions, and address the systematic displacement of local business by foreigners.
5. More health professionals need to be funded in the Provincial Health Authorities in all 22 provinces.
6. The Department of Education and the PNG Higher Education, Research, Science, and Technology Secretariat should introduce academic syllabus on Sustainable Development Goals (SDG) 1–4, 8, and 16, and the SDC Debating Competition at public universities to help the populace understand the importance of SDGs, how to move towards compliance, and how to initiate the required changes in mindset and behaviour.

Very few people in the world had knowledge of how to manage a pandemic of this nature. Indeed, all pandemic plans have failed, since they were originally written in response to the emergence of bird flu. The governments of the world tried many different approaches to control COVID-19 virus, and none of those are to blame for getting less than positive outcomes. Leaders who argued against the science or who were complacent were and are negligent. The situation is different now with, for the most part, excellent plans in place for regular variants of Covid and vaccination roll-outs; however, they may not be working for the Delta variant, or future variants. The learning process is continuous as the threat continually evolves.

14.7 Smart power and epidemic control

Difficult questions remain on territorial considerations: Should a hostile nation with an epidemic outbreak necessarily face infringements on national integrity? Or would sovereignty from other countries intend to prevent an international transboundary pandemic, the doctrine of nonpermissive international defensive (or preemptive) humanitarian intervention? If so, should soft and hard power be applied and deployed in this context; how far must the powers extend, and what are the associated national and international stability risks? Finally, would such globally protective but sovereignty-threatening steps—even if United Nations-sanctioned—be considered acts of war, or would the *vox populi*[d] in afflicted nations understand and support the intervention?

One of the factors contributing to the eventual successful containment of the 2014 West African Ebola outbreak was cohesive international relations and diplomacy, combined with military assistance [15]. Of course, the military may often have a logistical role in such situations. Still, soft-power and international relations are just as effective as hard power when addressing health security, or when combined military and diplomatic humanitarian engagements take the form of 'smart power'.

14.8 A need for national and supranational epidemic governance evolution?

Logically, future epidemic and pandemic control efforts necessitate the existence and funding of competent international or supranational bodies or initiatives that are completely detached from political bias, favouritism, and corruption. For example, the World Health Organization (WHO) may yet require a significant overhaul to rid itself of accusations of biased approaches and limited leadership capacity throughout 2020 and 2021. Still, to warrant continued support from all nations—though many would claim WHO lands in

[d] Voice of the people.

a no-win situation through a combination of current and prior circumstances—(other than via proportionate funding), how can such supranational bodies regain credibility and make sure of their roles as impartial arbiters of health?

The extent to which other related global relationships, national entities, and supranational organizations have performed in the 2020–21 pandemic situation will only be apparent in retrospect. It will quite possibly emerge that a combination of right- and left-wing policies and perhaps cherry-picking the elements of travel and visa restrictions were the clear winners. Yet, greater health security and diplomacy investments, and other defensive and protective policy aspects from highly divergent sets of agendas will be the guiding paradigm for decades to come [16].

Still, are public health arguments enough, in themselves, to justify neo-isolationist policies and associated connotations or accusations of regressiveness? In reality, can either pandemics or globalization be controlled by human efforts? How do we balance the risks associated with vector mobility versus global equality and freedom of movement—or are the two issues only spuriously connected, anyway? In the same way, if greater political authoritarianism resulted in better epidemic control (arguably not, in many aspects), we must explore the governments of South Africa, Hungary, Poland, and many other countries to less extensive or implicit extents during 2020. Can this authoritarian view be considered acceptable if such steps create international tension?

14.9 Guest essay: When do pandemics end?

Maisie M. Jones

As the COVID-19 pandemic enters its third year, with a string of lockdowns behind us, and the constant threat of more restrictions being implemented as case numbers remain steadily high despite growing vaccine uptake in numerous countries (in the Global North at least), many people are asking the same question: When will the pandemic end?

There are different ways to view the end of pandemics. There is the medical or epidemiological end, measured by the decline of disease cases and deaths. There is also the social end, which occurs when fear of the disease dissipates and normal life resumes. The ending of a pandemic is not always the same as the end of the disease, and there can be multiple endings to a pandemic, not only across the world, but even within individual societies [17,18]. Far more attention is paid to the beginning of pandemics than to their end; however, when we do focus on endings, we can see that they are not solely a biological phenomenon, but also have political, cultural, and social influences.

There have been several epidemics and pandemics throughout history, and these are often presented in a linear manner, as if they all had a defined beginning and end, and occurred one after the other often with disease-free gaps in between. The reality is less clear cut. While there is often a distinct beginning, defined using scientific metrics such as increased incidence, case numbers, and transmission rate, the end is often more ambiguous: a more long, drawn out, and often disputed process. One medical historian, Dr Erica Charters, uses the example of the Ebola epidemic in West Africa, which was officially declared over in 2016, to demonstrate this point. In her paper, she describes how the World Health Organisation (WHO) actually declared this epidemic over several times. They first declared that Liberia was Ebola-free in May of 2015, and the second declaration in September of 2015 more solemnly reported 'an end to Ebola transmission in Liberia'. However, cases of Ebola were reported in the months after this. And so in January of 2016, the WHO cautiously announced that the latest Ebola outbreak in Liberia was over adding that 'more flare-ups are expected'. It then declared the official end of West Africa's 'public health emergency of international concern' for Ebola on 29 March 2016. As its declaration clarified, this was not the end of the disease. Instead, the WHO explained Ebola transmission in West Africa no longer constitutes an extraordinary event, that the risk of international spread is now low, and that countries currently have the capacity to respond rapidly to new virus emergencies [17,19]. We can see that with each of these declarations, the language changed, from a definitive end of disease, to a more ambiguous and cautious end of transmission and somewhat temporary end of the current outbreak.

This points to the reality that in fact pandemics seldom conclude due to the end of the disease. Rather, they are more commonly declared over when the disease becomes endemic, and the disease is considered an inevitable and acceptable part of life [17].

As we know, smallpox is the only human disease to have been eradicated, and although a small number of epidemics such as SARS have been declared ended when cases amongst humans have ceased, this is not often the case. The 2009 H1N1, or Swine flu, pandemic was declared by the WHO in August 2010, to be in a 'postpandemic period'. The WHO explained that in this postpandemic period, cases and outbreaks due to the H1N1 virus were expected to occur according to seasonal patterns of influenza. In the week when Queen Victoria declared a national day of thanksgiving upon the great decrease of cholera in 1849, there were still 11 deaths from cholera recorded in London alone [17,20]. The Spanish flu pandemic of 1918 infected an estimated 500 million people worldwide and was responsible for 20 million to 50 million deaths. Despite the notoriety of this pandemic, pinpointing its end is almost impossible. Although the Spanish flu pandemic is said to have ended in the summer of 1919, in 1920, several newspapers reported the reappearance of influenza. Approximately 5000 cases were reported in Chicago in a 6-day period, while 60 people died from influenza in Paris around the same time. For years following, the virus was reported in cities throughout Europe and North America [21,22].

So rather than the end of disease, pandemics are more often marked by a perceived control of disease, the lifting of public health regulations and the associated political, social, and economic restrictions, and a resumption of normal life and activities. It can be said that the end of a pandemic is when the drama and panic ceases, when societies' attention is no longer focused on the crisis of disease but shifts to reflecting on the social, political, and economic repercussions of the outbreak [17,18].

Another point to consider is that often epidemics or pandemics end at different times in different places. HIV and AIDS are still classified as a global epidemic, but the geographical setting of the disease has shifted. For example, while this epidemic may have ended for most in the Global North, in some ways, it simply moved and continued on as an epidemic for many in the Global South [17,22].

The plague never really went away. The second plague pandemic, which ran between approximately 1350 and 1830, receded from Europe, while most of the world did not witness an end to this disease. In many ways, plague simply died down in a few select countries for a short period of time, while remaining active within rodent populations where it remains endemic today, and with scattered human cases regularly reported throughout the world. In the United States, plague is endemic amongst prairie dogs and can be transmitted to people. Polio is a thing of the past for Europe, the Americas, and Australia, but remains a threat in small parts of Africa and South Asia [17,18,23].

Furthermore, rather than this linear pattern of pandemics, with one ending and another beginning, there is often more of an overlap. Many epidemics and/or pandemics happen simultaneously, with more than one disease present, cases fluctuating in waves, one disease peaking while another troughs, or indeed more than one peaking at a time. In 2006 in Papua, Indonesia, there were simultaneous outbreaks of measles and cholera, alongside the ongoing high incidence of malaria [24]. In Brazil, the COVID-19 pandemic may have contributed to increased outbreaks of dengue, chikungunya, and zika [25]. In Kenya, the current COVID-19 pandemic is ongoing alongside concurrent outbreaks of Ebola, Cholera, TB, and others, and these diseases constantly interact with each other and with society [26].

This image of a history of a long line of epidemics, all with a beginning, a middle, and an end, is an attractive one. It gives us the illusion that we have conquered disease and therefore can do so again, that we can end this epidemic like we have ended all others. Epidemiology formed itself on this optimistic belief in human ability to control disease. However, the control of disease is not only in the hands of humans, but also relies on ecology, climate, and an element of chance [17].

All this to say that pandemics often do not have a definitive endpoint. For all the drama and upheaval that the declaration of the start of a pandemic brings, when it comes to the end, we are not given the same assertion. In an anticlimactic way, pandemics often simply fade out, transform, shift geographically, and become someone else's problem [17,23].

The current COVID-19 pandemic will likely be declared over once the virus is mostly contained and when rates of infection are considered to be at an acceptable or manageable level [27]. What is considered an acceptable or manageable level is not clear. Although the WHO is responsible for determining the status of a disease, and declaring when pandemics begin and end, it does not have a published set of criteria for how a pandemic could be considered over [28]. Nevertheless, this containment of the virus could happen in several ways. One such way pandemics can be brought to a halt is by starving the virus of hosts. This can be done through a combination of quarantine, social distancing, and restricting travel. This approach has been taken in many countries during the COVID-19 pandemic, some to try to stop the spread of COVID-19 completely, and some to slow the spread enough to allow healthcare services to cope until a vaccine can be developed. But, this variation in policy and procedure may be the reason it has not succeeded in ending the pandemic completely. For this method to succeed, requires a synchronized response, all countries working together as one interconnected system [27].

Another approach is by allowing for uncontrolled virus transmission, leading to herd immunity. The theory is that a large number of people become infected with the virus, and while many will die, those who survive will possess some level of immunity. Even if they are not permanently immune, they may be immune long enough to outlast the rate at which the virus can infect organisms to survive. With this approach, large numbers of people would be infected at the same time, healthcare facilities would reach capacity very quickly, and huge numbers of people would likely die, either from the virus itself or the collapse of healthcare systems. Furthermore, in the case of COVID-19, frequency of reinfection has not yet been proven and so uncontrolled transmission could lead to constant waves of disease [27,28].

There is another way to reach herd immunity, without such a dramatic loss of life, and this is through mass vaccination of the population. In this scenario, quarantine and social-distancing measures are put in place in order to slow the spread and buy scientists time to develop an effective vaccine. Once this vaccine is developed, it is rolled out worldwide, and once a large enough proportion of the population is vaccinated, we reach herd immunity and the virus dies out. This method relies on several factors; that the vaccine is effective in preventing infection and/or transmission of the virus, that the vaccine is available to everyone all over the world, and that people are willing to take the vaccine when offered it. In the case of COVID-19, the rush to develop vaccines may be the downfall of this approach. The WHO only recommends a minimum of 50% efficacy of vaccines, while other agencies such as the Food and Drug Administration (FDA) in the United States have allowed for vaccines that merely decrease the severity of the disease rather than prevent infection. In some ways, these organizations have adopted a 'better than nothing' narrative. This narrative, along with low efficacy and reported side effects, risks irreparably damaging public confidence in the vaccine and vaccines in general [28,29]. Furthermore, the unequal global distribution of COVID-19 vaccines is not only leaving millions of people vulnerable to the virus, but also allowing even more deadly variants to emerge and spread globally. Dr Tedros Adhanom Ghebreyesus, Director-General of WHO, has said that 'vaccine inequity is the world's biggest obstacle to ending this pandemic and recovering from COVID-19' [30,31].

So what does all of this mean for the COVID-19 pandemic? Will COVID-19 become endemic worldwide? Will we soon enter a postpandemic phase, like with Swine flu, with scattered seasonal cases? Will we formally end the pandemic in the Global North through the mass vaccination of the population, while leaving those in the Global South to continue to endure COVID-19? The answers to these questions remain to be seen, but one way or another, with all its successes and failings, the COVID-19 pandemic will take its place amongst the long history of disease. While we are all looking for an end to the pandemic, and we may well need that glint of light at the end of the tunnel, we would be better served by reflecting on the past two years and how we can put lessons learned into practice. We understand disease differently at the end stage of pandemics, so if in the outbreak stage, we could channel this end-stage perception of disease, it might help us reframe, and rethink our long-term approach to managing disease. We must also prepare for the possibility that COVID-19 is here to stay, and that the end of this pandemic may come in the form of a social and political end, rather than the end of the virus [17,22,28].

PICTURE 4.14.2 **Social distancing in Killarney.** Maisie M. Jones.

Political implications

Politically, as in 2020 and 2021, international tensions naturally heighten if states blame each other for pandemic spread. Similarly, political leaders who become infected can disrupt political decision-making while also increasing uncertainty about the leaders' capabilities within the country. An example is Iran's leadership, severely compromised in this manner, and the US's approach of turning a sharp corner following the 2020 elections to commit to remaining close to the scientific advice.

Similarly, in the political-epidemic context, the risk of xenophobia amongst states may all the while continue to grow [32]. As nation-states blame each other, political entrepreneurs seeking power advantages within their own states can use epidemics to gain support within their national political arenas; thus, while Iran's leadership came under stress without revolution, other regimes in similar future situations may choose to manipulate circumstances to crackdown on dissent or engender nationalist sentiment.

At the domestic level, the inability of nation-states to curtail future epidemics or pandemics effectively can also lead to perceptions amongst the public that the ruling regime may be ineffective or unable to confront the challenges they pose. Such problems are not unfounded in recent crises: throughout the Ebola outbreak in West Africa, the limited capacity of all three governments—Guinea, Sierra Leone, and Liberia—to contain the epidemic led to violence and protests, eventually requiring a large-scale external intervention to help control spiralling instability in all three countries.

Whatever the answers, most likely there will have been, looking back, opportunities for different approaches that might have contained the 2020–21 pandemic more effectively—the subject, in many ways, of this book. Ironically, it is only through international relations, communications, and transparency that both health and security can be advanced, even if this diplomacy requires the unpleasant discussion of measures that further isolate countries from each other. Yet, in the interests of public health and—sometimes, but not always—international security and stability, how can we continue to strive to improve?

PICTURE 4.14.3 Empty beaches and open spaces were a big part of the pandemic. Sarah O Dwyer.

14.10 Key points

- Preexisting tensions between countries (e.g. the United States and China) contributed to challenges involved in reporting and sharing of early-stage pandemic information.
- Infectious disease control interventions should be mindful of the differing cultural, social, and religious aspects of the demands they make on societies.
- Conversely, well-designed and well-managed health security efforts can, if appropriately explained and imposed, help to avoid escalations in tensions associated with limitations on movement.

References

[1] D.W. Drezner, The song remains the same: international relations after COVID-19, Int. Organ. 74 (S1) (2020) E18–E35.

[2] T. Johnson, Ordinary patterns in an extraordinary crisis: how international relations makes sense of the COVID-19 pandemic, Int. Organ. 74 (S1) (2020) E148–E168.

[3] World Health Assembly, 2021. [cited 2021 October 6]. Available from: . https://www.who.int/about/governance/world-health-assembly.

[4] R.W. Copsom, T. Salaam, The global fund to fight AIDS, tuberculosis, and malaria: background and current issues, in: Library of Congress Washington DC Congressional Research Service, 2005, March.

[5] A.J. Blinken, A Foreign Policy for the American People, vol. 3, *US Department of State*, 2021.

[6] Global Fund Thanks the United States for US$3.5 Billion Emergency Investment to Fight COVID-19—News & Stories—The Global Fund to Fight AIDS, Tuberculosis and Malaria, 2021. [cited 2021 October 6]. Available from: . https://www.theglobalfund.org/en/news/2021-03-11-global-fund-thanks-the-united-states-for-usd3-5-billion-emergency-investment-to-fight-covid-19/.

[7] International Day of Multilateralism and Diplomacy for Peace—Message, United Nations, 2021. [cited 2021 October 6]. Available from: . https://www.un.org/en/observances/Multilateralism-for-Peace-day/messages.

[8] https://www.un.org/press/en/2021/sgsm20699.doc.htm.

[9] Reinforcing the U.S. Commitment to Multilateralism at the G20 Foreign Ministers' Meeting—United States Department of State, 2021. [cited 2021 October 6]. Available from: . https://www.state.gov/reinforcing-the-u-s-commitment-to-multilateralism-at-the-g20-foreign-ministers-meeting/.

[10] Does the World Need a New Global Health Organization?, Foreign Affairs, 2020. [cited 2021 October 6]. Available from: . https://www.foreignaffairs.com/ask-the-experts/2020-12-08/does-world-need-new-global-health-organization.

[11] S. Devi, Travel restrictions hampering COVID-19 response, Lancet 395 (10233) (2020) 1331–1332.

[12] M.B. Allen, M. Mirsaeidi, Health and economy in COVID-19 era: a plan for reconstituting long-term economic security, Public Health Front. 8 (2020) 235.

[13] S. Taghizade, V.K. Chattu, E. Jaafaripooyan, S. Kevany, COVID-19 pandemic as an excellent opportunity for Global Health Diplomacy, Front. Public Health 9 (2021).

[14] D. Devakumar, G. Shannon, S.S. Bhopal, I. Abubakar, Racism and discrimination in COVID-19 responses, Lancet 395 (10231) (2020) 1194.

[15] V.K. Chattu, Politics of Ebola and the critical role of global health diplomacy for the CARICOM, J. Fam. Med. and Prim. Care 6 (3) (2017) 463.
[16] S.L. Greer, E.J. King, E.M. da Fonseca, A. Peralta-Santos, The comparative politics of COVID-19: the need to understand government responses, Glob. Public Health 15 (9) (2020) 1413–1416.
[17] E. Charters, K. Heitman, How epidemics end, Centaurus 63 (1) (2021) 210–224.
[18] G. Kolata, How pandemics end, in: W.C. Cockerham, G.B. Cockerham (Eds.), The COVID-19 Reader, first ed., Routledge, New York, NY, 2021, pp. 252–256. 2020 [cited 2021 November 6]. Available from: https://www.taylorfrancis.com/books/9781000332506/chapters/10.4324/9781003141402-30.
[19] Universitetet i Bergen, Corona Day—How Do Epidemics End? 2021, [cited 2021 November 8]. Available from: https://www.youtube.com/watch?v=LCWoWQ7sEOs.
[20] h1n1_donor_032011.pdf, [cited 2021 November 8]. Available from: https://www.who.int/csr/resources/publications/swineflu/h1n1_donor_032011.pdf.
[21] Editors H com, Spanish Flu, History, 2021. [cited 2021 November 8]. Available from: https://www.history.com/topics/world-war-i/1918-flu-pandemic.
[22] A. Arnold-Forster, The End of the Pandemic Is Coming—Just Don't Set a Date for the Party, The Conversation, 2021. [cited 2021 November 8]. Available from: http://theconversation.com/the-end-of-the-pandemic-is-coming-just-dont-set-a-date-for-the-party-157205.
[23] A. Gross, From plague to polio: how do pandemics end? Financial Times (2021). Mar 12 [cited 2021 November 8]. Available from: https://www.ft.com/content/4eabdc7a-f8e1-48d5-9592-05441493f652.
[24] Papua Hit by Simultaneous Epidemics, Médecins Sans Frontières (MSF) International, 2021. [cited 2021 November 8]. Available from: https://www.msf.org/papua-hit-simultaneous-epidemics.
[25] C.R. Vicente, T.C.C. da Silva, L.D. Pereira, A.E. Miranda, Impact of concurrent epidemics of dengue, chikungunya, zika, and COVID-19, Rev. Soc. Bras. Med. Trop. 54 (2021). February 26 [cited 2021 November 8]. Available from: http://www.scielo.br/j/rsbmt/a/xLBTRGsz8fwzWXf433DMLrJ/?lang=en.
[26] P.W. Geissler, R.J. Prince, Layers of epidemy: present pasts during the first weeks of COVID-19 in western Kenya, Centaurus 62 (2) (2020) 248–256.
[27] A. Rosenthal, When Is a Pandemic Over? 1591031448 [cited 2021 November 10]. Available from: https://www.ted.com/talks/alex_rosenthal_when_is_a_pandemic_over.
[28] In the Age of COVID-19: How Do Pandemics End?, Imperial Bioscience Review, 2020. [cited 2021 November 9]. Available from: https://imperialbiosciencereview.com/2020/11/06/in-the-age-of-covid-19-how-do-pandemics-end/.
[29] E. Torreele, The rush to create a covid-19 vaccine may do more harm than good, BMJ (August 18) (2020) m3209.
[30] COVID vaccines: widening inequality and millions vulnerable, UN News (2021). [cited 2021 November 10]. Available from https://news.un.org/en/story/2021/09/1100192.
[31] Vaccine Inequity Undermining Global Economic Recovery, [cited 2021 November 10]. Available from: https://www.who.int/news/item/22-07-2021-vaccine-inequity-undermining-global-economic-recovery.
[32] T.T. Reny, M.A. Barreto, Xenophobia in the time of pandemic: othering, anti-Asian attitudes, and COVID-19. Politics Groups Identities 10 (2) (2022) 209–232.
[33] International Sanitary Regulations, World Health Organisation Regulations No 2. 21 May, World Health Organisation, 1951. Retrieved from https://apps.who.int/iris/bitstream/handle/10665/101391/WHA4_60_eng.pdf?sequence=1&isAllowed=y. (Accessed 16 July 2021).

[34] D.P. Fidler, From International Sanitary Conventions to Global Health Security: the New International Health Regulations, Chic. J. Int. Law 4 (2) (2005) 325–392, https://doi.org/10.1093/chinesejil/jmi029.

[35] https://www.cdc.gov/sars/about/fs-sars.html. Accessed 16 July 2021.

[36] https://www.cdc.gov/vhf/ebola/history/2014-2016-outbreak/index.html. (Accessed 16 July 2021).

[37] M.E. Kruk, M. Myers, S.T. Varpilah, B.T. Dahn, What is a resilient health system? Lessons from Ebola, Lancet 385 (9980) (2015) 1910–1912.

[38] H. Kluge, J.M. Martín-Moreno, N. Emiroglu, et al., Strengthening global health security by embedding the International Health Regulations requirements into national health systems, BMJ Glob. Health 3 (2018) e000656.

[39] D.L. Heymann, L. Chen, K. Takemi, D.P. Fidler, et al., Global health security: the wider lessons from the west African Ebola virus disease epidemic, Lancet 385 (9980) (2015) 1884–1901.

[40] World Health Organization, Triple Billion Dashboard, 2021. https://www.who.int/data/triple-billion-dashboard. (Accessed 4 July 2021).

[41] World Health Organization, UHC Service Coverage Index, 2021. https://www.who.int/data/gho/indicator-metadata-registry/imr-details/4834. (Accessed 4 July 2021).

[42] World Health Organization, World Health Statistics, 2021. https://www.who.int/data/gho/publications/world-health-statistics. (Accessed 4 July 2021).

[43] S. Lehtimaki, A. Reidy, K. Nishtar, S. Darehshori, et al., Independent Review and Investigation Mechanisms to Prevent Future Pandemics: A Proposed Way Forward, United Nations University International Institute for Global Health, April 2021.

[44] United Nations Human Rights Council, Universal Periodic Review, 2021. https://www.ohchr.org/en/hrbodies/upr/pages/uprmain.aspx. (Accessed 4 July 2021).

[45] World Trade Organization, Trade Policy Review Mechanism, 2021. https://www.wto.org/english/tratop_e/tpr_e/tp_int_e.htm. (Accessed 4 July 2021).

[46] United Nations Framework Convention on Climate Change, Paris Agreements, Nationally-Determined Contributions, 2021. https://unfccc.int/process-and-meetings/the-paris-agreement/nationally-determined-contributions-ndcs/nationally-determined-contributions-ndcs. (Accessed 4 July 2021).

Index

Note: Page numbers followed by *f* indicate figures, *t* indicate tables, and *b* indicate boxes.

D

I

FSC
www.fsc.org
MIX
Papier | Fördert gute Waldnutzung
FSC® C083411

Kolibri 360 GmbH
Ferdinand-Jühlke-Straße 7
99095 Erfurt, Deutschland
produktsicherheit@kolibri360.de